ADAM AND EVOLUTION

By the same author:

Erik Bloodaxe: His Life and Times
A Royal Viking in his Historical and Geographical Settings

Beowulf the Jute; His Life and Times
Angles, Saxons and Doubts

Not Quite Trite
Poetry, Some Pleasant

ADAM AND EVOLUTION

A Look at Life and All Our Yesterdays

WILLIAM PEARSON

To order additional copies of this book, contact:
Xlibris
UK TFN: 0800 0148620 (Toll Free inside the UK)
UK Local: 02036 956328 (+44 20 3695 6328 from outside the UK)
www.Xlibrispublishing.co.uk
Orders@Xlibrispublishing.co.uk
516241

CONTENTS

PART 1
The World of Nature

FOREWORD

Many literary works have been produced that extol the pleasures to be found in studying natural history and exploring the intricacies of evolutionary theory. So is there any call for another? Well, yes, although the motivation needs to be strong, especially if the author embarking on such a project is an amateur naturalist and even more of a layman when it comes to biology. However, with the passage of time and the broadening of experience it has become apparent that certain aspects of the development of life - with its obscure origins in the appearance of the first microbes while eventually reaching the form regarded as the pinnacle achievement of the living world, i.e. us - provide plenty of room for speculation. Among the many arguments that have been presented over the years there are still those that are not rooted in the firmest of ground. In quite a number of areas alternative explanations seem possible that hitherto seem to have been ignored.

For us to get a firm grip on the origins of terrestrial life would seem to be an impossible task, since our intelligence – much vaunted as it is – has great difficulty even to understand what 'life' actually is. The one certainty is that the problem presented by the existence of 'life' cannot be approached without coming to terms with the manner of its beginning. With regard to this, three modes of approach can be recognized. These can be labelled as supernatural, cosmic and natural. The first mode attributes life on Earth to an extrasensory and omnipresent entity that lies behind all life and somehow sustains it, after having originally arranged its creation. The second mode maintains that life exists elsewhere in the Universe, where somehow it began and has since discovered a means to reach Earth and colonize it. The third mode assumes that life is unique to Earth and that it began here and only here. On the following pages it is assumed that the third mode is the most likely to be the true one. However, the intention is not to refute the others; acceptance of one does not necessarily exclude the other two. It is simply a question of putting the third mode to the test and using the exercise as a means to observe whether or not it is a workable proposition. Yet acceptance of this does bring with it the corollary that the other two modes might at very least be suspected to be the results of flawed reasoning.

In order to allow the work to be accessible to as wide a readership as possible the arguments are set against a background of uncontroversial biological and evolutionary knowledge gathered from a variety of sources. As and when appropriate the discourse diverges from the existing state of opinion to penetrate various unexplored zones of the intellectual jungle. These often result from random thoughts that lure one away from the generally well-trodden paths that criss-cross the areas where authenticity rules, often with apparently unassailable authority.

The work is presented in two parts. Part 1 deals with the living world without mankind. Here the nature of life in its various forms and the way they work are considered. Part 2 is concerned with mankind, how this species emerged from the rest and what it has achieved since in order to transform itself into a force that is so dominant and 'successful' that it now constitutes an overwhelming entity that threatens to destroy the very habitat that sustains it, which is the wonderful biosphere that clads planet Earth.

W. Pearson
Stockton-on-Tees
England
August 2021.

PART 1

The World of Nature

CHAPTER 1

The Forms of Life

■ Introduction

Let us start with the obvious. Living matter, having once emerged from obscure beginnings on this planet Earth - the home selected for it in the Universe - has continued to exist in countless different forms right to the present time. As a means to cope with the enormous diversity involved, natural scientists have created a comprehensive classification system. This enables one to get a grasp of the complexities of the chronological and biological relationships that all forms of life bear to one another. It is called the Linnaean System, named after Carl von Linné, its Swedish instigator. This is known more generally as taxonomy and the immediately following discourse is by way of a résumé to aid those to whom it is not entirely familiar. Yet is it ultimately a completely satisfactory method? Will it ever be able to explain all aspects of evolution? In later chapters there will be some discussion on this matter and suggestions made that are meant to clarify points that hitherto have been inexplicable.

■ Plants, Animals and Others

Apart from fungi and certain microbes, most forms of life - and essentially those most obvious to us - can be attributed to one or other of two groups, while including all those organisms that are readily perceived by our senses as 'plants' or 'animals'. In most cases the decision as to which of these any particular individual belongs is obvious, even to the casual eye. The external differences between plants and animals are however greatly diminished among the simpler forms of both, while some microscopic beings exhibit traits that seem to give them a foot in each camp. No-one would mistake a rhinoceros for a tree, but the tiny Bryozoa, those sedentary aquatic invertebrates otherwise known as the moss animals, might well be mistakenly taken for a kind of water weed, while the microscopic green flagellates display true characteristics from both 'kingdoms'. However, there are organisms in existence that are indeed neither plants nor animals and have had to be classified separately. The most obvious of these are the fungi.

The following pages will mainly be concerned with the animal kingdom and nearly all examples for investigation are to be taken from this.

■ Vertebrates and Invertebrates

Apart from that described immediately above, the largest subdivision of convenience of such kingdoms imposed upon life by human scholarship is the 'phylum'. The animal world comprises a score or so of such phyla, some of which consist of only a small number of non-conforming species. Below phylum level, further subdivision occurs to give class, order, family, genus, species and variety. The main concern in the following is for phylum and species, as well as for individual animals and their component parts.

The vertebrates constitute a group that comprises the mammals, birds, reptiles, amphibians and fishes, which are all animals whose bodies are supported by internal skeletons. Essentially these provide the vast bulk of the phylum called the Chordates. The remainder of the latter are some rather insignificant aquatic creatures that lack a skeleton and are hence technically invertebrates. Swimming forms do however possess a dorsal cord that is homologous to the backbone of the true vertebrates.

All other phyla of the animal kingdom contain invertebrates exclusively. The chief ones are the following.

The Arthropoda	- insects, centipedes, millipedes, spiders, scorpions, crabs, lobsters, etc.
The Annelida	- earthworms, lugworms, ragworms, leeches, etc.
The Mollusca	- slugs, snails, clams, mussels, squids, octopuses, etc.
The Nematoda	- various worms, i.e. roundworms, hook worms, etc.
The Platyhelminthes	- various flat worms, i.e. planarians, flukes, tape worms, etc.
The Echinodermata	- starfish, sea urchins, sea cucumbers, sea lilies, etc.
The Cnidaria	- coelenterates, i.e. polyps, jellyfish, sea anemones, hydras, etc.
The Porifera	- sponges
The Protozoa	- unicellular (or non-cellular) animals, i.e. amoebas, ciliates, etc.

Cells and Molecules

Except for the Protozoa, all animals are built up of numerous units known as cells. The 'cell' is the unit of construction of the living world and a group of basically identical cells when fused together form a tissue, in much the same way as a wall can consist of bricks mortared together. (With the sponges, by this analogy, the wall is not exactly dry-stone, but the mortar can be considered as poor quality.)

Where different tissues act together to form a functional whole the result is an organ, and this can be compared with a building. Pressing the analogy to its logical conclusion an individual animal is comparable to a village or town, depending on its size and organization.

Except for the Protozoa, which can selectively be regarded as single cells or creatures of non-cellular nature, all animals use the cellular method of body building. It is however a fact that many Protozoa do group themselves into colonies. These however can be seen in the badly mortared wall analogy in that they are usually merely connected individuals, being neither tissues, nor incorporating the differentiated and well-co-ordinated functions characteristic of multicellular animals.

The chief micro-ingredients of all living matter are minute combinations that incorporate, promote and allow life to exist. Yet such organic arrangements are very large and complex in comparison with all inorganic molecules. The chief constituent materials of living matter are the proteins, carbohydrates and fats, with the actual 'living' being primarily instigated by the second. The proteins are nitrogenous organic compounds that form structural components of tissues. Carbohydrates are organic compounds containing carbon, hydrogen and oxygen, which when broken down release energy into the containing organism and include sugar, starch and cellulose, the last two being more particularly associated with plants.

The various minute and elusive viruses would seem to be specific and active particles of nucleic acid coated in protein that, like the fungi, are dependent on the presence of other living matter. Lack of an independent sexual phase might put them outside of consideration as life, yet their ability to respond to circumstances by rapid evolution, i.e. mutation, none the less makes them discrete from mere chemical compounds. They are apparently fixed in some sort of intermediate parasitic stage between lifeless and living matter that cannot reproduce themselves except in association with living matter.

There will have been at a very early stage in the evolutionary story a variety of organisms that were more than chemical compounds but not in a state that can be called 'living'. They were groups of molecules whose instability had initially come under the control of outside circumstances with a certain degree of regular change. These can be regarded as 'infra-life' entities.

▧ Cells, Individuals and Colonies

The cells of most tissues normally cannot live as individuals. They are bound to their complete organism and if separated from it will die. However, one tissue is well known for its ability to survive disintegration. After being broken down to single cells by pressing through filters, the cells of sponge tissue can stay lively enough to be able to move together into clusters and with the latent ability to grow into new sponges.

Yet strictly speaking, a tissue ought to consist of cells that cannot exist independently, while a colony ought to be a conglomeration of some which can and, if need be, do. Otherwise a true tissue can be regarded as a fused colony of cells where the individuals have become quite interdependent and inseparable. At the cellular level, the body of a sponge apparently comes somewhere between the two clearly defined and idealised cases of what constitutes a tissue and a colony.

The path of life down through the ages has apparently involved a succession of groupings of like individuals into loose colonies that have later led to fused ones. In the latter there has been a tendency for the individual members to specialise; in fact evolutionary pressures have determined that they do so in order to justify themselves as part of the whole. (Nevertheless, within any organ, cells of any particular type have always kept company with those of their own sort, namely in the construction of tissues.) In this way, not only has a great divergence of function developed among the differentiated cells of an organism, but they have also diverged with regard to importance, so that there came to be the vital and the non-vital, the essential and the expendable and, by human analogy, the aristocrats and the commoners. Thus would it appear that class distinction has indeed long been a force among living matter, but the caveat must also be added that in the biological sense privilege does go hand in hand with actual worth.

Examination of the living cell shows that it is not a homogeneous piece of matter, but at the very least can be divided into the nucleus and the surrounding cytoplasm. These again can be subdivided into non-repetitive units. A cell is a fused colony at a lower level that consists of diverse units and the same system of specialised parts applies throughout the concept we call 'life' in all its larger and more complex forms.

▦ Organs and Individuals

The step described above as having been taken by individual cells, whereby they became differentiated within tissues to transform these into organs, has also been made by organs themselves in a very similar manner. This has eventually led to the most complex creatures in existence. The next step for animals that consisted of a single organ was to collect together in repetitive colonies, in a similar way to that in which polyps and Bryozoa now congregate with their own kind. Such earlier colonies of animals may not always have been connected to each other by common living tissue in the manner of the above two examples.

Given that the sharing of connective tissue existed, after this stage the individuals of such loose colonies of connected and repetitive organisms would be under pressure to differentiate and specialise. In the bulk of complex animals today the degree of fusion and specialisation of the original separate and virtually identical organisms has been so great in the course of time that their original repetitive nature is by no means apparent. This process demands further examination on following pages.

■ Polymorphism

This term is used to describe the conditions whereby individuals within a distinct species can appear in various forms to perform different functions. It is common among polyps, which can be designated as feeding, reproductive or protective members. This is indeed a case of specialisation of the individuals in colonies for the benefit of the whole. The next stage in this process does of course lead to the fused colonies of animals described above that have tended to become complex multi-part and multi-functional individuals, even though this may be far from obvious. More detailed discussion of this particular aspect will be undertaken later in this work.

CHAPTER 2

The Activities of Life

◼ General Activities

As a general rule living matter performs certain compulsory functions that we know by the terms feeding, respiring, growing, moving and reproducing; when one or more of these activities is permanently missing life can be judged to be finite in that the other activities will also ultimately fail. The labels attached to functions may make them seem to be more distinct from each other than they actually are, and on occasions one or other of them may be temporarily suppressed without life itself being eliminated. Encysted microbes seem to lose all contact with the outside world for prolonged periods, yet under the right conditions they can come forth and begin active life processes again.

◼ Feeding

The state known as 'living' involves an exchange of chemicals. Feeding is the taking in of suitable solid and liquid matter to this end. The equivalent wastes of metabolism are removed from the body by the complementary process known as excretion.

■ Respiring

Gaseous exchange also occurs to keep life's 'batteries' fully charged. As far as animals are concerned this primarily involves the absorption of oxygen and the release of carbon dioxide and can occur by breathing in air or - as do fish with their gills - by extracting oxygen contained in a liquid environment,.

■ Growing

Living matter exists in a series of unit lifetimes in which each individual starts off in a small way and, as time passes, develops to greater size and usually to greater complexity. The full specification of the mature animal is always embodied in the seed from which it springs. However, the scope can become wider and includes possibilities for change in development if circumstances demand it.

■ Moving

To the casual observer movement is the most prominent characteristic of life. The degree of movement varies enormously throughout life's different forms; but all life does move, either in a change in position with respect to its environment by using external organs, or by transporting matter embodied in its own tissues as a function of its internal organs.

■ Reproducing

Nothing lives forever, because living matter is basically discontinuous. For life as it appears to human senses to carry on, it must embody means of renewing its own forms. Thus, from out of their own tissues, the old forms create new ones that are referred to as being young. This reproduction might be regarded as the means by which life is kept going; or, conversely, one might even argue that life is actually the means by which reproduction is enabled to carry on. This aspect will be referred to again and more explicitly on later pages.

◼ Matter and Energy

Life can be understood as having come into being as a special alliance between matter and energy. These are still the two recognisable commodities out of which the world we live in is composed, at least as viewed at its simplest level. One of the ever growing preoccupations of the developing thinker known as 'man' has been a search for understanding of the Universe and this has led to two particular quests concerning matter and energy to be undertaken, the one seeking truth about the infinitely large limits of his environment and the other about the infinitely small ones. Yet as one hurdle has been jumped, another has always loomed ahead, making final success continually a distant objective. With every step forward in knowledge, the seeker is still confronted with the two infinities, either the seemingly boundless extent of space or the seemingly inscrutable minuteness of lack of space.

Matter and energy are essential to each other and form a forced and uneasy partnership. Matter apparently cannot exist without energy to hold it together; nor is energy evident until it has matter on which to act. This can be regarded as a game of football that is known as existence. 'Energy' does the kicking about of the ball known as 'matter' in the limitless 'park' that is the universe. But the ball is not immutable. Whereas matter is conservative, energy is restless. When matter seems to say, "Let's stay as we are", energy retorts, "Come on, let's change!" Indeed, matter only seems to exist because certain forms of energy have somehow been tamed and trained to behave regularly.

Life does indeed seem to be a sort of victory for matter, for within the scope of any particular environment or habitat it may have come to occupy, and in spite of many casualties, it has been able to resist the demands of energy for violent change by allowing itself to go through a long series of lesser changes, each one ready to meet a specific hostile circumstance, should it arise. And it is the same today as it always has been. This is the special characteristic of living matter, which can normally cope with small unpredictable changes within and outside of itself, but abrupt and larger changes result in its destruction. Such cause the end of any affected entities that bear that elusive commodity we call life, at least in the form they hitherto have taken.

The diverse component matter that embodies life is all highly unstable and 'living' involves the collective and delicate utilisation by certain special kinds of matter of the right amount of the available and suitable energy to preserve itself in the unstable state in which it exists. This occurs against the opposition of energy as a whole.

■ Self-Preservation of Matter

The state of all matter at any one time is dependent upon the uneasy alliance between itself and energy. The condition is maintained by the existence of a rough balance between the energy state inside and outside of the matter. With an increase in external energy the matter must be able to absorb a deal of it to preserve this balance, or alternatively to release some when the level of external energy is reduced, if it is to preserve its identity. Within certain limits, these adjustments result in changes in the matter itself that are neither permanent nor arbitrary. For example, with increased heat a metal bar expands, but a return to the original temperature shrinks it and it returns to its original size. However, excessive changes in external energy can induce radical changes in matter. Heat the iron bar even further and it will glow red, then white and finally melt. Cooling will then result in metal of quite different shape and structure. Apply heat to ice at normal atmospheric pressure and it will melt to water at 0°C. After all the ice has melted the continued application of heat will eventually boil the water off at 100°C. Here cooling back to a solid state results in ice of different shape from the original. Matter can survive limited external energy changes, for these induce only limited internal ones which it can reverse, but beyond certain limits it is forced to change itself materially.

The above examples only refer to the application of heat to specific substances, but where chemical compounds are to be formed, the circumstances of the relevant reaction can be much more complex and rare, and hence less likely to be repeated. At the very least the constituents of any compound need to be brought together.

The above remarks also apply to living matter, but in this case there is an important difference. Whereas lifeless matter is restricted to the passive exchange of energy with its surroundings, living matter takes two additional steps to try to buffer itself against damaging change.

1. It initiates unpredictable small changes at random within its structures to prepare itself in all directions, so that it might be able to stay in harmony with slowly changing surroundings.
2. It has a cyclic existence that at some point in the cycle provides the power to roam and thereby seek out more suitable environments.

Living matter is so complex that it can indulge in minuscule changes without either being severely provoked initially or necessarily losing its basic identity, for such can appear small against its whole structure and bring about comparatively insignificant alterations to its nature. This is not like, say, sodium chloride (common salt) and other chemical compounds of simple formula, where any change to their molecules results in them becoming quite different materials.

The Balance of Energy

For matter to remain unchanged internal energy must relate to external energy. This state is necessary for stability, as violent differences induce drastic changes. With living matter the actual energy turnover is greatly enhanced in frequency and it takes in and releases energy to a much greater degree than is necessary simply to maintain its energy balance. This means that normally it always has a large bank of accessible energy present and ready to perform all the functions and respond to all the reasonable demands of living.

The energy balance of living matter must always be maintained, but the amount of energy involved can be very flexible as long as this balance is observed. Yet even though energy turnover may indeed be quite flexible, for maximum efficiency it must on the whole stay within reasonable limits, so that the standard state of the energy bank of the particular organism concerned is maintained. Life forms absorb energy; but in a healthy state they always retain an adequate accessible store of it within themselves. They use this store to transform other kinds of matter into their own specific organic requirements, i.e. within themselves and throughout themselves.

CHAPTER 3

Reproduction of Plants and Animals

■ The Forms of Reproduction

1. Fission of Cells

The normal form of reproduction found among cells, whether as individuals or forming tissues (mitosis), is similar to that of the protozoan amoeba. (Among bacteria and archaea the form of fission is simpler.) The fission commences in the nucleus, where the chromosomes replicate themselves. This initiates duality throughout the whole cell, which results in separation into twins that are theoretically identical to the original entity.

2. Fusion of Cells and Microbes

Sometimes cells fuse, but this can normally only be related to reproduction when it occurs with cells in the special reproductive organs. However, as an act of reproduction certain protozoa, under particular circumstances, indulge in a form of temporary fusion known as conjugation. As an example, the paramecium is an altogether more complex protozoan than the amoeba, with a greater variety of functions built into its structure, including the possession of two nuclei – a larger and a smaller. Normally a paramecium reproduces by fission in a similar way to an amoeba, but with conjugation activity takes place that has aspects that resemble sex.

The conjugation cycle has replaced simple fission as the immortalizing phase of the paramecium and this occurs when the animalcules find themselves in an extraordinary energy situation. Among multicellular organisms the sex cells are the only immortal form, while, with regard to these, the female ones may hasten growth and reach their ripe form by a process involving the preliminary fusion of several individuals with one another.

3.   Asexual Branching

In many of the simpler forms of multicellular animals reproduction takes place by sending out 'branches' that grow to be new individuals, except that they remain integral parts of the colony, sharing the food supply with each other by means of common tissue. This is true of the sponges and the more primitive coelenterates such as the polyps. The distinction between this form of reproduction and growth is not always clear.

4.   Asexual Budding

Sometimes organs called 'buds' form on animals and these can grow to become new individuals that eventually break free from the parent body to lead separate existences. This again is a feature of the coelenterates and it is by this method that the sexually mature medusa forms, resembling small jellyfish, are produced by polyps.

5.   Asexual Division

Some multicellular animals can reproduce by 'pulling' themselves apart into two or more pieces, followed by reconstitution as new wholes. This is found among sponges, coelenterates and flatworms. As might be expected, such animals also retain a high capacity for regeneration of damaged parts.

6.   Hermaphrodite Self-Fertilisation

In animals where sex cells are produced it is necessary for a male cell to fuse with a female one for sexual reproduction to occur. Female cells are usually comparatively large and passive, while their male counterparts are much smaller and motile; they must find and enter the females. Once a female cell has been fertilised other males are inhibited from entering. All multicellular animals have the means to reproduce by sexual means. With hermaphrodite self-fertilisation both male and female reproductive organs are contained in the same individual and reproduction normally occurs without physical contact with others of their kind.

7.   Hermaphrodite Cross-Fertilisation

This happens when each individual has both male and female organs, but they can only fertilise or can only be fertilised by other members of their species. This occurs in many phyla, but not among the vertebrates.

8.   Heterosexual Cross-Fertilisation

Here each individual has male or female reproductive organs, but not both. It is widespread among the phyla and normally is a universal characteristic in the case of the vertebrates

■ Comments on the Above Eight Forms of Reproduction

It is to be noted that both plants and animals use all of the above methods. However, with plants it is more often hard to distinguish what is meant by 'individual' even among the highest sexual forms. Yet the implication here is that sex itself arose in a common ancestor of plants and animals, but that animals represent a group that primitively broke away from direct reliance on photosynthesis. At various times this was also achieved by other groups, including some bacteria and the fungi, and ultimately by viruses.

Among animals the groups can be arranged as follows:

Unicellular Animals	1	2					7	
Multicellular Animals			3	4	5	6	7	8
Asexual Methods	1	2	3	4	5			
Sexual Methods		2				6	7	8

The term 'asexual' does not only suggest that no differentiated sex cells take part in reproduction, but also implies that out of one individual, alone and unaided, there come forth others. It is hard to determine what specific conditions arise within a cell or tissue to initiate this, but it does seem probable that some form of extension on the process of growth is involved.

When a cell or asexual organism reaches its full size further normal growth becomes undesirable, if not impossible, so, as already mentioned above, something has to be done to maintain the energy turnover at this point and asexual reproduction can be a means to this end. In this way destructive ageing of member organisms of a species is forestalled.

Unicellular sex occurs among paramecia, but the simple ability to divide is also retained; division is indeed an immediate consequence of the 'sexual' act they perform. In this context the more elaborate reproductive method of paramecia is 'sexual' in that it involves elements that go (male) and those that stay (female). Paramecia can indeed be said to utilise hermaphrodite cross-fertilisation (type 7 above) at the unicellular level.

From all this it would seem that ageing and death are not inevitable in the animal world – nor for that matter in the plant world either – if only asexual reproduction be used. Yet not only mice and men, but also elephants, whales, sharks, crocodiles, turtles, squids, etc. (and even mighty oak trees) are doomed as individuals through using sexual methods exclusively. Why? The harsh reality of the case must be that life and evolution do not work for the preservation of the individual as such, but only insofar as this assists in the perpetuation of the type. This would seem to suggest, that as far as everlasting life on Earth is concerned, the basic unit is the species and that individuals are the expendable component parts of such, just as cells are expendable in the individual organism. However this does not imply that individual members of a species can be immutable. The governing force behind the reproduction methods is evolution. It always has been and still is an evolutionary advantage to have sexual differentiation. A greater sensitivity towards evolution than that of animals reproducing asexually is necessary, not only to be able to survive in a more changeable habitat, but also greatly to improve the ability to break into previously unconquered environments. However, rapid evolutionary changes can be achieved by some single cell organisms by means of mutation. This particularly applies to viruses and makes them so dangerous.

It would seem that the present limit on the evolution of reproductive systems themselves has been reached with heterosexual fertilisation and that all other forms are more primitive. Yet how did sex originate in the first place? It is found with all its methods in both the animal and plant worlds, yet the only apparent bridge between these two great streams of life is to be found at the unicellular level among the green flagellates. This appears to imply that sex itself originated at unicellular level, which is indeed suggested by the example of the paramecium. This animal could not of course have constituted the common sexual ancestor of the multicellular animals and plants (for such a creature has presumably long ago itself evolved to something quite different and perhaps even to extinction, or at least to being unrecognizable). Yet, even so, the paramecium might serve as a living example of the type on the animal side of life, but after some evolution specific to its kind having occurred.

How could the sexual system evolve from supposed unicellular beginnings and start to control the reproduction of multicellular animals, culminating in complex creatures such as the elephant and man? Firstly there is a requirement to assume that sexual reproduction always must follow in the wake of the asexual, both at the unicellular and multicellular stages. Let reproduction at the unicellular level first be considered in further detail.

When paramecia conjugate, they do so with their own mating types. In other words, if there are types A, B, C and D present, an A will always mate with a B and a C with a D, but an A will never mate with another A, and so on. This may not in itself be a demonstrably sexual differentiation, but it leads to the possibility of the eventual formation of such, for evolutionary trends may favour the A-type to specialise in the direction of maleness and the B-type likewise towards femaleness. Of course this would mean that the A-types ultimately no longer reproduced themselves and the post-conjugation fission of B-types would then have to produce both A- and B-types. If dense populations of similar microbes are considered, then the real difference from a tissue is that they are not in permanent and direct contact with each other. Consider theoretically then an increase in density of such a population until they are thus in contact. This would resemble a tissue where most of the member cells were reproducing by means of ordinary fission, but perhaps under special circumstances using a form of conjugation similar to that described above. It has been shown with paramecia that if they pass a certain age without conjugation, individuals lose the power and such a colony degenerates and dies. It apparently follows that the acquisition of immortality of the species by way of conjugation has somehow deprived it of the same as generally achieved among microbes by way of simple fission. A colony of paramecia must use conjugation as part of its reproductive activity or die out.

Imagine the hypothetical proto-tissue at the stage where it had to employ a form of conjugation or be doomed. In tissues the specialisation of cells or the permanent gathering of cells into groups were inevitable trends, and in our example it can be assumed that conjugation became selectively the specialized function of some groups. At first they would be hermaphrodite cells exchanging male and female elements, but the evolutionary pressure to specialize would eventually lead to separate male and female functions (as suggested with the A- and B-types above). Thus might one organ only produce male conjugating cells and another only female ones. Thereby has been reached the stage that can be referred to as hermaphrodite self-fertilisation.

Progress from self- to cross-fertilisation was the result of further evolutionary pressure. Once all animal life was found in water and the renewal of multicellular ones by cellular conjugation therein could better accomplish the spread of sessile species if it took place externally, with the distribution of the primitive young thus formed being allowed by means of currents. If the male and female cells could be shed into the water first, then each resulting combination or 'larva' might drift some distance before settling on the bottom. Eventually such could even evolve swimming mechanisms to further the ability to wander.

However, when male and female cells were shed by members of sufficiently compact colonies of aquatic animals, each stood a good chance of conjugation (or fusion) with a partner emanating from a different adult to its own parent. The evolutionary advantage of this would encourage a tendency for this sort of union to be preferred and eventually mandatory. Hermaphrodite cross-fertilisation would thus first become established and then consolidated.

It is more difficult to give a good evolutionary reason for the development of heterosexual reproduction. While it is true to say that the most 'advanced' creatures, such as the vertebrates and the arthropods, use it exclusively, it is also present in such 'primitive' phyla as the coelenterates, sponges, roundworms and rotifers.

In the case of the vertebrates division into males and females has led to distinct advantages, one being a corresponding division of activities and, eventually, duties. Vertebrate males and females tend both to look and act differently from each other. Such differentiation hardly applies to 'lower' forms like jellyfish. The disadvantage of halving in this way the breeding potential of species at this level of development must have been offset by some advantage directly connected with reproduction. Such sexual differentiation perhaps owes its origin to the fact that primitive animals tend to shed their eggs and sperms separately into the water. An advanced means to prevent self-fertilisation was by segregation into individual males and females, for cross-fertilisation is a distinct advantage where evolutionary progress is concerned.

Where quite evolved terrestrial species preserve hermaphrodite cross-fertilisation, while heterosexual aquatic species from the same phyla also occur, the observation can be made that the former went ashore at a time when hermaphrodite cross-fertilisation was widespread. Since once ashore the shedding of eggs and sperms to find each other, would no longer be feasible, direct contact with each other must have been a means already used before the conquest of land could be embarked upon. Yet it was those that remained aquatic gamete shedders that came under greater pressure to become heterosexual in order to obviate self-fertilisation. The annelids illustrate this with the hermaphrodite earthworms and heterosexual polychaetes, as do also the hermaphrodite slugs and snails on land, which compare with the general heterosexual nature of aquatic molluscs. This all suggests an earlier phase of the invasion of land by hermaphrodite invertebrates, which crawled legless ashore.

Later on, in the water, other heterosexual creatures developed whose females also came to be fertilised by direct implantation by males and some of these subsequently became ready to leave the water as a result of acquiring this facility. They all had appendages of some sort that were capable of evolving into legs. Firstly this involved certain arthropods, the group that culminated in the insects, followed eventually and much later by vertebrates. Some vertebrates have attempted life on the shoreline without evolving the necessary implanting organs; they remained aquatic breeders and are represented by the modern amphibians.

A compounding influence in all this was the incidence of parasitism, both internal and external, for the rules governing such creatures were clearly somewhat different.

■ Parthenogenesis

This is "virgin birth" and features in several phyla, including the arthropods. It generally occurs where heterosexual reproduction is the rule and is hence a repression of the male sex. A typical and prominent example is found with the green fly, an aphid. Females usually produce nothing but female young from fertile cells without mating having taken place. It is only at the end of summer that any males emerge from eggs produced, have their fling with the females and soon perish.

A somewhat similar situation occurs with social insects such as ants and bees, except that in such cases all the breeding potential of the specific coherent social group is concentrated in one special female – the queen. All the other females of the colony, while technically sisters, are actually sexless slaves of the reproductive unit.

What advantage could have led to parthenogenesis? With aphids loss of the advantages of evolutionary flexibility must have been outweighed by the need to produce great numbers quickly. To some extent this must also apply to the social insects, but, in addition here, the concentration of reproduction in the super-female has freed the energies of the huge multitude of sterile females for specialisation in other directions, i.e. workers and warriors.

■ The Reason for Reproduction

While the apparent aspects of reproduction have been discussed above, no attempt has really been made yet to get to the root of the problem. Why should it be a necessary function of all living matter? However humble it may be, every unit of life is ultimately the result of some form of reproduction. Hence, where reproduction is absent, life can also be deemed to be absent.

The apparent function of reproduction is to keep life going. The method is to replace old parts or units with new versions of themselves before they are completely worn out. But in view of the above claim that reproduction is integral with life and has been its companion from the very outset, can this view be upheld?

As will be discussed below, it can be claimed that life began as a response to cycles already existing on the lifeless Earth, in particular the day-night one. There was fusion during the day under the influence of the Sun's rays and fission at night when this source of energy was absent. There was a rhythm about the process, with this at first being in time with that of the Earth, but so much energy was gathered and stored by such matter exhibiting 'infra-life' (as we might call it) that it was able to have an effect on additional extraneous matter, and extend the period beyond the day-night swing. Yet the energy supply available was still always above the requirements of any such body of molecules, which would become over-energised and forced to break itself up. The subsequent state reached, here dubbed 'proto-life', shared with infra-life the trait of always being impelled to find and utilize material to form new combinations for its ingredients to absorb the excess energy it was continually building up and thus to save itself from destruction by way of resultant disintegration.

At a primitive stage the period of the rhythm would become various and may even have been shortened from the day-night alternation. Each distinct unit would have a limit as to how much energy it could absorb before it must split itself in a controlled way to relieve the excess. Before such behaviour could be achieved, destructive fission would usually be put off till the last moment by growth. When disintegration finally took place as the limit was reached, the energy released would escape into the environment and no immortal chain set up. This latter would come into existence only when the energy released by destruction was sufficient to allow the same material to reform itself immediately into two or more new – but lesser – wholes. Later development would enable the matter to take the precaution of specifically assembling itself essentially into two new wholes before final division occurred; this is a primitive and simple version of what happens during the fission process of a single cell. This unlikely procedure happened because it gave an advantage when it came to survival. It was evolution in action. Indeed, if it had not happened reproduction would have become a stalled process and progress towards life in its higher forms would have ceased. By pure chance some matter always behaved in a way that resulted in survival advantage. This was incipient evolution.

Immortality would seem to be the prerogative of the primitive unit cell due to its capacity for binary fission. But it has already been noted that more advanced units, such as the paramecium and the individual cells of plants and animals, are not immortal by themselves and some form of sexual process must be employed to ensure the continuity of life. The limitation on the lifetime of non-reproductive cells of a multicellular animal (and indeed of a plant) seems essentially to be set by a limitation on the size attainable by the species concerned, but perhaps also by other considerations. Mortality among paramecia, for example, may be induced by adverse conditions such as harmful chemicals in the water, wrong temperature, or volumetric restriction of the colony, through these causing the inhibition of the rejuvenating form of reproduction.

At its very start life had to escape from its primitive regular rhythm. Should one accept this, then it seems probable that the first life really consisted of repeated acts of reproduction and that further embellishments that led to life as we understand it were merely evolutionary developments to maintain any unstable combinations that had been reached between acts of reproduction. Life is hence the means originally created by evolution to span the rhythmic periods of the matter involved. In this sense life is the horse placed between the shafts in order continually to pull the cart of reproduction. However, in the course of time the horse itself seems to have grown into such a fine animal in the case of many species that the cart now spends much of its time in the shed, awaiting special occasions, while the horse is usually taken out on its own for the pleasure of the ride. *Joie de vivre*? Looking at it another way, it may be said that the cart of 'reproduction' was created first, but was unable to go anywhere until its driver 'evolution' harnessed the horse 'life' to it. Of course this is the complete reverse of what we learn from our senses, our culture and our upbringing. Based on these experiences our intelligence tells us that life came first. As discussed above, reasoning and imagination can argue the opposite, i.e. that reproduction came first and life was merely a build-up of activity that enabled matter to preserve itself in its specific form between such events.

So, cyclic conditions in the external world initiated reactions among infra-life forms. These repetitions were the forerunners of reproduction. Life merely arose as a recognizable reality due to the coordination of survival activities that developed during the intervals required to accommodate the ever growing nature of such necessities.

CHAPTER 4

Heredity and Reproduction

■ Inheritance of Characters

Throughout many millions of years living matter has existed, reproduced and evolved on the planet called Earth. Evolutionary changes have been passed on continuously from one generation to the next and this system of inheritance of characters is called heredity. Along with the act of reproduction, heredity is one of the major tools of evolution, for not only can it transmit characters *apparently* lost in the past, but also those that seem quite new and spontaneous can be fed into the loom of heredity, to get woven into the ever-lengthening cloth of evolutionary development. Lost characters can re-emerge to be used again, because they can linger on in the depths of the genetic pool (apparent evolution - say), but brand new characters that can crop up represent unprecedented variations in the genetic structure and are known as mutations (real evolution - say).

■ The Advantage of Sex

Sexual forms of reproduction provide a great advantage over the asexual in that heritable characters form a common pool, which may be dipped into in the course of time by all the breeding members of a species. In other words the lines of inheritance criss-cross down through the generations and a proportion of the characters thus secreted can provide considerable steps forward by not being eliminated, being selected together as just right for contemporary conditions. On the other hand, quite unfortunate combinations can also be created, but these tend to be eradicated from a species quite quickly by natural selection.

Where reproduction is only by asexual fission, no future genetic crossing of paths can occur among the descendants of the individuals concerned and any evolution can hence only take place through individual mutation. Yet one large evolutionary step might seem to have been taken by asexual living matter; sexual beings have indeed somehow evolved from the sexless. On later pages this important development will be dealt with in greater detail.

◼ Genetics

As discussed above real evolution could not occur without mutation, for otherwise offspring would simply go on forming exact replicas of their parents' genetic structures down through time. Yet mutation itself can be real or apparent, and the two forms proceed together along the path of evolutionary change. The real introduces brand new characters, while the apparent is the re-emergence of lost characters from the reservoir of the past; any of these might prove useful in a favourable phase of conditions and such occurrence is the basis for the procedure known as genetics. In genetically simple organisms it is possible to predict the appearance or non-appearance of recurrent characters from the past and also the proportion of any generation they may represent. It was the Austrian monk Gregor Mendel who first performed breeding experiments with plants to demonstrate this and thereby the science now known as genetics was incidentally founded.

Mendel worked with pea plants, a species of which may, for example, exist in two forms – a tall and a short. Peas are sexual and if one starts off with a tall plant (T) and a short plant (s) of the same species, then the first generation (F1) will all be tall. If peas from F1 are then mated, the second generation (F2) will average out 75% tall and 25% short. In such species, which fall into two clearly defined forms, the characters concerned can be called 'either-or'.

In sexual organisms the genes that govern characters occur in pairs that are located identically on the corresponding chromosomes. Such gene pairs are called alleles. In cases where the alleles are the same (e.g. TT or ss) the form is known as homozygote, but when differing (e.g. Ts) heterozygote is used.

The descent of the peas occurs in the following manner.

Alleles	Genotypes
TT – ss	(homozygotes)
Ts Ts Ts Ts	F1 (heterozygotes)

F1 are all tall since, if T is present at all, it takes precedence over s, a condition known as dominance. T is a dominant gene and s recessive. Hence s can only influence the height of the peas when T is absent, which has only occurred above with the homozygote ss.

Alleles	Genotypes
Ts – Ts	(heterozygotes)
TT Ts Ts ss	F2 (mixed)

Thus, because T is dominant, in the second generation F2 75% are tall and 25% short.

◼ More Complex Patterns

The above system is relatively simple and only applies where a specific character is controlled by a small number of genes that are strongly linked together. If a character is determined by a greater number of more easily dispersed genes, which is frequently the case with more complex organisms, then the patterns are much more confused than those illustrated above and the actual descent more difficult to demonstrate experimentally, while tabulation becomes so much more involved. Such characters as these can be described as 'more-or-less' and, as a generalisation, the height of human beings comes into this category. Apart from the difference in average height between males and females (which has a different cause), human beings do not exist in well-defined size groups, but in graduations between rather vague upper and lower limits, the greatest numbers being around the mid-point of height measurement.

The Division of Cells

Single living cells, whether independent or parts of larger organisms, reproduce by division. Normal self-copying of cells is called mitosis and simply stated it involves the longitudinal parting of the chromosomes themselves followed by replacement of the lost companion strands by copying so that a full complement will be retained in each resultant daughter after cellular division takes place. In non-sexual animals the chromosomes are single, all being different from each other. This condition is called haploid. With sexual animals the chromosomes are duplicated (i.e. diploid), but they too use mitosis to reproduce their cells. However, when sexual animals produce their own reproduction cells they do so by a process known as meiosis, in which cells divide without the chromosomes present being split. Any sex cell thus formed by selecting one undivided chromosome from each relevant pair is then in a haploid state.

Male or Female?

In general sexual animals have a mechanism that, at the very outset, decides the sex of each of the young. Among the paired chromosomes one pair governs this choice. Diploid paired chromosomes are normally identical, but with that pair that specifically governs sex there is a marked difference between those of the male and those of the female. It is conventional to refer to the paired sex-governing chromosomes of the female as XX, with the male pair being XY. After meiosis there is a random chance of embryos still being XX and XY in equal numbers, but with shuffled constituents.

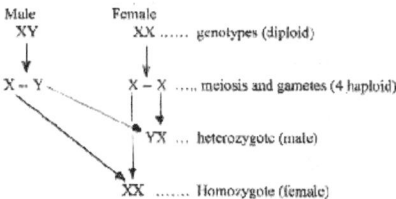

Maleness (Y) in the zygote acts as though it were controlled by a dominant gene, while femaleness (X) is recessive. If Y is present at all the result is male, while XX is necessary for a female. The combination YY is not naturally possible due to the very nature of sexual reproduction. The new zygotes differ from the originals in that inherent genetic material has been transferred. Occasionally human beings are born with a faulty genetic structure, such as with an additional sex chromosome to give XXY, with the result being a person of mixed sexual characteristics who is sterile.

■ Sex-Linked Characters

The Y chromosome in humans is smaller than the X and certain characters can appear more often in males because of this. Colour blindness is caused by a recessive gene (r) and one would expect it to occur equally often in males and females, i.e. when the genotype is DD. However, although this expectation is valid for females where the sex chromosomes are XX, the pertinent allele is missing from the Y side of the male chromosome.

Male	Chromosomes	XY	XY	XY	XY
	Genes (alleles)	D-	D-	r-	r-
	Colour blind	No	No	Yes	Yes
Female	Chromosomes	XX	XX	XX	XX
	Genes (alleles)	DD	Dr	rD	rr.

	Colour blind	No	No	No	Yes

Like most recessive features, colour blindness is a defect that natural selection tries to eradicate. It is shielded from this by the very nature of Mendelian succession. Apart from their furtive nature, the persistence of such unwanted traits is also assisted in the case of modern man by the sheltered nature of his existence. In actuality the incidence of colour blindness indicates a more complex situation than that shown above.

■ Secondary Sexual Characters

These are those features that are characteristic of each sex, but play no direct part in sexual reproduction. There are some startling examples among animals, especially the birds, and in man the different amount of facial hair and difference in stature between men and women can be indicated. The male features of hairiness, largeness, etc., appear as though due to dominant genes found upon the XY chromosome.

Male	Chromosomes	XY
	Genes	rD
	Description	Bigger, hairier, etc. (due to D)
Female	Chromosomes	XX
	Genes	rr
	Description	Smaller, less hairy, etc. (due to r)

There is of course a big overlap in human sizes between men and women and a slight one as regards facial hair. These features are not overall distinctions, but apply to averages, for the average male will always be bigger and hairier about the face than the average female. Secondary sexual characters have a psychological basis and because of this are not restricted to physical features; differences of attitude, manner and behaviour between the sexes can also come under the same heading. In this light the argument can ultimately be extended to include differences in apparel between men and women as the result of heritable behaviour.

Primary Sexual Characters

These are the organs concerned directly with sexual reproduction and they must be selected in the same way as those characters described in the section above. The genes covering both primary and secondary characters must be tightly bonded to their chromosome, so that there is little chance during division of mixing, as can be seen to happen with many other genes. Even these primary characters are to some extent 'secondary' in that certain primitive sexual animals have no specific organs for implanting or receiving male cells.

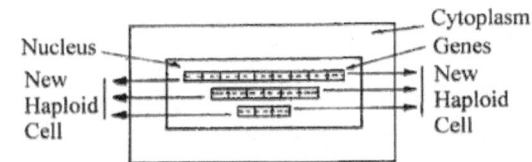

A haploid cell showing 3 chromosomes preparing for mitosis

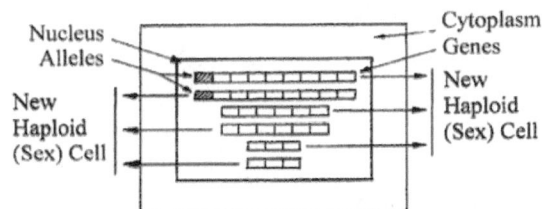

A diploid cell showing 6 chromosomes preparing for meiosis

**Diagrammatic and Simplified
Representation of Cells**

CHAPTER 5

Evolution

■ The Inevitability of Change

All matter is subject to change even if the nature and rate of change may vary depending upon composition and circumstances. On Earth changes are induced by the very composition of the planet itself, with its molten interior, solid crust and surrounding gaseous spheres. Also important is its position relative to the other members of the Solar System, in particular the vital central star called the Sun. Also relevant is Earth's satellite, the Moon.

■ Sensitivity to Change

The sensitivity of lifeless matter towards change is predictable and relatively simple. The converse is true of the living, and the ultimate fineness of the accompanying sensitivity is roughly proportional to the degree of complexity of the life unit. Fine and active reaction to changing conditions is an intrinsic part of life itself. Living matter has a much more complicated physical and chemical structure and a far greater diversity of outlets for its inherent energies than does the lifeless. The two essential channels into which these energies are released by any unit or individual grouping of life forms are self-preservation and species perpetuation.

In order to achieve these twin ends, the peril to be overcome by the earliest forms of life was the hostility of the surrounding lifeless world, with its possible excesses of heat and cold, drought, starvation and other environmental sources of danger and death. In times of shortage of space and nutrients, living units have also always been liable to come into direct competition with others whose needs were very similar, not the least being their own kind, the so-called competitive exclusion principle. In such adverse circumstances it would normally be the best fitted to meet the challenge that would survive. Finally the state would arise wherein danger existed from the active hostility of other forms of life (predation and parasitism) and life units would be competing against each other to survive this threat too, on top of those imposed directly by their fellows and the lifeless world.

■ Change under Control

For living matter, structural change - except for that being its own characteristic - is liable to be damaging and in the long run fatal. Yet activities of living matter, such as feeding and breeding, allow a continuous and characteristic amount of change to occur within itself that the unit endeavours to keep within desirable limits, while at the same time resisting all deforming change that outside influences or internal foreign bodies try to impose on it. From this it may be argued that the *raison d'être* of living matter is to resist change and that all its energies are devoted to this end. Yet change has occurred among life forms to an enormous degree, for are there not living in the world today life-units consisting of little more than enlarged molecules and at the same time others comprising millions of differentiated cells, while simply bristling with various organs and sensitive functions and representing the latest models within the scope of living matter?

Although it was asserted above that living matter has always resisted change that was of external origin and imposition, this in itself was accomplished by producing controlled changes from the inside that were not of such a violent nature and some of which eventually enabled life to survive. These changes meant the creation of new features to perform additional functions that became imperative; but to what extent they involved improvisation of an existing organ is at this stage not clear. Or the change might simply have entailed an alteration in the existing pattern of activity. Such changes must not be thought of as having taken place in individual animals, but in species over long periods of time and the main agent for their perpetuation was – and still is – the reproductive system. Such change is known as evolution.

Lasting physical changes in species, i.e. those with evolutionary causes, have all been initiated among the smallest items of matter recognisable as being associated with life. These are the elements that make up DNA and which govern the nature of the genes within which it is housed. Structural evolution can be seen as one of the two forms in which it occurs, the other being behavioural. In the beginning the latter could not take place without the former having occurred. Eventually nervous systems developed among some animals and these made it possible for changes in behaviour to evolve without causative physical change, except among the nerves themselves.

■ Natural Selection

There are two possibilities in which evolution may be thought to occur.

1. An organism can alter in some way during its lifetime and pass the acquired change on to its offspring.
2. An organism can give rise to offspring that differ to a degree from itself and this change can then be perpetuated in their successors.

Of these two methods the first has long been out of favour as against the second. Charles Darwin is the man who is accredited with having made sure of that.

Evolution by natural selection works by a tendency to retain changes that are advantageous and to eliminate those that are not. The organisms that survive are liable to transmit to their offspring the favourable traits thus acquired, while in the same way faults tend to be bred out through them interfering with reproduction itself, either by damaging its efficiency or preventing it completely. In the end it all comes down to the survival of the fittest to reproduce.

Sexual Selection

This is a special and important projection of natural selection. With those animals that reproduce by sexual methods, selection often occurs through the acceptance or rejection of a member of one sex by a member of the other. This may depend on physical features, scents, patterns of behaviour and so on, choice being governed by the comparative strength or the presence or absence of such qualities. In any one instance any rejected suitor obviously has lost the chance of passing on his (or her) characters to the next generation. This deliberate form of selection tends to concentrate reproduction among a reduced number of chosen ones and can increase the rate of evolution, as well as sometimes leading to quite bizarre forms.

The Aim of Evolution

An abstract concept cannot have any aim that the human mind can recognise. However, should one arbitrarily ascribed one to evolution, then it can hardly be change itself, but rather the preservation of the identity and intactness of any particular kind of organism by the provision of means by which suitable change can take place. Such change as does occur arises out of necessity in order to accomplish this. Evolution selects from the innumerable changes that occur randomly within a species in order to combat and possibly overcome diverse hostile forces that are always at large or may arise and which threaten to injure or extinguish it. It is always good to bear in mind the earlier proposition that the unit of life that is destined to be protected and preserved by evolution is not the individual but the species. Hence, if evolution does indeed have any aim at all, it is to facilitate survival of each and every species involved in the business of living and reproducing.

However, the fact that a species exists in the form of numerous member units is important to its survival for various reasons. There is certainly safety in numbers and a species can usually afford to lose many of its individuals in the course of the struggle against its environment without becoming extinct itself. Nevertheless, a reduction to too few in number is always a danger sign, as evidenced by many species in quite recent times, which through excessive reduction in numbers have reached the end of the line or are now in danger of doing so.

■ Paths of Evolution

This phrase is meant to cover the routes taken by single characters of species down through the ages. Owing to the continuous nature of life these paths, except where leading to extinction, all stretch right from the very beginning of life to the present day, but with very frequent forks along the way. This is not to say that any feature of a modern animal is recognisably the same in appearance, structure or function as one on a remote predecessor of, say, 300 million years ago. Recognisable steps of evolution will always be of much shorter duration. One would, for example, look in vain for horns on the earliest ancestors of any of the ungulates; the earliest forms of these adornments were simply modifications of the skin and bones of the head.

■ The Crossroads of Evolution

The 'paths' described above are the routes taken by the characters possessed by all types of living matter from the birth of terrestrial life to the present. Any such path is crossed at countless places by lateral tracks that represent the lifetime of individual creatures from conception to adulthood and death. Thus any character has not only evolved during the existence of its ever-changing specific type, but has also changed as the lifetime of each individual has unfolded, i.e. involving growth and ageing. The important point to remark here is that any character has not just been subject to evolution of its finished form on what is regarded as the adult creature, but evolution has also been occurring at every stage of the character's development through the lifetime of each individual.

◼ Evolution and the Adult

Adulthood is best defined as the attainment of sexual maturity, the ability to reproduce. On an earlier page it was claimed that preservation of the species is the main aim of evolution, with the perpetuation of reproduction itself - rather than life – being the ultimate objective. In this sense the species, rather than the individual being, can be regarded as the reproductive unit.

The protozoan adult is mainly distinguishable from its young by its greater size, but with multicellular animals there is a vast difference between the adult and the seed from which it sprang. Here adulthood is only reached by passing through a variety of intermediate stages. Complexity of form (phenotype) is not necessary for reproduction, but only to facilitate the survival of any individual of a species evolved to such a high level until sexual maturity is attained and reproduction can take place. If circumstances change in the course of time and make it unnecessary for the existing adult form to be reached, natural selection will ensure that the species can begin to reproduce at an ever earlier stage and thus incidentally (by a selective process of elimination) evolve its own adult form to a state of superfluity and extinction. Later on – or even during the same time – further positive evolution may occur, but it is impossible for the new selective pressure to lead back exactly to the old adult form again, even though faint clues as to its former identity may linger on in the resulting animal.

CHAPTER 6

Some Aspects of Evolution

■ The Rate of Evolution

Our method of using fixed names to label the various animals in existence induces unwittingly the attitude that their forms are also fixed and that, though they may have greatly changed in the past, they are now quite steady. Living beings are always subject to evolution due to the ever-presence of its agent natural selection, even though the rate of change may have varied considerably at different times during Earth's history. The main quickening forces for evolution, as it sails along on the ocean of time, are geological, geographical and climatic changes, for these tend to change the course - while also rocking the boat - of even the best-adjusted species. Though many may survive, others can fall overboard and be lost.

■ Divergent Evolution

Evolution is indeed normally thought of as working on living matter in a divergent manner. The observation has already been made that there is a tendency for living beings, in the course of time, to diverge from their fellows and their ancestors owing to the appearance of mutations. The rapid divergence that could well ensue is controlled by the existence of the interbreeding community that is provided by the species. The mutated genes are in this way kept within the common pool of heritable material and similarity of members preserved. Yet the chance always remains that any advantageous change can be spread throughout the whole species and this process is hastened by natural selection.

So, in the past species have diverged and broken up to form numerous new species. How did this come about? The most important factor at work has been geographical isolation. Any species of animal might at one time have a large and continuous range, but at last the formation of a physical barrier, such as a desert, might split the species, with no further interbreeding being possible between the two groups thus formed. Differing conditions on either side of this barrier could then lead to divergent evolution. If after a great deal of time the desert became fertile again the two groups might then mingle once more, but by this time it might be virtually impossible for them to interbreed for a variety of reasons. They might no longer stimulate each other sexually owing to changes in appearance or behaviour. They may no longer be able to recognise each other's signals. Even if in some instances they actually succeeded in mating, their offspring might prove to be sub-fertile or infertile, or the two genetic structures might have become so incompatible that the production of young at all could be highly unlikely or impossible.

This state of affairs would drive a wedge into a species and, as time passed by, evolution would normally serve only to widen this rift in order that both of the new subspecies might survive. The closer that such still resembled each other in structure and function, the greater the competition for the available space and food. If too close and they continued to share the same area, one would eventually gain an advantage and pose a threat to the very survival of the other. The only chance to avoid the risk of elimination of the one in such coexisting circumstances is for both to diverge sufficiently to be able to occupy two different ecological niches, whereby they would cut down on the areas where they competed directly for the necessities of life.

◼ Convergent Evolution

In spite of the above observations, some species do seem to have moved closer together. This can indeed involve members of different genera and even different families. Take, for example, the ungulates (or hoofed animals), which can be largely divided into two groups – the odd-toed and the even-toed.

The zebra and the white rhinoceros are examples of the odd-toed ungulates, while the wildebeest and hippopotamus are even-toed. Although the toe arrangement reflects a true evolutionary relationship, it might be thought that a zebra and a wildebeest, although not very alike, at least resemble each other more than do their mentioned closer relatives. In truth, all grass eating herding species tend to have a superficial resemblance to each other, due to their way of life; but this is superficial and restricted to the general outline of individuals. In spite of the close association of these two species, wildebeests (or gnus) remain antelopes without trace of horse, while zebras remain horses without trace of antelope. Any resemblance between white rhino and hippo is even more coincidental, both being bulky and with skin largely lacking pattern and hair. These features happen to be independently suitable for both, despite their different ways of life.

Animals are in fact subject to both convergent and divergent pressures. Between species the genetic structure tends to drive them ever further apart, but habitat and general way of life may tend to hold them together; they are indeed held in some sort of balance between these two influences. They must not diverge too far from the ideal standard so as to be unsuitable for the habitat obtaining, but neither should they converge too much while occupying it so that excessive interspecific competition takes place. This statement applies to every form of life.

◼ The Limits of Convergence

Although many similarities between otherwise dissimilar species must certainly be due to convergent evolution, there is a great danger for this term to be overworked. The possibility has to be considered that in some cases similarities may be attributed to convergence simply because the actual relationships are hidden from us. The hair on some insects and that on mammals would seem to be a good example of convergence, for it is an adaptation for living in air, while their ancestors clearly left their original watery habitat quite independently.

A good test case for convergent evolution is the remarkable resemblance in structure between the eyes of cephalopods, which are molluscs, and those found on many vertebrates. This is all the more remarkable when one considers that no other branch of the molluscs, such as the slugs, snails, bivalves, etc., have eyes remotely resembling those of the squids, octopuses and cuttlefish. So is there any other way in which this surprising structural resemblance between creatures in different phyla can be explained, except by convergent evolution?

The grouping of animals into different phyla does not necessarily indicate that these major divisions are equally remote from each other, for this must be regarded as a wrong impression created by the classification system itself and be guarded against. That the Annelids are closer to the Arthropods than they are to the Molluscs seems obvious. It must also be borne in mind that primitive and rare intermediate forms that have become extinct may not occur in the geological record. Such information could perhaps have provided knowledge of lost links between the phyla.

The eyes are not the only features that the cephalopods have that provide resemblances with the vertebrates and that are generally lacking (or are less comparative) in the case of other molluscs. They also:

1. Grow internal bony or cartilaginous structures;
2. Are active in the water, being able to swim freely;
3. Nurse their eggs and young;
4. Possess teeth (the radula);
5. Are provided with a compact central nervous system (brain) in a braincase (of cartilage) and hence have a relatively elevated level of intelligence;
6. May have light producing organs in the skin (bioluminescence);
7. Have a circulatory system with properly lined vessels.

Some of these features are possessed to some degree by other groups of the molluscs, but on the whole it would seem that in certain aspects of structure and function Cephalopods resemble Vertebrates far more closely than they do their own supposedly much nearer relatives. Convergent evolution may have contributed to these apparent coincidences, but to attribute every resemblance to this cause seems to be stretching its possibilities too far. This poses the question as to whether there can possibly be an area of contact between two such phyla of which we have no knowledge. The finding of a possible solution of this problem will be endeavoured on later pages.

■ The Role of the Individual

In species that are sexual the cross-fertilisation that occurs gives a greatly enhanced diversity of material for evolution to draw upon. Yet individuals are specifically important because they house the reproductive functions that keep the species going. The more of them there are the more diversity there is likely to be within that species and hence the more material available for the purpose of evolutionary change at a time of need.

It is the continual struggle of living matter against its total environment that causes the single life unit to be subservient to the requirements of the species, just as in human affairs the needs of the single citizen are ultimately subservient to those of the state. At no time is this more apparent than when his country is under threat, especially of warfare.

■ Combination of Assets

The frequent association of zebras and wildebeests occurs too often to be due to mere chance. Can there be advantages in this arrangement, especially when they seem to be competing for the available grass? Well it is not the same grass because the one species likes it longer and actually prepares it for the other, which prefers it shorter. Another advantage may be better protection against predators. It is certain that two such different species will have the senses developed at rather different levels. If the species with better smell gets wind of a predator, its uneasiness will forewarn the other. The same will apply with sight and hearing. Any mixed herds, or lesser groups, thus have advantages over unmixed ones. A further advantage for the zebras is that they almost invariably form a small minority when accompanying wildebeests, thus reducing the chance of being selected by impartial predators. This statement implies that zebras can never hope to compete in numbers with wildebeests, yet, nonetheless, the zebra is the only kind of wild horse to survive in numbers in the wild in Africa and indeed elsewhere.

Safety is to be found in numbers. The greater the herd the safer it is for the individual. On a world-wide scale the already sparse population of odd-toed ungulates is still diminishing, mainly due to the comparatively recent human pressure on wildlife as a whole, but also because they have long been losing the survival struggle with their even-toed competitors. (It might almost seem that the zebras, unable to beat the wildebeests, decided to join them. The reality of course is an illustration of an evolutionary principle that if when under pressure any available escape route is not taken, then doom will loom.)

■ Escapism

Where a species is losing the struggle for survival against its competitors, it may escape elimination by retiring to a harsher habitat and adapting for it, a place where its competitors are loath to follow. Once the necessary transition has been made, the new abode might even become increasingly advantageous. Once established in the harsher environment any worsening conditions, if not fatal in themselves, would strengthen the position of the adapting species by making it more secure from competition.

The camels are related to the llamas, guanacos and vicunas of South America, but are more obviously adapted to desert conditions. The camel tribe originated in North America, but eventually failed there totally, presumably due to a prolonged loss of a suitable habitat in the past and failure to compete under the changed conditions with better adapted animals, such as ungulates. Camelids survive in restricted areas of the Old and New Worlds, places where ungulates that are successful elsewhere, such as wild cattle, sheep, goats, antelopes and deer are not found. Wild sheep and goats are mountain dwellers and one might well surmise that their ancestors were forced into these inhospitable regions for similar reasons, even though they can be considered as successful species in their chosen habitats

Another form of geographical escapism is the worsening of habitat due to climate change. This works in conjunction with interspecific competition. A big question arises as to what spread of circumstances could have driven established but diverse land animals like the ancestors of the turtles, whales, dugongs, seals, otters, etc. back into the water.

■ The Evolutionary Spearhead

When a type is escaping into a new habitat, unless it stumbles into an area that actually eases the pressure, it is to be expected that the most evolved forms will be created at the extremity of such migrations. Thus are camels animals of American origin that once crossed the former land bridge to Asia, but both surviving Asiatic species appear to be more evolved creatures than are their South American cousins. The horse too originated in America and native wild ones still roamed the plains until comparatively recently. Perhaps the arrival of man there finally tipped the balance right against them. Wild horses failed in their American ancestral homeland and yet survived under what appear to have been similar conditions in Asia to those in North America. Presumably the vanguard of the horse tribe, moving westwards across Asia, grew gradually and selectively stronger as it leaped difficult hurdles encountered in its path. However, the horses remaining in America had provided a lesser 'hurdle' that was easily surmounted and trampled underfoot by even-toed ungulates from Asia with good spearhead qualities. The American horse never really recovered from this shock, could not compete and had no adequate escape route left. It could only move towards and reach extinction. When the Spaniards eventually re-introduced the horse to America, this animal had then completed its east-to-west encirclement of the Globe.

It might be noted that when domesticated horses and camels have escaped in America and Australia, they have found habitats where they can survive well. Among their evolved characters is large size. This, coupled with maintenance of fleetness of foot, may or may not allow them to cope better with the adopted environment, but it certainly enables them better to resist the limited predation pressure now obtaining in the areas where they have thus come to thrive.

CHAPTER 7

Multicellular Animals and Their Early Evolution

■ Volvox – A Colonial Protozoan

This is a flagellated Protozoan that lives in colonies. Thousands of individuals are assembled spherically around the surface of a jelly ball. They each resemble the non-colonial flagellate Euglena in so much as they also contain chlorophyll bodies. However, the many constituent cells of a Volvox are colonially connected together by a network of protoplasmic matter.

Sexual reproduction at a stage prior to the differentiation of the sexes has been observed among some of the relatives of the Volvox. The latter itself produces male and female gametes. The egg is produced inside the spherical colony and grows on the nutrition provided there by the whole collective body. It is fertilised by a sperm from the same or another colony.

Although the colony is spherical, it exhibits signs of polarity in that the same part of its surface always faces forwards as it moves about. Swimming is accomplished by the pair of flagella with which each component microbe is equipped, while the fact of the distinct orientation in the direction of movement indicates that the travel is deliberate and that there is co-operation between cells.

As an entity Volvox also retains the ability to reproduce itself by asexual means, but starting at the cellular level. A cell can be produced that is held within the sphere while it divides over and over again and eventually comes to resemble a small version of the whole colony. There can be several of these 'juveniles' developing within a sphere and at the appropriate time this will burst open to release them.

■ Porifera – The Sponges

Sponges are to a degree amorphous, but even so, different species can be seen to develop in their own ways. Yet there are several characteristics most of them have in common. They are colonial, rather than being clearly defined separate individuals: they have two layers of cells, one facing out to the exterior and the other inwards to a body cavity; between the two layers is a jelly-filled gap in which are found 'mesenchyme' cells; the body is supported by structures built up of specific hard materials secreted by the animal and called 'spicules'; the walls are traversed by certain cells, each of which provides a pore allowing water to be drawn into the body cavity from the outside; the body cavity is lined with collared flagellated cells.

Sponges may be male or female, or bisexual. In any case the gametes are produced within the mesenchyme, either for self-fertilisation or combination with the equivalent cells of other sponges.

As a larva, the sponge is motile. After sexual reproduction, this swims about before settling down to found a new colony. The body of the larva is covered with flagella to drive it along.

The incurrent pore cells, with their "openings right through them", show a trait not found among Protozoa. Non-cellular animals like the Paramecium each have a gullet leading to a separable and renewable food 'vacuole'. The postulate is attractive that the 'pore' has been formed by a fusion of equals in the past and that the two gullets have thus been combined to form a 'thoroughfare'. This union may also be linked with the creation of the diploid cells of sponges. The pore cells, like the gametes, are produced from mesenchyme cells.

The collar cells of sponges are very similar to certain flagellated Protozoa, such as 'Codosiga', which form colonies of stalked individuals as a result of sexual reproduction. It seems possible that sponges, with their greater integration, are more advanced forms of such colonies. However, this concept alone does not explain the cellular polymorphism of the sponge, nor the way in which the different parts are put together to a plan that in varying degrees is predetermined, dependant on species.

The reproductive cells of a sponge, whether sexual or asexual, stem from the mesenchyme. It is in the genes of these that the predetermined plan of a whole new colonial animal is to be found, because in the cellular colony this has become their specialised role. Other kinds of sponge cells have apparently lost this capacity to produce new sponges through specialising in other ways. For these other duties their diploid nature seems superfluous; they might indeed be just as effective if they were haploid cells.

The mesenchyme of sponges can also produce capsules of cells called 'gemmules', which can survive hard times (such as drought). From gemmules new colonies can spring.

Sponges feed like Protozoa, having no common digestive capacity. The loose-knit nature of their cells has been indicated by experiments in which a sponge was typically broken down to its individual cells by being pressed through a filter. The social and interdependent nature of the cells was then made evident by their subsequent ability to reintegrate themselves into a sponge.

As an animal with no motility, a completely predetermined shape or size is not essential for a sponge, although each species has its characteristics, governed historically by the specific environment and the selective evolutionary pressures caused by this. There is theoretically no limit to the number of recognisably individual sponge bodies that can be joined together as a colony; the primary restrictive factors would seem to be food supply and available space.

A sponge body is a fine-and-whisk feeder and achieves this by using its flagellated cells, although eventual absorption is amoeba-like, the nutrition being taken through the cell walls, as is also oxygen.

The simpler sponge bodies often seem to have more clearly defined shapes than the more complex. The various purse sponges can look like little vases joined together at their bases.

Common fossils in Cambrian levels are 'Archaeocyathids'. Their skeletal remains suggest the hard parts of individual animals that were once vase-shaped and sponge-like. However, they had internal partitions; this is not sponge-like and is reminiscent of the partitions found in coral polyps and some other coelenterates.

Sponges have some cells that act like muscles to narrow their body openings when adverse conditions obtain, but in the absence of nerves such action is limited and uncoordinated.

■ Coelenterates

These comprise various groups, each with its own characteristics. All have tentacles for seizing passing prey, with this being overcome by means of stinging capsules that inject poison and it is then passed into a digestive cavity by way of a centrally placed mouth. Wastes are ejected by the same route. The animals may be solitary or colonial, while in certain cases the colonies are polymorphic, i.e. with parts discrete in structure and function. Colonial members are linked by protoplasm through which run communal passages uniting the individual body cavities. Colonies are mainly found on the seabed, or attached to rocks or seaweed, but others float on the surface, being, as it were, "upside down". Mouths may point in any direction, but for convenience here are regarded as though "at the top". Free swimming individuals are produced by some species, which in certain cases dominate the life cycle. All unitary animals, whether individual or one of a colony, exhibit radial symmetry.

Coelenterates can move some of their parts. Animal movement in general can occur without nerves, but only very slowly, as can indeed plants too, to an even lesser degree. But polyps, medusas and jellyfish can move their tentacles quite rapidly by means of a nervous system usually concentrated at the basal ring of these, while the free swimming forms can also open and shut the bell to move about.

The shapes of coelenterates are more specific than those of sponges, while the ability to change shape is greater and exercised at greater speed.

Hydrozoa These are colonial animals, the feeding heads of which are small. A basic trait is to build up plant-like stems and branches encased and supported by a horny covering, with the individual animals being emergent from cups along the branches, into which some species can retract.

Polymorphism is present in certain species, in that breeding is confined to particular polyps, while others specialise in feeding, using their tentacles, and others have a specialised protective role using their stinging capsules.

In the case of Obelia the reproductive polyps use a version of asexual budding, which occurs as a branch of the colony that specialises in the internal production of a series of buds: these are going to turn into free-swimming animals resembling a juvenile jellyfish, i.e. at the tiny 'medusa' stage. The reproductive branch is vase-shaped and the ripening medusas represent the coming sexual phase: as they reach the top opening they then break free, each in its turn. Sperms and eggs are soon produced by these to unite externally and the zygote develops into a free-swimming larva that in time settles down to found a new colony.

The fresh water hydra is non-colonial, but reproduces by asexual budding, with buds forming on the outside of the body, where they develop into smaller replicas of the original, finally being severed from it, after which separate existence occurs. A hydra can also reproduce sexually, with some species being hermaphrodite.

Siphonophora These resemble the Hydrozoa in their organisation, but the colony is not attached and floats on the surface of the sea. A well-known dangerous species is the Portuguese man-of-war.

Scyphozoa These are coelenterates whose medusa stage is dominant over the sessile one. Such larger, radially symmetrical free swimmers are generally referred to as 'jellyfish'. In the swimming mode the central mouth is normally directed downwards and its edge is extended into lobes that in some species can be of great length and be interpreted as 'tentacles'. The main body consists of the round swimming bell, the fringe of which is also provided with tentacles, although numerous and of different design. The sexes are separate and the ovaries or testes are situated under the bell, from which position eggs or sperms can easily be shed into the water.

Anthozoa These are all sessile creatures and have no medusa stage. They comprise the various sea anemones and corals. The colonial corals resemble the hydrozoa, but their hard housings are not always on delicately branching structures. Their tentacles are more regular, being in multiples of four or six. Solitary corals are like sea anemones, except that they have but one ring of tentacles and construct hard cups for themselves to retract into. Anemones are also solitary, but do not produce hard parts and usually have multiple rings of tentacles.

In General The unique feature of all coelenterates is the provision of thread capsules that are embodied in special exterior cells and which, when triggered off, can be everted to penetrate an adjacent body and inject poison. However, not all capsules work in this manner; some have a sticky thread, while yet others have a wrap-round one, all of which are selectively useful, depending on the species of both coelenterate and prey. The very fact that all coelenterates possess the basic feature represented by thread capsules, while this is apparently lacking in all other phyla, suggests that it came into existence on some discrete common ancestor of the jellyfish and polyps.

Just as the sponges may have developed originally from colonies of collared flagellates, so is it possible that the coelenterates too consisted originally of colonies of sessile ciliates, which were each able to whisk small prey towards themselves to transfix it with some device and then to pull it back into the main tissues. The device can tentatively be compared to the trichocysts of the paramecium, which this can fire outwards, apparently as a defensive measure.

The hypothetical microbe in mind would certainly be after bigger prey than the particles whisked in by the paramecium and which might even have approached its own size. As an example, the didinium is a free-swimming ciliate that moves about until it transfixes a paramecium with a protruding spike and pulls it into itself.

This hypothetical protozoan ancestor of the coelenterates is now apparently absent, presumably having evolved earlier forms of itself to extinction. The thread cells on present animals are diversified, which is presumably subsequent to the onset of the metazoan condition, while they are also so specialised that they now sacrifice themselves, for the capsules can only fire once.

Coelenterates all exhibit radial symmetry and the question presents itself as to how this came about. Did it occur as a natural result of the way of life? Likewise one might wonder as to whether the primitive ancestor was a swimmer or sessile. Although coelenterates are carnivorous, their feeding can hardly be described as 'hunting'; they and the prey usually blunder into each other, thus stimulating the tentacles and their weaponry into action.

Assuming that the primitive members of the phylum were bottom dwellers, one can imagine an earlier colonial stage during which cell specialisation took place, some catching the prey, while others absorbed it. The 'prey' would basically be protozoa, which may have been inducted by currents caused by cilia and seized by primitive versions of the thread capsules. The food absorbing areas would better be served by becoming hollows, while the food catching parts would work better if they were on protrusions. The progressive development of such advantages would lead to gastro-vascular cavities and tentacles.

In colonial organisms it can be difficult to decide what is the actual individual. One might argue that, if an organism consists of cells that are in continuous and intimate contact, then the limit of the individual ends where the body mass becomes definitely recognisable as connective tissue or protoplasm. Under such circumstances polyp-like protrusions could be thought of as 'organs', even if their activities were uncoordinated. However, this view breaks down if nerve cells are present, for if these pervade the whole of the protoplasm, then it is truly an individual and with organs that are co-ordinated. Yet, if what might otherwise be considered to be organs possess their own isolated nervous systems, then the whole protoplasm is then clearly a colony of individuals that are co-ordinated only within themselves, but act as a community through an internally shared food supply and common interests.

New colonies are usually created through sexual reproduction, with each resultant larva aiming to settle down and multiply its cells. The hydrozoan branching colonies result from subsequent asexual budding, with the buds becoming polyps that remain part of the common protoplasm. In the hydra the buds produce young that detach themselves. The polyps of Scyphozoan jellyfish likewise do not develop colonially. Anthozoan corals of some species form branching colonies reminiscent of those of hydrozoans, while other colonies are not like branching plants, but do have cup openings and, as they live and die, limestone reefs are built up.

Solitary corals also exist, while the coelenterates called sea anemones are without cups and are always individual. The strange "dead man's fingers" are obviously colonial, but are also without hard parts; they comprise a mass of protoplasm supporting numerous tiny polyp heads.

The sessile origins of the coelenterates may be inferred from there being many polyp species without medusas, but few medusas exist without a polyp phase.

A curious fact about the radial symmetry of coelenterates is that their radial parts can generally be shown to be in multiples of four or six, as with tentacles, mouth lobes, sexual organs, etc. This geometrical arrangement might be inferred as being due to asexual reproduction at an earlier stage of evolution, indeed, to a process one might well dub "arrested fission". A yet prevailing example is the binary fission still employed by sea anemones, and one might cast one's mind back to the very early history of the phylum, before any of them were producing horny branching structures or limestone cups. Each would consist basically of a cavity whose single opening was surrounded by protuberances (or proto-tentacles). A new colony would grow and reproduce asexually. The shallow seas were so full of life that growth was rapid. Three forms of asexual reproduction were available to absorb the available energy: fission, budding and branching. The last implies a spread of the protoplasm into new areas and, with the vast amounts of energy about, it was eventually to provide a solution to overcrowding by allowing the colony to grow upwards away from its base into open water. Before hard parts were available, the first may have been an earlier kind of solution, in that a way was found to create a more efficient animal.

Asexual reproduction by fission is a persistent occurrence and hence is heritable. This implies that the instruction to do so must have arisen through a chance mutation or series of mutations, should one take the strict Darwinian view. The pressure for space would encourage the wider cylindrical animal to pucker and this would have been in a regular 'doubling' form on most occasions. Thus could a round cavity acquire 2, 4, 8, etc, alcoves. Should these happen to become sealed off both from each other and the original cavity, then a group of separate animals could eventually be created. Yet should such fission be arrested, then an animal with alcoves in its gastro-vascular cavity had evolved. This is essentially the sea anemone as known today. The gene mutation behind such physical manifestation must eventually have been such as to prevent the final splitting up of the original cylindrical body into virtually independent segments and this mutation must have been selected because it provided incidentally an improvement in the animal's powers of feeding and digestion.

The regular radial division of all the polyps, medusas and jellyfish would then owe its origins to this arrested fission state among primitive sessile coelenterates. It would seem to be among the anemones that signs of this development are best preserved and their multiplicity of tentacles must not blind one to their otherwise primitive status even among coelenterates.

The asexual budding of a creature like the hydra would seem to be between fission and branching. Were the formation of hard parts possible, one could well imagine a hydra bud being the source of a branch and hence of a colony. Any link with fission is less obvious. In the case of an anemone this procedure now starts at the top and works its way downward, while budding of hydra type occurs below the tentacle head.

The partitions in the gastro-vascular cavities of today's anemones can be thought to represent the modified walls of the alcoves of their ancestors. This itself would be like a simpler form of the coral polyp, but without hard parts.

The question remains as to why an anemone has so many rings of tentacles. The answer would seem to lie in linear reproduction by fission, but arrested and followed by retro-fusion. The formation of the jellyfish Aurelia provides an example in another group. In this the polyp develops into a column (like a pile of saucers), with the topmost, when ready, breaking away as a medusa. Among the ancestors of the anemones, with their multiple rings of tentacles), this fission to provide medusas will have been arrested and the matter retro-fused to give the rings of tentacles so characteristic of anemones. Once again one has to assume a genetic change (mutation) that satisfied needs created by evolutionary pressures.

The radial division of coelenterates into multiples of 4 or 6 demands explanation. The first could easily have been promoted by progressive and arrested binary fission, to give 2 followed by 4, 8, 16, etc, with such progress being arrested selectively in different species. The second case presupposes an initial division into 3, followed by 6, 12, 24, etc. The initial 3 would be the result when further division of members of a created pair - to start with - happened to be sufficiently out of synchronisation.

The medusa or jellyfish can be regarded as an upside-down swimming polyp with two rows of tentacles, one around the mouth and the other around the bell. Transferring this notion from the anemone, each thus represents two individuals resulting from a form of arrested linear fission, followed by different evolutionary development of the two longitudinal sections.

The swimming action of a jellyfish can be assumed as being secondary to the feeding action, but derived from it. Apart from the movable tentacles, ingestion by coelenterates is assisted by ciliary action along tentacles and gullet, with the produced current being reversible for the expulsion of wastes. Assuming further that the earliest swimmers were tiny and with a density near to that of water, the action of feeding may have been sufficient to lift such polyps off the bottom and take them into areas that represented pastures new. In order for this form of activity to become normal for the species, it again, of necessity, would have to coincide with any fortuitous genetic changes that made it permanent.

Floating up and swimming would relate to certain conditions. Jellyfish may have originally developed from polyps that easily tore. Even present day anemones, when moving over rocks, occasionally leave part of the base behind, which then may grow into a new animal. At a very primitive stage detached polyp parts might simply have floated away to do likewise, but eventually the animal came to anticipate the move and could pre-empt it by creating developing asexual offspring before the event. Eventually this would lead to the "pile of saucers" effect, a form of linear fission that, when arrested, leads to segmentation.

Among the coelenterates a fully segmented and integrated animal was hindered from creation because of the lack of a through-gut. The basic animal among them is always represented by the segment with the mouth. Any other residual and rudimentary segments are subsidiary and contribute to the new whole they have come to form part of by means of arrested fission, because of the advantages they could still contribute. Thus in sea anemones, the animal proper, together with its gut, is really concentrated in one segment, while the others basically provide extra rings of tentacles. In jellyfish the animal proper supplies mouth, mouth lobes and gastro-vascular cavity, while one can attribute the outer row of tentacles and associated swimming bell to a single added linear segment.

The alcoves derived from arrested radial fission are represented among jellyfish by pouches lined with stinging cells. Aurelia thus has four of these, to go with the four mouth lobes and four sexual organs. The stinging cells are on tentacle-like structures known as "gastric filaments". The suggestion emerges that these are actually developed from the tentacles of the four would-be offspring represented by the four pouches, but whose evolution was changed by arrested radial fission.

The bell of a jellyfish is its swimming organ. In the case of Aurelia it is subdivided into eight lobes, this suggesting one subdivision more here than in the rest of the animal. The action of the bell was presumably derived originally from the feeding movements of the outer ring of tentacles. The swimming action is so important that in some species the outer tentacles have disappeared, with food catching being the sole preserve of the inner "mouth lobe" ring.

The action of the bell would appear to be counterproductive to food gathering and the inner tentacles are often very long, seemingly in order to achieve distance from the resultant ill effects. A tactic adopted by some is to swim upwards and then 'parachute' down, the latter being the feeding stage, i.e. while the bell keeps still.

The coelenterates have not developed elongated forms that can swim actively and laterally through the water. Their 'hunting' relies largely on chance. This may be the reason why they, and they alone, retain the stinging capsules; these have been to such advantage that they have helped the phylum to resist evolutionary pressures to change in other ways. On the other hand one might consider the alternative whereby other phyla have survived without them, or have abandoned them at an earlier stage, because their general evolution in other ways made the development or retention of such unnecessary.

The coelenterates do however exhibit traits that can be observed as forming part of the evolution of other phyla. One can mention the development of a ring of tentacles around the mouth, especially with the presence of ciliary grooves. There is also multilateral evolution; firstly radial formation of potential offspring, but with fission arrested and followed by retro-fusion; secondly the linear formation of potential offspring, again followed sometimes by arrested fission and retro-fusion and with some specialisation. This last is the basis of longitudinal segmentation, although, in the case of the coelenterates, a through-gut cannot be formed for such a 'colony'. Nonetheless, the "pile of saucers" is reminiscent of certain segmented worms. Even an Anthozoan such as "Peschia hastata" is worm-like in that it inhabits a burrow.

■ Bryozoa

These are sessile aquatic animals, passing their adult lives attached to rocks or plants, or within branching structures of their own making. Each inhabits a self-made hard cup that serves as a refuge when the animal is disturbed. Bryozoans are tiny and superficially resemble hydrozoan coelenterates, with each animal having a mouth surrounded by a ring of tentacles by means of which food is collected. However, bryozoans are never gastrically linked together; they are colonial in that the group grows by asexual budding, but all resultant individuals live separate – even if intimately gregarious – lives.

Sexual reproduction takes place to found new colonies. Marine bryozoans have ovaries and testes and the gametes form zygotes in the body cavity called the coelom and make their way to the outside as free-swimming larvae of the 'trochophore' type.

Only coelenterates have tentacles bearing stinging cells, these being essentially designed to subdue larger prey. Those of bryozoa are ciliated: microscopic particles are swept by this 'fine-and-whisk' method into the gastro-vascular cavity or stomach, this process being augmented by further cilia in the gullet. Wastes are ejected by a separate ciliated passage leading to an anus outside the ring of tentacles.

The bryozoans would seem to have remained tiny because they are only capable of dealing with minute prey and because they do not have any special organs for absorbing oxygen. There is muscle tissue to effect the withdrawal of the tentacles while a rudimentary 'brain' near their base facilitates the rapid accomplishment of this.

The coelenterates have a space between the inner and outer body walls ('endoderm' and 'ectoderm') that is filled and stiffened by a non-living jelly-like substance and wherein are produced the mesenchyme cells. With the bryozoa the equivalent space is itself lined with cells that form a covering called 'mesoderm'; this has a potential for producing various organs and cells, such as ovaries and nerves, in a manner reminiscent of that of the function of the mesenchyme of coelenterates and sponges. Such a space is called a 'coelom' and is filled with fluid rather than jelly.

The coelom is a feature of animals at a greater degree of organisation than the coelenterates. With it go other traits, such as a tendency towards a bilateral structure, the provision of an excretory system to clear wastes from the coelom and a through-gut having an exit called an 'anus'. It seems reasonable to suggest at this point that all these developments are interrelated.

The possession of a through-gut decisively distinguishes the bryozoans from the coelenterates and its existence might well be seen as connected with the 'fine-and-whisk' feeding method. By acquiring nourishment in this way the through-gut is clearly more efficient than any arrangement whereby incoming food and outgoing wastes must share the same opening, no matter how phased. However, its efficiency alone does not explain its existence. In view of the probability that the gut was initially blind, in the same way as found with a coelenterate polyp, one is faced with the problem of the evolution of the alimentary tract that possess two openings, an inlet and an outlet.

The explanation would seem to lie in an incidental result of asexual budding. One might consider a hypothetical earlier stage in the history of the bryozoans, before the development of any hard parts and with young being produced asexually by lateral budding, similar to that of the coelenterate called hydra. During the growth of such an offspring, it shares a common gut with the parent and lives on food captured by the latter. One must then imagine a genetic development that ensured that the common gut was retained after the mouth of the offspring was formed and came into use. It is clear that such blind-gut ancestor must have had a mechanism for expelling wastes through the mouth and this could have taken the form of cilia that could reverse their action. A change in the adult-young relationship would result in such a pair, one would have cilia that only wafted in, while the other (presumably the bud) only wafted out.

A colony of ancestral bryozoa can easily be pictured as a mass of soft-bodied pairs, each pair of which having an incurrent and an excurrent opening. However, these pairs as yet cannot be considered to be individual animals, but rather as being at an intermediate stage, namely a form of polymorphism. Individuality of the pairs would not arise without full integration and this could only occur by genetic means, whereby the pairing became basic, right down to the single cells and through the various embryonic and larval stages. In other words, the pairs were integrated as individuals once the necessary intercellular amalgamations had taken place. Pairs of cells would fuse, the chromosome count would double and gametes would hence produce zygotes that were bound to grow up into the fully integrated pairs that the bryozoans represent.

It may be that under these circumstances the member of the pair contributing the anus was otherwise atrophied. Even were this true, the fact is that the number of effective genetic units had been doubled, which allowed the possibility for duplication of parts and/or further specialisation of parts. This concept is very important with regard to the evolution of animals that were to be much more complex than the bryozoa and for which degrees of bilateralism were to feature strongly in their adult forms.

The genetic development of the through-gut can be compared with the lobe development of the coelenterates, in which lateral accretion might provide, for instance, four sex organs and eight tentacles in radial symmetry. The number of tentacles of coelenterates would seem to result from a stage of this sort of development and this should apply to bryozoans too. The chief difference is in the longitudinal development of coelenterates, which, as argued above, led to the swimming bells of jellyfish and the multiple rings of tentacles of sea anemones. It is in a kind of longitudinal development of an ancestral bryozoan that the through-gut came about and the process can be seen as incipient segmentation, the number of segments being two! The incipient bilateralism of bryozoans may be evidenced by some having tentacles arranged on a U-shaped ridge, rather than a circular one round the mouth.

There are some ostensibly related animals that superficially resemble bryozoa and are usually grouped with them. However, there is one fundamental difference; the anus with these opens within the row of tentacles. Hence their name 'endoprocta' (as opposed to the bryozoan 'ectoprocta'). This arrangement could not have the same history of development as that outlined just above. To illustrate how the endoprocta might have arisen one can use the example of the anemones, rather than of those polyps that reproduce by budding. Some anemones have differential areas of cilia to provide circulation. Most of the cilia have an inward action to draw water into the gut, for oxygenation rather than feeding. There is also a groove containing cilia whose action is outward, to provide the necessary evacuation. This groove has the potential to evolve into a separate opening by a process akin to asexual fission; the endoprocta may have taken this path. But this development was ultimately just as sterile from the evolutionary point of view as those characteristic of the coelenterates and the greater future lay with certain ancestral relatives of the ectoprocta

■ Bryozoa-like Worms

The phylum 'Phoronidea' embraces some elongated creatures that hence have acquired the description 'worms'. They do in fact resemble the bryozoa in that they have a U-shaped through-gut and feed by means of a U-shaped lophophore bearing ciliated tentacles. These worms are generally not so intimately gregarious as bryozoans and, rather than cups, they inhabit individual tubes into which they can withdraw.

The U-shaped lophophore, with its spiral ends, is reminiscent of some of those on bryozoa and likewise can be thought of as reflecting the incipient bilateralism resulting from the fused pair, which itself can be inferred from the presence of the through-gut. The apparent single lophophore of each animal hence represents a pair derived from the two fused and integrated ancestral individuals.

■ Segmented Worms

These are collected into the phylum 'Annelida', which has two main classes, the 'Polychaeta' – "many bristled" and 'Oligochaeta' – 'few bristled'. The polychaetes are essentially marine or shore-dwelling and among them are the fan worms and feather worms, animals that are tube-dwelling and equipped with tentacles, often very numerous, which they can withdraw into the tube when disturbed or exposed to the ebbing tide. The tentacles also serve as gills, although in some species protrusions further along the body serve specifically as oxygen absorbing agents. The tentacles are often in two clearly divided groups, indicating incipient bilateralism.

The bodies of polychaete worms are characteristically divided up externally into distinct segments. Frequently found on both sides of each segment is a protrusion called a 'parapod'; this is provided with bristles. On many species, particularly those that swim, the parapods are not only used for this activity, but act as gills. Even some polychaetes that are sedentary in tubes do not have 'tentacles'; instead they pass water through the burrow by means of the parapods and in this way feed, respire and get rid of wastes.

The segments can be regarded as individuals whose separation has been prevented by fission arrest. The parapods, seen in this light, are derived from the tentacles that would have graced the individuals of former times. In the case of the bilateral segments, these have been fused to form the paired parapods, as well as the more specialised sensory extensions and feeding jaws at the head or other feeler-like extensions at the rear.

Even the more active polychaetes tend to lurk in cavities, perhaps relating to an original tube-dwelling life-style for the whole class, but those living off larger prey (coarse-and-grasp feeders) eventually evolved into forms that could roam more freely.

The fine-and-whisk feeding annelids are clearly related in function to the bryozoa. However, a clear difference can be seen in the segmentation itself. It was stated earlier that the through-gut indicated that the bryozoans really had two segments, although this was not readily observable; but, in any case and if true, the principle of segmentation was already about and also obtained among the coelenterates. The difference with the annelids is that the segments remained externally visible, with retro-fusion apparently restricted in its extent. Only at the head – especially - and the rear has this last process occurred to a large degree to obscure the segments and produce a mass of organs.

Segmentation is derived from the asexual means of reproduction whereby, in sessile animals, a "pile of saucers" develops on which the top one can keep breaking away as a free swimmer. The young did not just appear at the top, but the whole animal formed a column of developing individuals, with individuality, at its most rudimentary, first appearing at the bottom. The larva of a polychaete shows the segmentation developing in a similar manner. In the event the process is used, not to produce young forms, but segmented adults. The fission of the individuals has been arrested, followed by integration and retro-fusion. All the segments, however modified, have the latent ability to produce 'tentacles', which often appear in specialised forms, such as parapods, feelers, antennae, etc.

As described above, the coelenterates have followed two avenues of asexual evolution: multilateral (or radial) and longitudinal, leading to corresponding types of segmentation. But the through-gut added a distinct 'two-ness' or duality to animal species, leading to the evolutionary advantage of bilateralism. The bryozoa show little sign of this advantage, this being indicative of their primitive nature and it is much more in evidence among the segmented worms, but not existing to the same degree in each species, nor even along the same animal.

Longitudinal segmentation allowed animals to grow larger and, in particular, longer. However, the greatest advantage was in the repetition of organs along the length. When retro-fusion took place the repeated organs could be amalgamated and integrated and then become specialised within the new enlarged organ, to produce something that was far better than the sum of its parts. Nowhere is this clearer than with the more advanced nervous system that grew into ganglia in the segments and into a brain at the head, where retro-fusion was much more extensive.

Possession of a through-gut means that the animal concerned basically represents two former ones, fused together. This last process has instigated bilateralism as a tendency with regard to the ectoderm and the coelom, but not the endoderm. However, with a segmental animal, different segments might exhibit a mixture of the new bilateralism and an older condition, i.e. radial multilateralism, depending on their position along the animal.

The heads and tails of polychaetes are so specialised that new segments clearly could never be created there, but somewhere between. Logically they ought to appear at the point along the body that represents the original 'joint' of the pair co-operating to provide the through-gut.

The free-swimming polychaetes often have an eversible pharynx, equipped with hard 'jaws', which is used to seize prey. This apparatus need not be thought of as having evolved on its own, but as being homologous to the ability to extend and retract tentacles, as seen among the coelenterates, bryozoa and sessile polychaetes, even if any relationship between the jaws themselves and hypothetical inner tentacles is highly tentative.

Those polychaetes that can swim are carnivorous, but prey on creatures smaller than themselves. There would seem to be no sign of an adult annelid ancestor of sedentary habit. It would appear that predation was first undertaken by an animal at the larval stage; once this kind of juvenile had arrogated for itself the role of sexual maturity it was only a matter of time before the sedentary adult became extinct through obsolescence.

The other class of annelids – the oligochaetes – is largely found on land or in fresh water. It comprises mainly the various earthworms, but also those ectoparasites known as leeches. The earthworms need to keep their bodies moist and this restricts the occasions when they can comfortably leave their burrows. Their underground life-style suggests an origin in common with the tube-dwelling polychaetes, from which they have broken away to exploit a different habitat. Only aquatic animals can be fine-and-whisk feeders. The earthworm moves about, eating earth as it goes and expelling it through the anus, but minus nutrients. Tentacles are obviously not required for such a life-style, nor are any derived sensory organs of any elaboration. Since the worms do not swim, parapods are not required either, although vestiges of these remain in the form of bristles.

The two classes are clearly distinguished in that the polychaetes have separate sexes, while the oligochaetes are hermaphrodite. This indicates a fundamental split of great antiquity. It was suggested earlier that the original sexual condition was hermaphrodite, with unisexuality being a later specialisation. If this can be upheld, then the example of the annelids suggests that the change to the unisexual condition occurred in the marine environment. The animals that became terrestrial could theoretically have taken the bisexual condition with them, while retaining the evolutionary option of going unisexual; but any that happened to be unisexual before leaving the water (i.e. with distinct male and female forms), would have no chance of appearing in terrestrial bisexual forms, for such reversion would seem to be biologically impossible, even if it ever became desirable. The oligochaetes are hence bisexual because their ancestors were already in this condition when they abandoned their aquatic home.

Polychaete reproduction gives rise to larvae of trochophore type (or one evolved from such as often as not). This super trochophore has polar bristles and an equatorial band of cilia. Most important to note here is the U-shaped through-gut, with both mouth and anus pointing downwards. This is essentially the same as the digestive tract of a bryozoan, except that fine-and-whisk feeding is still at the 'protozoan' ciliated stage, rather than the tentacled stage of the Metazoa. Such an advanced trochophore could readily be imagined as settling down as a bryozoan, except that this is in fact pre-empted by the occurrence of segmentation. Polychaetes are promoted segmented juveniles that have superseded their earlier adult form.

Proboscis Worms

These are also known as "ribbon worms" and constitute the phylum 'Nemertea'. It contains comparatively few species, which are mainly marine, although a few are to be found in fresh water and even damp soil. They are carnivorous and catch prey by means of a very long 'proboscis' that can be extended from a cavity situated in the head above the mouth and retracted by means of a muscle attached to the cavity wall. In some species the proboscis end is equipped with a needle that can be regenerated if necessary.

This cavity beside the mouth may indicate its descent from a degree of polymorphism. So, imagining some earlier colonial form, the proboscis could represent a member that specialised in catching, while others did the subsequent absorbing (feeding). The retraction of the proboscis could well be a form of the tentacular withdrawal so apparent among coelenterates, bryozoa and annelids. Indeed, bryozoans still exhibit a form of polymorphism that may throw light on the proboscis of the phylum Nemertea and even on the jaws found on some polychaete annelids. Thus do some bryozoan colonies bear curious members that are reminiscent of the curved beaks of some birds and are hence called 'avicularia'. However, they are not now used in connection with feeding and seem rather to play a defensive role.

The head of a nemertean is defined from the body, but the rest of the animal shows no external sign of segmentation. The shape of the body can also be varied considerably by the animal, shortened and fattened, or lengthened and thinned. Length difference between species is also great, some being but a few inches, while others can reach many feet.

The worms are provided with a through-gut that has lateral bulges at intervals. This suggests incipient bilateralism, which concept is also reinforced by the presence of twin longitudinal nerve chords and the sexual organs (these occurring between the intestinal bulges), as well as the generally flattened shape. This bilateralism and the 'worm-like' length are related to the through-gut. Yet one might propose that both bilateralism and segmentation can manifest themselves selectively in any species, while vestiges of other and earlier organisation are also retained. The lateral gut-bulges of a nemertean may thus indicate bilateralism, yet at the same time be related to the 'alcoves' found in the guts of coelenterates. Likewise their repetition along the worm's body seems to indicate some internal segmentation that is not evident from the exterior, but has given rise to the repetitive sex organs and lateral connections by which the paired longitudinal nerve chords and blood vessels form unified systems.

Thus might one claim that internally the Nemertea are quite advanced, while externally they have remained much more primitive. They are slow swimmers and their skins retain the covering of cilia one has seen among the Protozoa and other primitive Metazoa.

The nemerteans enjoy the obvious advantages conferred by the through-gut, in that the digestive cavity can specialise in function along its length. Food is moved along it by muscular activity that ripples from the head to the tail end (peristalsis) and, in the absence of any heart, this also brings about blood circulation.

The larva of a nemertean is not a trochophore, it being called a 'pilidium'. It has a ventral mouth, but no anus. This is almost as though it has emerged from the egg in a form that reflects the animal in a more primitive state than does a trochophore; in other words the animal's larval form itself is yet destined to pass through the stage of the phylogeny during which the through-gut was acquired.

◾ Roundworms

These constitute the phylum 'Nematoda'. They are all very similar in structure and are to be found just about everywhere that other animal life exists, many of them being indeed endoparasites.

The nematodes are completely free from appendages. Their bodies are simply cylinders that taper at each end, with a mouth at the front and an anus at the back. Coupled with this through-gut there is a rudimentary bilateral symmetry. At the mouth there are inconspicuous lobes that constitute sense organs and can be provisionally identified as vestiges of the production of tentacles. The exterior of nematodes is in the form of a cuticle secreted by a skin layer beneath it called a 'syncytium'. This layer has nuclei, but no cell walls, and hence is reminiscent of the condition of the non-cellular protozoans.

These worms are devoid of cilia, both externally and internally, while the movements they accomplish are apparently pointless. They cannot swim as such, but can make progress when in contact with solid matter.

The nematodes exhibit no sign of segmentation, either externally or internally. They would simply seem to represent the original pair that was necessary to give the through-gut, but elongated to give the advantages of specialisation along the intestine. Indeed, one might suspect that the round worms are derived from the larval forms of some bryozoan-like creatures that came of age, and thus eliminated their original adult forms.

Males and females are separate and the reproductive system is elaborate, the ovaries and uteruses constituting long tubes that fold up and down the body several times, this itself telling against any hidden segmentation. The female system consists of two of these long ducts that come together eventually near the head to form a 'vagina' opening to the exterior. The single duct of the male opens towards its rear and sperms are transferred by contact.

■ The Flatworms

These constitute the phylum 'Platyhelminthes'. They inhabit both salt and fresh water, as well as following numerous parasitic life-styles. There are even land-dwellers. Virtually all are free of appendages. The flatworm body is more flattened than that of a nemertean, but like that has a surface covering of cilia. The non-parasitic flatworms are divided into groups, e.g. the triclads and the polyclads, this reflecting whether their gastro-vascular cavities are three- or many-branched. The common fresh water planarian is a triclad.

Externally a planarian has all the appearance of having bilateral symmetry; this also extends to the nervous, excretory and reproductive systems, The first consists of two ventral nerve chords, joined together in ladder fashion along the whole length and with extensions towards the outer edges of the body. It is to be noted that the two longitudinal chords are well apart, unlike those of the annelids, which run side by side. The annelid system would seem to give better integration and hence to be more advanced. At its head end the planarian has its nerves thickened into two ganglia, these being associated with the positioning of the two eyes and the pair of sensory protuberances, one on each side of the head.

Planaria and other flatworms are hermaphrodite. The male and female systems are arranged with regularly spaced organs and linking ducts down each side of the animal, the male system higher than the gut and the female lower. Both systems have access to the exterior through a common genital pore, situated ventrally and more towards the posterior end. The animals are not self-fertilising; two worms place their ventral surfaces together, whereby sperms are transferred to the partner's genital sack using a penis that can be protruded through the genital pore.

Planaria can also reproduce asexually, whereby the rear portion holds fast while the front pulls away from it. Both pieces can then grow into whole animals. This is reminiscent of the asexual division of sea anemones.

There is an excretory system down each side of a planarian and these are connected to the exterior by pores.

Thus, from a description of its nervous, reproductive and excretory systems could one hold that a planarian was a bilaterally symmetrical and segmented animal, just like an annelid. However, the description breaks down with reference to the digestive system; there is no through-gut and the mouth is not at the head end. The gut has three main branches, one leading forward and two backward, and these have further small branches on them, while the mouth is more or less centrally placed on the ventral surface.

In polyclads the worms tend to be shorter in proportion to width and the gut's many branches lead from the centrally placed mouth. It is just as though these 'turbellaria' still retain certain characteristics of a sessile adult ancestor, which have only been partly eradicated by a bilateral larva coming of age. This seems to suggest that even pronounced adult bilateralism is not of necessity accompanied by the adoption of a through-gut, even if the option to form one may have arisen.

A more primitive form of flatworm is found in the 'rhabdocoels', which have a straight gut, but still no anus. Sometimes these will reproduce asexually to become 'colonial', in that they are still fused together longitudinally and share a common gut, while each 'link' of such a chain has its own opening to the exterior. Such worms apparently represent an earlier stage of flatworm phylogeny than do triclads.

Even more primitive are the 'acoels', which have no gut at all. The fact that the flatworms, even the more advanced ones, have no through-gut, does not necessarily mean that the evolutionary opportunity to have one was never presented to them. Those bilateral aspects that are present suggest the dualism generally associated with the through-gut. This could have been rejected by the phylum, which has made evolutionary progress otherwise, namely by increasing the area of the gut by its branching characteristics and by developing more sophisticated reproductive, excretory and nervous systems.

The centrally placed mouth of a planarian suggests a feature retained from a radially symmetrical extinct adult form. The branching gut would seem to be a relic of the type of multilateral asexual reproduction hypothetically ascribed to the coelenterates, while the bilateral arrangements are a sign of segmentation that has suffered external retro-fusion to reconcile it with the retained aspects of multilateral development. When adulthood is adopted by an erstwhile juvenile form that is then subjected to further evolution beginning from that stage, the process involved is the so-called 'paedomorphosis' or 'neoteny'. Most phyla show signs of this, although there tends to be little token of it in those with radially symmetrical adults, such as coelenterates and bryozoans. The original radially symmetrical adult ancestor of the flatworms may have left a trace of its ring of tentacles in the pair of feeler-like projections occurring near the mouth in some species.

■ Brachiopods

These are exclusively marine creatures and the number of both species and individuals is now greatly reduced from what it was in the remote past. Their way of life is very similar to that of the bivalve molluscs and it may be that the evolution of the latter contributed to the extinction of so many types of brachiopod.

Present day species fall into two groups. They have in common the way of feeding, which is fine-and-whisk using a pair of arms, known as 'lophophores', which bear a great number of ciliated tentacles that direct food along a ciliated groove to the mouth. Oxygen is also conveyed to the tissues by the same organs. Each animal has a foot that is used to anchor it to a rock or other solid matter. Their way of life is also similar to that of bryozoa and they share with them a true coelom, i.e. a cell-lined body cavity between the gut and outer layers.

Of the two groups into which the phylum is divided, in one the shells are equal and horny and the mouth leads to a through-gut. This group is usually considered to be more primitive than the other. The species 'lingula' is unchanged from fossil forms. It dwells in burrows and protrudes its lophophore to feed, but can withdraw underground by contracting its foot. The life-style hence resembles that of the tube-dwelling bristle worms and bryozoa-like worms. The foot protrudes between the twin shells, which are held together by muscles that control their opening.

In the other group the shells are unequal and calcareous, while the mouth leads to a blind-gut. The foot protrudes through an aperture near the edge of one shell. As with flatworms, the loss of (or lack of) anus might be explained by them concentrating on evolution in other ways, with it being the supposed "more primitive" group containing lingula that has concentrated more on the digestive system and hence has acquired or preserved the through-gut.

One might postulate that the calcareous brachiopods are related to the calcareous bryozoans, while the horny ones are similarly related to the horny bryozoans. The two shells would then represent the 'arrested' development of the cups of the twin individuals that are the basis of the through-gut. The ensuing bilateralism is evidenced by the twin lophophores. However, relative to the lophophores the shells are not bilateral in the same plane and seem to have moved through 90°.

Molluscs

These have various features in common, but take their name from their 'soft' fleshy bodies. Apart from certain aquatic arthropods, they are the only other kind of invertebrate to form a substantial food source for humans. All have a through-gut and bilateral symmetry has earlier been more thorough-going, although in certain cases this has been subsequently distorted. There are five classes of mollusc.

1. *Amphineura* (or 'Loricata') – Chitons. These are bilaterally symmetrical and have eight shell plates along the back, suggesting a degree of segmentation. However, there is the fleshy foot that is like those possessed by most other kinds of mollusc. The life-style is limpet-like.

Head appendages are absent. The impression given is of an animal that once passed through a segmented immature stage, before settling down as a radially symmetrical adult. The chiton is a resultant "compromise form"; this is indeed a convenient term to apply to all molluscs and its value can indeed also be extended into the membership of other animal classes.

2. *Gastropoda* – Univalves – such as limpets, whelks, snails, slugs, etc. The 'foot' of such is homologous to that of a chiton, but instead of the eight plates on the back, a gastropod has but a single shell (in some cases obviously spirally coiled), or vestiges of such. The head usually bears paired appendages, but the rest of the body (visceral mass) lacks the organs from one side and normally is twisted upwards, with the anus pointing forwards above the head. The shell grows out of the resultant 'half-body' and this general asymmetry has initiated the spiral shell development, although some species have resisted this tendency and even possess - or have reverted to - simple cones (e.g. limpets). The sacrifice of one side of a bilaterally symmetrical body would seem disastrous; it has in fact been just the opposite, for the gastropods are the most numerous and wide-ranging class of mollusc, while being the only one to invade land.

The original dualism is maintained by the through-gut and the head, while the foot may be thought to reflect an original radial feature. The slugs are gastropods that look on the outside bilaterally symmetrical, but internally have characteristically lost one each of the paired organs. Gastropod gills are external and normally found between mantle and foot (as also found with chitons). In some sea slugs these can be seen to be retained, but of course only developed on one side. On others the gills are on the back, towards the rear and clustering around the anus. There can be many of these, in both odd and even numbers. It would seem that the side and anal gills are of somewhat different origins, with the former being parapodial and hence representing segmentation, while the latter are tentacular and reflecting the earlier radial stage. Some species can retract their anal gills.

Some sea slugs can swim using extensions of the mantle (confusingly referred to as 'parapodia').

3. *Bivalvia* (earlier 'Lamellibranchiata') – Bivalves – The foot here is not used as a means to glide along, as is the case with the foregoing, but serves to anchor the animal against current and turbulence and, by using it intermittently as an extensible hauling device, to effect locomotion. The bivalves can characteristically defend themselves by closing the paired shells. Apart from defence against predation, this also allows some species to survive exposure to air when left between the tide lines. The bodies retain bilateral symmetry. The gills are permanently inside the shells and are also used to direct food particles that have been trapped in them into the mouth.

4. *Cephalopoda* – Nautilus, squids, cuttlefish, octopuses. 'Cephalopod' means 'head-foot'; this is derived from the surprising assertion that the tentacles of these, which surround the mouth, are homologous to the 'foot' found on other classes of mollusc. It would seem more just to claim that these tentacles are actually homologous to those found on other molluscs and indeed elsewhere, whereas the 'foot' of the cephalopod is now no longer immediately evident, having been either made redundant or converted to something else long ago by the active life-style.

All are predatory and swimmers. The body plan exhibits strong bilateral symmetry, except that the tentacles round the mouth may be thought of as vestiges of a former radial body pattern. Now enclosed in the mantle, a vestigial shell still remains within the squids and cuttlefish. The small cephalopod known as Nautilus retains an external coiled shell into which it can retire. Those 'octopods' grouped under the name 'argonauts' produce external shells of quite different nature to those of other cephalopods.

5. *Scaphopoda* – Toothed shells. This is the smallest class of molluscs. The foot serves as a burrowing organ, with the animal dwelling underground. Its long tubular shell protrudes into the water. It has a rudimentary head provided with thread-like 'tentacles'. A created current draws food down the tube.

....

In consideration of the molluscs in general, one can apply the profound principles extensively explored on these pages and concerning remoter and more concealed evolutionary links with other phyla, as well as phenomena related to paedomorphosis. The molluscan ancestral adult should be considered as being a radially symmetrical entity, whose cavity was open at the top for the ingress of food and with an asymmetrical exit, both openings being equipped with tentacles, or vestiges of such. The body may have been protected by some form of shell or shells, but not completely enclosing the base or foot, for the animal could perhaps move like a sea anemone or a limpet. To reach such an adult stage the animal went through juvenile forms that were bilaterally symmetrical and swimmers.

Using the "plastic view" of phylogeny, with characters in the course of time being able to move selectively along the ontogenetic scale in both directions (i.e. nearer to the egg or nearer to the adult), one can attribute the different classes of mollusc to origins at different stages of the lifetime of the ancestor, but at all times including borrowings from other stages. This way of approaching the problem necessitates that the ancestral adult was a sessile, radial, fine-and-whisk feeder, while its juvenile was an active bilateral coarse-and-grasp one.

The chitons have retained the bilateralism of the juvenile, although the foot represents the original base of the adult. The pronounced segmentation still evident in the mantle suggests a relationship with the annelid worms, particularly those polychaetes whose bodies are short and wide.

The gastropods resemble the chitons, but they have even lost the residual segmentation shown by these; bilateralism has also been selectively abandoned and they are altogether more complex and diverse creatures. On the whole they give the impression of chitons whose evolution has gone wrong; yet they are the most diverse and numerous class of molluscs. The loss of organs down one side of the body is their great peculiarity, yet the lack of one of each pair of such seems to have been to some evolutionary advantage, even though the reason for this is far from obvious. One might tentatively imagine that, in growth toward the radial sessile adult, the juvenile was a bilaterally symmetrical swimming animal with through-gut. As evidenced by other phyla, this is a successful combination, yet the very success would increase competition for those adopting this form. Some molluscs compromised by going only part of the way towards the original radial adult condition, by becoming poorer swimmers and by achieving sexual maturity at this very stage.

The radial adulthood originated from an animal prior to the state of duality associated with the through-gut and, in partially reverting to this ancient form, the organs down one side were lost. The animal then could approach the radial form during its lifetime by twisting its rear forward about its middle. This did not affect the foot, which was already of radial origin, while duality was always retained at the head.

As adulthood was transferred into gastropods of even more juvenile form, the lost organs did not reappear, but ostensible bilateralism emerged instead; shells became less coiled, as on limpets and the like, or tended to dwindle and vanish, as in slugs. Thus can one postulate that snails, winkles and whelks represent adulthood reached at an earlier stage than do limpets, while slugs are of even later origin than these. Yet the fact that sea slugs often can swim suggests that they somehow at the same time represent an earlier phase of immaturity than do the other gastropods, which cannot do so.

The loss of organs and accompanying loss of symmetry would originally have taken place at a larval stage known as a 'veliger', itself grown out of a trochophore. The onset of this peculiar arrangement has hence slipped to near the beginning of the ontogenetic scale of the gastropods.

Gastropods are descended from creatures that were comparatively low in the number of segments. Their subsequent evolution has been towards disguise of segmentation, partial loss of bilateralism through a tendency to partially retain the radial symmetry of the ancient adult and loss of one each of certain paired organs in the course of these developments. However, the duality was retained in the nervous system and the sense organs of the head, while slugs have reverted to or retained an apparent external bilateralism.

The loss of the gills from one side does not seem to have seriously hampered gastropods and one can attribute their tolerance of a 50% loss to their generally 'sluggish' behaviour. Ironically, it is the sea slugs, which are swimmers, that have elaborate gills and, if anal, numerous too. The reduced requirement for gills, due to low activity, has allowed certain of the land snails - which have all lost them - eventually to re-colonise the water by the adaptation of the mantle itself as a sort of lung.

The bivalves, being fine-and-whisk feeders, are of necessity aquatic. Typically, they too are sluggish movers, which leads one to the question as to why they have retained full bilateralism and two elaborate sets of gills. The primary purpose of the gills here is not for gaseous exchange, but for filtering out of food. The gills appear to be 'inside' the animal, but in reality are external, being merely enclosed by both shells and the mantle, yet lying before the mouth. Their duality suggests lophophore-like origins, the ciliary action of which is retained. Otherwise they are much modified to give a system of enclosed water channels and minute entrance pores that exclude larger particles. Presumably the gills were originally grooved ciliated tentacles that could be retracted within the shells in a way similar to that in which barnacles still manage their 'legs' now.

The cephalopods do not creep, so they have lost the 'foot' that is so characteristic of other molluscs. With them a trace of earlier radial symmetry has been preserved in the case of the mouth and its ring of tentacles; otherwise they are bilaterally symmetrical. The pair of gills are laterally placed and as with most molluscs lie within the cavity of the mantle. This can be sealed by a collar behind the head, thus enabling water drawn into it to be ejected rapidly through a tube formed of the mantle material, this 'funnel' being analogous to the excurrent 'siphons' of gastropods and bivalves, but enabling cephalopods to swim off rapidly.

Extinct cephalopods are often recognised as fossils by their shells, some being of giant proportions. Some of these were straight (e.g. belemnites), while others were coiled (e.g. ammonites), as is also that of the small surviving Nautilus. Unlike those of gastropods, these shells were generally bilaterally symmetrical, being coiled in the vertical plane (although spiral ones are known). The full retention of bilateral symmetry among cephalopods relates to their life-style; active predators could not afford to sacrifice organs in the way of their sluggish gastropod relatives. The retention of shells by such swimmers implies enemies. However, subsequent evolution has indicated that the swimming itself has proved a more important character for survival than the ownership of a shell, the weight and shape of which must have been a drag. Lack of shell is now complete among most true octopuses and they generally offset the increased vulnerability by skulking in rocky holes.

The scaphopods, although burrowing like worms, have retained many of the specific molluscan features, i.e. shell, foot, tentacles and radula. One can recall that some bivalves use the foot to haul themselves underground. Yet scaphopods are purely fine-and-whisk feeders; so why do they need a radula? One must presume that, lacking the efficient filtering mechanism of the bivalves, they retain it to deal with the larger drawn in particles.

....

As has already been proposed, the different phyla can be seen as representing different stages in the ontogenetic development of a common ancestor, with characters, in the course of time, having also been able to pass up and down the ontogenetic scale in a surprisingly free manner. This has contributed to great change and some breaking up of species, but is a process that has not itself alone resulted in the "branching off" of new species.

This last has had its cause in phylogenetic change, with its roots in geographical isolation, which has allowed evolution to take place using the changes resulting from ontogenetic shift and/or those arising from brand new phenomena, i.e. mutations. Indeed, in earlier forms of life it would seem that evolution had much more plastic material to work on, starting with colonial development. This led initially to radially and then longitudinally segmented forms, and indeed combinations, together with the changes brought about by ontogenetic shift. A major aspect of the latter is paedomorphosis. In later times animal life has become more and more stable, so that evolution has had to rely more and more on selection from the chance irruption of mutations. It remains to be added that animal life was already firmly in the stage of the ascendance of "mutation evolution" by the end of the Pre-Cambrian Era.

Regarding the molluscs, one might wonder how it is that the bivalves are fine-and-whisk feeders, while the other three major classes are coarse-and-grasp. The reasons are due to "ontogenetic evolution". Reproductively the bivalves represent the most adult form. The cephalopods, gastropods and chitons have retained forms of coarse-and-grasp feeding methods used by various juvenile forms of a common ancestor.

◼ Echinoderms

These take their name from the spines that protrude through the skin and are most apparent with sea urchins and some starfish. All have a through-gut; but completely absent is the obvious bilateral symmetry in association with longitudinal segmentation as evident in some other phyla.

Echinoderms comprise the crinoids (and their feather star derivatives), the sea urchins, the starfish, the brittle stars and the sea cucumbers. The phylum is notable for its consistent radial symmetry, which may be based on an odd or even number of segments. However, there are most commonly five (which is also the normal minimum) and well known in the shape of the common starfish.

In the case of the stars, starfish and crinoids, the segments are marked by the presence of apparent tentacles (the common starfish hence normally having five such), while the sea urchins and sea cucumbers indicate the segmentation by clear radial divisions of the body, the characteristic number again being five.

Studies of the starfish larva have shown this to represent a bilateral phase. But the eventual 'rays' can be seen to develop within the larva on one side only and strung along a strip of curved protoplasm. This asymmetry has evoked the conclusion that this string of developing 'rays' represents a former bilaterally symmetrical and longitudinally segmented adult animal. However, there is nothing in any of the present living forms to suggest that this was ever the case. In accordance with the ideas that are being presented here, one can rather infer an erstwhile animal that was bilaterally symmetrical when immature yet later became radially segmented. As already described, the gastropod molluscs have resulted from the partial phylogenetic atrophy of one side of a longitudinal forebear: the starfish similarly represents the loss of one half of the organs of a former sessile radial animal, while in a form even nearer to that of the primitive adult. As in most of these cases, one can express the result in a different way. Hence the starfish might alternatively be thought to represent a bilateral animal on which have been conferred the benefits of the through-gut, but which otherwise has reverted to a form redolent of a primitive radial blind-gut ancestor, this development presumably with advantages that are not immediately obvious, as has also been observed in the case of the gastropods. One can of course plausibly argue that in these cases evolution took what was needed from ontogenetic developments and discarded that found to be superfluous for the particular life-style being followed.

One might add that in following their special line of evolution, the echinoderms lost all chance of developing degrees of complex behaviour and any ensuing intelligence that were possible for animals which maintained full or near full bilateralism, as evidenced by the cephalopods, insects and vertebrates.

The echinoderms have certain features unique to them, this giving rise to the question as to how such traits fail to be represented in species found in other phyla, given the type of unity of origin and since obscured links that once existed between the many through-gut animals as claimed in this work. The most clear-cut feature unique to the echinoderms is the water vascular system that operates the external "tube feet", which themselves show modifications of design and purpose between the various species. The reason why tube feet do not feature in other phyla is that the water system arose on an archaic adult form, of which the echinoderms are the modern representatives, while the equivalents of such have either evolved to extinction in the most closely related phyla by the onset of degrees of paedomorphosis or they are unrecognised through being transformed and converted to other use.

The many separate plates built into the outer layers, and on which are based the protruding spines, are also unique, but here one might compare the shells of molluscs, even if these are coalesced into continuous masses. Even so, one can observe that the plates of a sea urchin are not isolated, but form a continuous shell protecting the entire animal except those organs that protrude through it.

The 'pedicellaria', small protective 'jaws' found on the skin of a starfish or sea urchin, have been compared earlier with the 'avicularia' found on bryozoans. One can explain them both as polymorphic forms of a very early colonial ancestor of the modern forms. This ancestor has retained its primitive colonial relationship with these surviving organisms, despite them having eventually evolved into phyla of otherwise fully integrated individuals. The skin 'jaws' have been retained as very junior partners of the main amalgamations of assets.

There are other skin features that appear to be unique to the echinoderms and which may also have polymorphic origin, e.g. the skin gills, which are difficult to relate to the gills of other through-gut phyla.

'Tentacles' are a conservative feature retained by the echinoderms, only the sea urchins being without them. The word, as used idiosyncratically here, describes organs that are ostensibly of different form, but, even so, represent different versions of the same kind of structure. The basic 'tentacle' of an evolved through-gut animal is a fleshy protuberance that may be subdivided. Ultimately the final fine-and-whisk effect may be achieved by cilia. Coarse-and-grasp feeders have little use for such whisking and their ultimate food gathering organs may be suckers (e.g. cephalopods), tube feet (e.g. starfish) or adhesive matter (e.g. sea cucumbers).

Tentacles can usually be claimed to share a common ancestry, but intermediately may be understood as being of three kinds. Mouth tentacles are arranged round the mouth of the animal (as in cephalopods, crinoids, etc.), which was the position on the archaic adult form. However, such positioning on animals in recent times does not imply direct descent from the original state, for various evolutionary processes may have had effect, especially bilateralism. The twin lophophores of animals such as bryozoans and brachiopods represent fleshy tentacles that have been reduced to two by the effects of bilateralism. In the case of the starfish, the present ring of fleshy tentacles actually represents the result of bilateralism further affected by unilateral phyletic atrophy.

Side 'tentacles' occur on the segments of longitudinally segmented animals (e.g. annelids). Their original tentacular nature can be postulated from the standpoint of longitudinal segmentation having evolved from an earlier form of asexual reproduction, with arrested fission. The tentacles on such segments can then not normally function as such, but have evolved to serve other purposes, resulting initially in their conversion into parapodia.

Anal tentacles represent the former mouth tentacles of one of the pair originally providing the through-gut. Since they obviously could no longer feature as feeding organs they have usually disappeared, although adaptation to various other purposes has occurred in some phyla.

Hence only head tentacles usually retain the capture of food as a prime function, while side 'tentacles' primarily are found with locomotor functions; although not always, nor exclusively so. Yet all can have acquired or retained sensory uses and the same applies to gaseous exchange, with their conversion into gills. Head tentacles can be seen in a highly evolved state in the case of the bivalve molluscs, where the functions of fine-and-whisk feeding and gaseous absorption are well combined. Side 'tentacles', as parapodia, are used to absorb oxygen on the polychaete worms and can be seen in more elaborate form as the gills on such molluscs as cephalopods and gastropods. Anal 'tentacles' are especially featured as the gills on some sea slugs, but also persist as 'features' on the rear ends of many annelids and arthropods.

The most archaic living echinoderms are the crinoids, which have ciliated tentacles leading food into an upward turned mouth, while wastes escape through an upward turned anal tube alongside the latter, indicating their through-gut nature. The animals are sedentary (except when assuming the released "feather star" phase) and are attached to the bottom by a long column of their own making. In the Palaeozoic there existed other classes of attached echinoderm that are now extinct. The 'edrioasteroids' and 'cystoids' were similar to crinoids, but the body of the animal was contained in a plated box, as are now sea urchins, but situated on top of an attachment stem in a similar manner to a crinoid.

The stems of attached echinoderms join them in positions that appear analogous to the attachment of shells on gastropod and cephalopod molluscs. This suggests the possibility that the shells that existed on extinct cephalopods were in some way homologous to those of edrioasteroids and cystoids, even though the motile molluscs actually had to carry theirs about. This argument is tenuous, but would at least point to one area of contact with another phylum for adult echinoderms, which otherwise is hard to observe. Should this postulate be true, the implication is that the 'mineral' attachment stalks of crinoids and their extinct relatives is of a different origin to the organic attachment feet of molluscs. This matter is the subject of further attention on a later page.

◼ Chordates

The invertebrate chordates are recognised by the presence of one or more of three features – a dorsal tubular nerve chord, a stiff 'notochord' below it and gill slits in the pharynx. These features have reference to creatures that are elongated and bilaterally symmetrical.

The lancelet is the only creature that completely satisfies all three requirements. The acorn worms are variant in having a second shorter ventral nerve chord, while the short feature thought to represent the notochord on them is held in some doubt as such. The most common creatures considered to be chordates are the many species of sea squirt, even though the adults have only the gill slits to support such a claim. However, the larvae are bilateral creatures and fulfil all other requirements to be dubbed chordates.

Sea squirts can exist as individuals or in colonies, the latter having fused bodies that surround a common excurrent opening. The solitary sea squirt is a good example of the type of creature earlier postulated as resultant from the formation of the through-gut. While the ectoproct bryozoan is also characteristic of this proposal, it is more primitive, being consistently tiny and with the mouth provided with tentacles, these however tending towards a bilateral lophophore arrangement. A sea squirt has no tentacles, the mouth being merely fringed by lobes of small definition.

The life-style of sea squirts is reminiscent of that of bivalve molluscs, in that water is drawn into the gut together with food and ejected with wastes by means of an excurrent opening. However, with the bivalves the gills are basically a filter that prevents unwanted particles from reaching the mouth; the absorption of oxygen is secondary to this. With the sea squirts the gill slits eject any unwanted particles from the flow through them, while oxygen absorption will again be secondary for such sessile creatures. Both systems also have the advantage of obviating the necessity of passing excessive quantities of waste and water through all or much of the gut.

Both bivalves and sea squirts have body cavities that are not coeloms (i.e. spaces lined with mesoderm and which they both also possess), but are formed externally. Such a mantle cavity of many molluscs is often an enclosed space with clearly defined openings, although it is generally most loosely organised and open among the gastropods. The equivalent 'atrium' of chordates is a more perfectly enclosed space with a very specific excurrent opening, the "incurrent opening" in this case being the creature's mouth.

The gills of bivalve molluscs are hence used to receive water currents from outside themselves, large particles being thereby excluded and the finer ones, comprising the food range, being transferred internally to 'palps' and then to the mouth. With chordates a somewhat opposite effect obtains; water currents enter by the mouth and unwanted particles are ejected with the bulk of the water stream by way of the pharyngeal gill slits. The bivalve gills are hence before the mouth and the chordate ones after it. The latter can hence be deemed of parapodial origin, rather than tentacular. This differential origin of gills can be observed with the primitive chordate lancelet, which has both. It retains tentacles at the mouth but also has gill slits after the mouth, these being homologous to the gills on fish.

The functions of gills are basically twofold; to trap food and to absorb oxygen. These functions are also performed more primitively by tentacles and derived features like parapods. It is thus inferable that, generally speaking, gills are modified tentacles. In the case of bivalve molluscs the gills can be imagined as being derived from mouth tentacles, which once formed a screen over the mouth and only admitted desirable elements. Eventually these tentacles fused into the finely perforated organs that constitute bivalve gills. Sea squirt gills, being behind the mouth, cannot be of exactly the same origin and this also applies to the homologous gill slits of other chordates. These can be proposed as resulting from longitudinal segmentation of 'colonial' origin, but with arrested fission that left gaps between the segments as prototypes of gill slits. However, the shapes of gill slits tell against this being the whole story. It seems desirable to consider also the contribution of parapods - i.e. earlier tentacles that have become redundant as such due to intermediate positioning - in the formation of the rather different forms that chordate gill slits display (as well as those of the derived vertebrates).

Returning to the lancelet ('amphioxus'), this creature has mouth and through-gut, gill slits and atrium. It is also bilaterally symmetrical like sea squirt juveniles and acorn worms and has all the signs of many segments, now externally disguised (selective de-segmentation). Around the mouth is a hood with a ring of 'tentacles' that are believed to serve a sensory function. At the mouth itself are "inner tentacles", which combine to form a screen that excludes larger particles. In this function it would seem to emulate the tentacles earlier postulated as being ancestral to the gills found on bivalve molluscs.

The lancelet is fish-like in its shape, but has no side fins and swims about by using lateral movements of its tail. However, most of its time is spent buried in mud, except for the head, and using its tentacles, mouth and gills to feed and respire. This creature would seem to be of some importance regarding relationships between phyla in that it has all the typical chordate structures and it retains a life-style and tentacular appendages that are reminiscent of several other invertebrate phyla. Yet it is this chordate that by its shape and swimming capabilities resembles the vertebrates, the earliest forms of which were fishes that could swim by lateral movement of the tail. This swimming organ, protruding well beyond the anus, is a particular vertebrate trait that also proved to be advantageous, and was hence retained when certain species took to the land.

■ Arthropods

This was the phylum that first comprehensively conquered the land environment, including dry circumstances. The most successful of terrestrial species can run fast, withstand desiccation and even fly. The secrets of their success include a hard outer covering (chitinous exoskeleton), jointed legs and air tubes opening through the body wall (trachea).

The Crustacea remain largely aquatic arthropods. The segmentation of their bodies, as also applies with the arachnids and insects, has been split up into discrete sections, generally consisting of head, thorax and abdomen, with each section comprising a number of segments; this is often disguised by thorough-going retro-fusion. While insects generally retain the distinction between head and thorax, the two are fused into a single mass among crustaceans and arachnids.

The vast bulk of crustaceans are small creatures of bilaterally symmetrical form and a life-style based on swimming. This description also applies to the larval forms of some of the larger ones. Lobsters and crabs have appendages on the rear part of the cephalo-thorax that have evolved to serve as legs for walking on the sea bed, while those appendages on the front part serve feeding or sensory functions. Adult lobsters can swim as well as walk and to do so their abdomens are flexible and provided with appendages to assist with this activity. In general terms adult crabs of most species only walk and with them the abdomen is shrunken into insignificance (especially with the male), being tucked away under the cephalo-thorax. Some more primitive creatures are still known as 'crabs' yet retain some abdominal appendages with which they can swim.

While crabs share the same continental shelf as lobsters, they have also extended their range into areas where lobsters cannot follow, some of them inhabiting the tideline itself, while others are completely terrestrial in their adult form.

As evolutionary developments on fully aquatic creatures, the limbs of lobsters and crabs do not seem a possibility as original functioning 'legs'. They seem most likely to have developed as such as a result of an existence on or near the shoreline. The lobsters were created among certain of these creatures that became thoroughly aquatic again, even regressing somewhat towards a juvenile form, except that the "land legs" were retained as a convenient means of locomotion across the sea bed, where a generous supply of food was concentrated. Lobsters and crabs are omnivorous, being prepared either to kill or to scavenge for their food.

Crustaceans have also spread from the marine environment into fresh water, where are found, for example, crayfish and certain shrimps.

A fused carapace largely protects from above the cephalo-thorax of a lobster. The gills are also protected by this, for it extends downwards on each side, but otherwise they can be considered as external protrusions on the segments. The carapace thus serves to form a pseudo-internal space in a way redolent of the shells of bivalve molluscs and even the atrium of chordates. The ostracods are tiny fresh water crustaceans that are entirely enclosed in a carapace of bivalve form, and bring to mind the twin shells of molluscs and out of which they can protrude their appendages in order to swim.

The barnacles are of great interest in that their life-style is sedentary and they use a fine-and-whisk feeding method. The adult is protected by its hard plates from which it regularly protrudes and then withdraws the whisking appendages on its segments in order to lead food into its mouth. Should one envisage that the ancient crustacean adult form was sessile and fed by means of a mouth ringed with tentacles, then the barnacle clearly approaches it most closely. The four or six plates that protect the adult reflect a tendency towards the retention of radial symmetry of the mantle components that was resisted by the rest of the animal. The barnacles can hence be adduced as exemplary of the crustaceans that represent most closely the ancient bottom dwelling adult form, although they retain to a great degree the general body design of a free-swimming, longitudinally segmented juvenile. All the other crustaceans represent evolutionary variations on the free-swimming juvenile form, with the segments frequently differentiated into discrete functional groups.

The Arachnida are terrestrial arthropods of which the most important group is that of the spiders. The class also comprises the scorpions, mites, ticks, harvestmen and others. The head and thorax are fused and all possess eight legs.

The scorpions share with crustaceans the possession of pincers, but also have in common with some insects a poisonous sting in the tail, although the upward delivery of this is unique to them and makes them readily recognisable. Spiders, on the other hand, have a poisonous 'bite' delivered by paired 'fangs'. They can also in most species produce a sticky filament from the abdomen that can be used to weave traps called 'webs' or provide a 'trapeze' whereby the animal can traverse a space. The arachnids are carnivorous, with a number of smaller species being parasitic.

The insects are also terrestrial arthropods, although some go through a larval stage in fresh water. It has been argued that they constitute the most successful class of animal because of the numbers of both species and individuals, their social as opposed to their physical adaptability and their mastery of flight. The insect head, thorax and abdomen are distinct. The head specialises in feeding and sensing, the thorax in locomotion and the abdomen in reproduction and certain other functions.

The centipedes and millipedes are also terrestrial arthropods, whose bodies comprise many undifferentiated segments, except for some at front and rear. They are hence reminiscent of many of the marine segmented worms. They are generally creatures that shun the sunlight, unlike many of the insects.

Except for the many smaller crustaceans, the arthropods are generally speaking terrestrial animals with legs and segmented body construction. The latter trait suggests overwhelmingly that they are an offshoot of the annelids, having most kinship with the polychaetes. This is most clearly evident with the centipedes and millipedes, which one can clearly envisage as 'polychaete' derivatives that have adapted to the terrestrial way of life. The life-style of arthropods makes it highly unlikely that they owe anything to the oligochaetes or the tube-dwelling polychaetes, but indicates much greater affinity with the free-swimming worms like 'nereis'.

One may find the proposition attractive that the arthropods originally evolved on the shoreline. Here certain of them gradually improved the hard casing so that it provided better protection from the Sun's rays and evolved the jointed legs that enabled them to move about freely on the land surface, as well as developing the necessary musculature for the latter purpose.

At first the larval and juvenile forms all remained aquatic. In the case of the crustaceans this remained the way with those cases within the phylum where the adults developed a shoreline lifestyle. Yet these remained fully or partially tied to the water in that they must continue to lay their eggs in it, a procedure indeed used by dragonflies today, while others have reverted to it. As an example, water spiders use it as their hunting zone. However, the majority of them, including those adjudged to be the most successful, have a lifecycle that takes place entirely on land (and including the air above it), although this involved constraints with regard to the attainment of size.

CHAPTER 8

Metamorphosis and Fusion

■ Metamorphosis

This term applies to radical physical changes that take place as phases in the course of growth and maturation. It is among the insects that it commonly manifests itself, often with spectacular results.

The change between egg and larva is so common that it is generally not referred to as metamorphosis, with the term being restricted to intermediate transformations on the way from egg to adulthood.

Metamorphosis can be 'gradual', as is the case with grasshoppers, the young (or 'nymph') of which resembles a small adult, except that the head is disproportionately large and wings are absent. As the animal grows it passes through several moults, after each of which it resembles the adult a little more.

Metamorphosis can be 'incomplete', as is the case with the dragonfly, the nymph of which is an insect in its own right, but is aquatic and differs radically from the adult. This emerges from the final drastic moult as a fully functional flying insect.

Metamorphosis can be 'complete', as is the case with the butterfly, the larva of which is structurally imperfect as an insect and becomes finally a rigid 'pupa' from which the 'perfect' insect eventually emerges.

The evidence suggests that insects arose in fresh water, for many of them even now start their lives in this habitat. While one can well imagine that the crustaceans took steps to becoming terrestrial by climbing out onto some seashore, with the insects it seems more likely that their forerunners clambered up onto swamp-standing vegetation, using parapodial projections to do so. At this stage they would have no wings and were segmented like annelids or centipedes. The two most amazing aspects of complete metamorphosis are the degree of change and the appearance of wings. The wings are so special, being all of the same membranous character and always found on the same segments, that it seems a certainty that, from out of various primitive species of wingless insects, there arose just one interrelated winged group. From this has evolved the vast range that now exists.

The characteristic division into head, thorax and abdomen is basic to all insects. The primitive wingless insects exhibit it, but without the clear-cut differentiation shown by advanced insects, such as ants and wasps, and show no sign of metamorphosis. Thus, while metamorphosis is not essential to be an insect, it does appear in some form or other with all the fliers and the winged phase is only found in a special final adult form.

Metamorphosis clearly has some connection with moulting, but is also derived from a separate ability to go into a quiescent state within a self-made casing. The latter was probably originally a device for surviving adverse climatic conditions, in particular aridity.

The original insects were arthropods with regular segments, which in the adult form could survive out of water, but whose eggs and immature stages were aquatic. Any incipient tendency to specialise among the segments was enhanced by the needs of living on the land, while the chance existed for changes to be transmitted back to the various immature stages. At the same time growth was accompanied by moulting and the water-to-land transition came to be facilitated by a special moult during which, as evolutionary time passed, ever more radical changes could be accomplished, especially if the land environment was becoming drier.

With many insects the main moult began at an ever-earlier stage, so that the degree of change during it must increase. At this stage the form at the main moult had become a pupa. Pupation first involved a quiescent stage wherein the removal of all unnecessary features occurred, the material released absorbed, and transformation into the adult status achieved. Eventually even the most advanced form of the larva was different from the adult in every respect. Once this happened the only way that pupation could work was by the complete reversion of the enclosed larva into a primordial mass of cells. Changes then took place within these that allowed them to be reassembled as the perfect insect. All the insects that could pupate were fliers and proto-wings had evolved at a much earlier stage. All flying insects once had larval forms that were nymphs. These developed four movable organs on their backs, to cut a way out at the end of the pupal stage and which were destined to become one or two pairs of wings. In the case of the flies one pair of these does not provide wings, but movable stabiliser knobs instead. With the beetles the anterior pair have been modified to provide hard wing cases.

The vast majority of insects eventually evolved to a state that enabled their entire life cycle to be on land. In achieving this they acquired many complex means of ensuring the survival of their larvae, either by laying the eggs carefully so that a food supply for the larvae was at hand, or in providing aftercare as do the social insects.

The insects are so basically uniform that their single ancestor must have been omnivorous, with the arachnids being their first predators, having followed them from the water. However, the insect form was so successful that it came to encompass every manner of feeding, including following the fully predatory lifestyle.

In the case of the dragonfly, whose terrestrial adult and aquatic nymphal forms are both carnivorous, one might then presume that it was the adult who first became a hunter, with the change being transmitted back to the young; yet it was probably the reverse! In all such cases some form of transitional omnivorous stage would be involved. A further development of the opportunist insects allowed some of them to parasitize other insects and other inverebrates, as well as vertebrates when these eventually made their appearance on land. Insect endoparasites are only present in the larval stages; the emergent adult flies off. Ectoparasites like fleas and lice are permanent attachments; they are characteristically without wings, but are larger than ticks and mites, which are arachnids.

▒ Introduction to Fusion

Unicellular and non-cellular microbes have been able to fuse with their fellows deep in the past. Such fusion may have been temporary in most cases, but when conditions were right it could have led to permanency of the condition and the creation of new species by accretion. Extremely early fusion may have taken the form of steady build-up, as seems to account for the strings of genes forming the chromosomes, as well as the severalty of the latter themselves within the cell; or it may eventually have happened as a simple combination of equals, as might explain the existence of diploid cells.

The probability presents itself that this fusion, in both its dual aspects of accretion and combination, may be a fundamental property of living matter. Hence it might be thought to transcend the small worlds of cells and microbes, as well as their infra-life predecessors, and be applicable to the larger beings that are more familiar to our senses, those known as multicellular animals and comprising the Metazoa.

▒ Multicellular Fusion

If agglomerate fusion is to be recognised, repetitive units of an animal must feature as additions to the original. These are often disguised in later forms through being differentiated by specialisation. In multicellular fusion there are two forms of agglomerate function that are clearly discernible: there is colony forming, in which adjoined individuals may be created volumetrically, as well as radially or linearly, and there is segmentation, which basically results in radial or linear addition. It is to be appreciated that in both of these cases the 'fusion' leads to the production of a prospective new individual, which then fails to detach itself. In other words, its ancestors would have broken away to become separately functional creatures, even if not physically apart. The evolutionary advantage of avoiding separation is in the prospect of specialisation. Among companions, this is made possible by the sharing of a common food supply collected by selected feeding members. Of the two forms of agglomerative fusion, it is clear that colony forming is more primitive than segmentation. Yet, in a way, both are extensions of the same process at multicellular level as the "replica forming" of genes at cellular level.

Regarding the "combination of equals" among multicellular animals, the evidence is much more obscure. However, the way in which it may have occurred was at the times when one was reproducing by asexual budding, but the offspring failed to separate and the pair carried on as a single animal. Once again one can indicate that parent and offspring in such cases were then free to differentiate within the new whole, whereby specialised parts could be acquired.

A form of specialisation can be dubbed 'retro-fusion'; this goes beyond the failure to separate and represents a degree of reintegration so that signs of former individual identity are masked or eradicated. The term is more applicable to segmentation and combination of equals than to colony forming.

CHAPTER 9

Pre-Cambrian Evolution from Microbes to Complex Organisms

◼ Pre-Cambrian Life

Geological history is divided into several eras, the most recent being the Cenozoic (modern life), preceded by the Mesozoic (middle life), the Palaeozoic (ancient life) and the Cryptozoic (hidden life). The first period of the Palaeozoic began some 600 million years ago and is known as the Cambrian; time before that is hence also called the Pre-Cambrian.

Cambrian life has left abundant, if incomplete, evidence in the rocks for the existence of most of the phyla into which animal life is divided. Even those for which there is apparently no evidence, e.g. the ancestors of the vertebrates and the protozoa, can be assumed to be present, but without leaving a discernible trace, because of either minuteness or bodies with no hard parts; or because they are not recognised. The Pre-Cambrian rocks were long considered to bear no evidence of life, but subsequently traces have been discovered of what have been claimed to be life forms. Everywhere in the World where the Pre-Cambrian rocks appear they show this general dearth of life, while the subsequent Cambrian layers indicate the presence of shallow seas teeming with life. This apparently abrupt appearance or burgeoning of life is not consistent with evolutionary theory. In the context of this work the apparent sudden onset should be recognised as being due to a gap in the evidence, rather than to any event verging on the miraculous.

Humans share with many other creatures the state of being land-dwelling vertebrates, and in considering these earlier periods of life's history we must beware that our thinking is not contaminated by the general experiences of our own lives. The life forms of the early Palaeozoic were all aquatic and all invertebrate. Life on land then was impossible and creatures able to secrete hard body parts were probably in a distinct minority. However, it is from the skeletons and housings of the latter that we are mainly able to recognise life in the fossil record. From this point of view the evolutionary spearhead produced at the start of the Cambrian is due to conditions arising that enabled living forms to absorb mineral matter into their tissues and to transform it into protective and stiffening structures. This development implies some general change in the environment has allowed it to take place.

However, the initiative may have first occurred as a result of minerals being incidentally absorbed into bodies by those using a certain way of feeding and then evolving so as to find a use for otherwise superfluous matter. The apparently rapid spread to other species would then be due to the predatory habits of these, by which they would pick up minerals in increasing quantity from their prey and start similarly to turn them into hard parts, e.g. skeletal use. This need to develop protective and supportive structures would accelerate their complexity and diversity through causing evolutionary pressure.

Both plants and animals emerge from the Pre-Cambrian in forms that exhibit the distinctions between these two major kingdoms as still recognised. There are however three bridges uniting them. For one, the fungi have to be considered, which do not contain chlorophyll and to some extent behave like animals in having a scavenging and sometimes pseudo-predatory role. They feed by initiating or accelerating decay and hence can be supposed to digest their food externally before absorbing it. Then there is the unicellular form *Euglena*, which, though basically a flagellated protozoon, yet secretes the light-absorbent substance chlorophyll, this being otherwise the prerogative of the green plants. Finally reproduction by 'sexual' means needs to be considered, forms of which are found in both main kingdoms.

Evidence for plant life in the Pre-Cambrian is more substantial than for animal life and is largely in the form of calcareous algae, which were plant forms that could incorporate calcium into their tissues and which have contributed to the building up of certain kinds of limestone. These were presumably the prime source from which filter-feeding animals initially obtained the materials for building up their 'hard parts'. The development that accelerated the apparent sudden emergence of animal life (and indeed most life as we know it) at the start of the Cambrian was an increasing turning away by members of the animal world from the 'fine-and-whisk' method to adopt 'coarse-and-grasp' feeding. In addition there was a need for defence against increasingly voracious feeders, for even the hunter itself could become the hunted as the scale of predation on other life forms accelerated.

Despite the scanty fossil record, it is clear that in Pre-Cambrian times life not only emerged out of the lifeless, but also had made considerable progress from its far simpler physical and chemical beginnings. Sexual reproduction became the major tool of evolution and it is evident that even in Pre-Cambrian times it accelerated the appearance of new kinds of plants and animals.

It was argued earlier that reproduction was not new ground that life at one time started to break, but had indeed been its companion from the start. It will be argued later that living began by the effect of light from the Sun on the behaviour of magnesium ions. Such chlorophyll (magnesium + nitrogen) gave rise to carbohydrates (carbon + hydrogen + others), which had the potential for an apparently endless number of forms. Chlorophyll is fairly stable, with its contribution being purely chemical; infra-life only occurred in the unstable carbohydrates it produced. While chlorophyll operated as a result of the alternatives of light and darkness due to the Earth's rotation, the carbohydrates were not sensitive to light, but to heat, which itself tended to fluctuate in a like manner, but was subject to retention plus selective variability.

The production of carbohydrates by the action of light on chlorophyll was irreversible, but the products themselves were unstable. The processes whereby ever more complex compounds were unmade, as well as made, occurred among the carbohydrates and these were the beginnings of infra-life very deep indeed in the Pre-Cambrian.

The processes of infra-life, proto-life and evolution began among the carbohydrates and came to be increasingly affected by changing environmental conditions, but initially by variable temperatures. A regularity caused by day-night alternation of light led to reproduction, but carry-overs and irregularities occurred that interfered with this, eventually to initiate infra-life and this evolved to proto-life.

Yet early sex and life can hardly have been of the same order as those processes we know now. In order to get some provisional grasp of what the earlier forms were like and the reasons for them, it seems apt to look in some detail at the reproduction methods of the ciliate protozoon called *Paramecium*.

■ Conjugation of the Paramecium

1. Two paramecia approach each other, each having a large and a small nucleus.
2. The two make contact and the small nucleus of each divides into two.
3. The small nuclei divide again to make a total of four such in each microbe.
4. Three small nuclei in each degenerate, while the fourth divides again into two.
5. One of each pair of such small nuclei migrates into the other microbe.
6. The migrant nucleus fuses with the remaining one it encounters in each case.
7. The microbes separate and the original large nucleus of each degenerates.
8. The other nuclear body divides into eight.
9. These 8 parts develop into 4 large and 4 small nuclei, 3 of which latter degenerate.
10. The remaining small nucleus divides into two
11. The whole microbe divides, each half receiving 2 large and 1 small nucleus.
12. The small nucleus divides into two

13. The whole microbe again divides into two, each half receiving one large and one small nucleus.
14. The resultant microbes may then continue to reproduce by normal cell fission, with division of both nuclei (vegetative reproduction).

At stage eight one can postulate that each recent conjugant actually represents four potential individuals, as the result of the division of its small nucleus and exchange with the other animalcule. The exchange has ensured that the resultant paramecia embody a combination of characteristics of both original partners who, though virtually alike, would not be exactly so. This gives at least a potential for evolutionary advantage in the same way as sexual reproduction does in multicellular organisms. The crux of the matter lies in the reason for conjugation to arise in the first place. The answer is obviously buried somewhere in the obscure earlier history of the paramecium.

The evidence indicates that it is the small nucleus of the paramecium which solely governs reproduction by conjugation and the consequent transference of hereditary material. The large nucleus keeps the being functioning until a certain stage is reached. When animalcules come together thus, it is just as though they were trying to fuse into a new whole, a feat that their present complexity prevents. However, it is difficult to imagine how paramecia could have evolved to their comparatively complex state (in comparison with amoeboid protozoa and other unicellular types), plus their proclivity to conjugation, without some actual permanent fusion of more primitive ancestors, under conditions that have favoured such an event.

Experiments have suggested that paramecia under less than perfect conditions cannot survive indefinitely by means of vegetative reproduction (binary fission). They pass through a mature phase during which conjugation not only becomes possible, but also necessary for ultimate survival by renewal of the nuclear material.

Paramecia are divided into strains within which conjugation does not occur. If an animalcule has the urge to conjugate, but not the opportunity, it has the alternative of autogamy (self-fertilisation), in which stages 5 and 6 above are replaced by a fusion of its own nuclei within the individual. One presumes that this has some value with regard to rejuvenation, but clearly does not provide the same evolutionary possibilities, however slight these may be.

The original fusion seems to be evidenced by the presence of the two different nuclei. In the past, conditions would seem once to have been right for certain primitive ciliates to fuse. The way in which this came about may have been a failure to part after amoeboid style binary fission. This could have occurred during an earlier lost lifestyle stage. The two adjoined creatures must be assumed at first to have worked individually, but with shared assets; each would have its own nucleus. Yet where binary fission took place simultaneously, each hemi-nucleus could have combined with one produced by the other partner. This was sufficient to rejuvenate the nuclear material and the newly paired creature would now have greater integration, because the two nuclei were now theoretically exactly alike and hence "knew each other". The next step was the specialisation of the nuclei, once traces of the former paired structure grew obscure.

With the passage of time, the evidence for the members of the species being indeed fused individuals would be totally obscured, with the parts becoming integrated and co-ordinated to form a stable new kind of individual. When such reproduced by binary fission it would be to create like copies of themselves, not of their unicellular mononuclear ancestor.

With more parts available the chance for specialisation was increased. The large nucleus took over those functions proper to a unicellular microbe, such as an amoeba, while the small one undertook a specialised function relating to reproduction by conjugation. It is almost as though the large one was in charge of fission, while the small one took over fusion. The process produces surplus nuclei, which, being superfluous, disintegrate. The evidence of conjugation suggests a one-time total of eight nuclei, i.e. eight ancestral ciliates, or four of the first fused versions. With the small nucleus specialising in fusion, it would seem that there have been two more acts of retained fusion to produce the species we know today.

During the course of this evolution, the paramecium abandoned its erstwhile life-style to become a free and directionally orientated swimmer. The advantage to the species of the original fusion was that the duplicated parts thus resulting were open for diversification and specialisation. The disadvantage seems to have been the partial loss of the immortality that is enjoyed by those simpler protozoa whose only option is to reproduce vegetatively. The small nucleus is an unstable component that is sensitive to a worsening environment and initiates conjugation: during this operation it is dominant over the large nucleus. Thus does the small nucleus assume temporary control over the whole creature. The main factor determining the onset of conjugation is an increased awareness of difference between internal and external energy.

The historical fusion of earlier simpler forms to make a single non-cellular animalcule in this way has not greatly increased it in size, but has led to a more complex creature, because parts duplicated by fusion have become available for differential evolution. The complexity achieved is admirable, but there is a structural weakness leading to instability under worsening conditions

■ Pre-Cambrian Reproduction

Conjugation of paramecia might appear to result from the urge to take fusion even one step further. This urge seems to be fuelled by the threatened onset of nuclear disintegration. With permanent complete fusion not being possible, it is replaced by the reception of a hemi-nucleus from the other partner, which itself is in a similar condition.

The development of certain protozoa into fused non-cellular organisms sets them apart from the simpler organisms in the phylum, which have evolved as best they can within the bounds of a single cell. However the particular development that led to creatures like the paramecium had one great disadvantage; without the separate building blocks of cells in close association, the potential for the sort of size attained by organisms made up of multitudes of independent cells was missing, and this likewise precluded the same sort of complexity being reached and other evolutionary achievement.

However, the sort of fission that occurs with paramecia during conjugation is very similar to that employed by multicellular organisms (meiosis) to produce sex cells (gametes). The difference is only that the specialised producers (gonads) of such sex cells can no longer conjugate and the gametes they produce must find other means to achieve fusion.

Regarding the gametes, in the case of females, the ovum gives the impression of being representative of the pre-fusion unicellular individual, while this would seem to be less so on the male side. The tiny spermatozoon is very numerous, but is not designed to last on its own; apart from the lack of bulk, the reduction achieved serves primarily the need to be motile and the carriage of the necessary quota of genetic material. The ovum is normally far less numerous, but far more durable, than the sperm. Once fertilised by the entry of a sperm, the ovum divides in a way that is reminiscent of the fission sequences of a paramecium after conjugation, except that the new cells always remain in contact with each other.

The phenomenon of parthenogenesis, in which females produce further females without repeated fertilisation by males, would seem to have something in common with the autogamy of paramecia. Indeed, with these animalcules not being divided into sexes, one can conceive both members of a conjugating pair as being 'female' entities, even though the transferred small nuclei can be perceived as playing the role of 'maleness'.

The reproduction processes of the paramecium suggest strongly that the contents of any nucleus can be rejuvenated by being disintegrated, followed by changed disposition and suitable reintegration. Additional advantage would seem to be derivable from an exchange of nuclear material with a sister cell, which not only has a possible further rejuvenating effect, but also gives prospects of evolutionary changes, which may embody improved prospects of survival.

As already described in detail, prior to ordinary binary fission of any cell, the stable nucleus is organised with its genetic material in the form of chromatin. Before division this matter is reintegrated and reformed into more definite structures known as chromosomes. It is only at this chromosomal stage that the cell can divide. One might postulate that the chromosomes represent original individual bodies that have resulted from successive acts of fusion in the past. In similar vein, one can regard the genes that arrange themselves along the chromosomes as being themselves the result of the accumulation of very complex molecules. It is in the very complexity and individuality of the molecules that the answers to the inherent problems must lie. It is in the enormous multiplicity and flexibility of the carbon compounds that were involved in the eventual emergence of genes that the apparently infinite possibilities of infra-life can be seen.

One might imagine a watery environment in which the ingredients were being built up from simpler beginnings. As already described, the chemicals were bathed daily in the Sun's rays and were able to build up by fusing forms that in turn became marginally more stable and hence more durable. The molecules of like nature would be grouped together in a specific way and in such gregarious co-operation were better able to dominate their immediate surroundings. Each group would be surrounded by matter that had not developed to the same degree of complexity and whose constituents were to a degree chemically distinct. (Brought to mind at this stage is the concept of a nucleus with its molecules representing the genes of a single chromosome.) The continuous bathing in energy meant that further change was possible, if not inevitable. Nuclei could fuse and the dependent matter would be shared by two virtually equal yet possibly ever so slightly variant groups of genes. Once such amalgamation occurred the groups could disintegrate internally and mingle with each other, so that a new 'nucleus' was formed, consisting basically of two 'chromosomes'.

By such processes of fusion 'chromosomal' groupings could build up within a unit called a 'nucleus', but this itself became surrounded by less evolved and chemically distinct matter constituting and defining the bounds of a cell. This matter depended on the nucleus for continuing existence and was also, to some extent, under its control.

The principle at stake here is the "fusion of equals", leading to duplication or multiplication of parts and hence eventually to the specialisation of function within an integrated individual. Once specialisation took place, exact reversion to original forms was impossible, while any inexact reversion that was coupled with enhancing the chances of survival, although unlikely, became desirable. The immense complexity of the molecules made it possible for an infinite variety of combinations, with subtle differences arising from gene to gene; and hence from chromosome to chromosome.

At the same time it is worth recalling the earlier and simpler days of each cell and the fact that, though advances could be made by fusion with one's peers, at every stage of evolution the ability to fuse with simpler compounds had not been lost. For the more complicated combinations, the absorption of the simpler forms became 'feeding' and this particular unequal form of 'fusion' involved the complete disintegration and loss of identity of the minor partner.

To recapitulate, absorption of otherwise excess energy continually caused the molecules of forms of matter involving infra-life to combine, with the fusion being a chemical process producing a new whole and an amount of discarded material. This was a developing process and, as the molecules of infra-life became more and more complex, they had to grow in size and maintain or regulate their immediate energy flow requirements by continuing to absorb suitable lesser molecules, thus sustaining their chemical processes and producing waste products. This can be dubbed the "fusion of unequals".

At the same time the evidence suggests that growth in complexity resulted from the "fusion of equals". This was a form of physical integration of molecules and compounds that occurred when circumstances were suitable and it produced no wastes. As this process advanced it resulted in entities that were more and more complex, arising from a long series of fusions. Each fusion of equals, however, meant that in such complex structures, albeit so tiny, there would always be slight variations that could open the door for further differentiation after fusion. This is the very stuff that evolution is made on. Once post-fusion differentiation had set in at any stage, the possibility for the molecule to revert to its former pre-fusion parts would become difficult and eventually eliminated.

However, in the case of genes, when reproduction is about to take place, one can observe that they are temporarily discernible as individuals along a chromosome. At this time the chromosomes have re-emerged as individuals within the nucleus. Binary fission seems to occur within a cell (or similar unit) when sufficient material has been assembled by the fusion of unequals, and then organised and fused by the nucleus to allow the basic material to reproduce itself. This is accomplished at the gene level by each gene making a copy of itself and, when the process is finished, original and copy part, which occurs as a separation process along each chromosome.

In a way this looks like a form of "de-fusion of equals". Here, however, the preceding fusion has been replaced by the artificial creation of a 'partner'. This is not easy to reconcile with the theory being expounded here, but one can regard it as going back to the behaviour of the historical components of the gene, but not of the chromosome. The progressively simpler molecules of the ancestors of genes act in retrospect, having previously fused with like partners at all stages of the process. At last the gene stage is reached, which results in the possibility for binary fission of the components still being retained. As the twinned parts progressively separate, they each maintain the original half structure of the gene, upon which is then built up an adjacent replica from suitable components already present in the nucleus.

The gist of this might be more easily appreciated by understanding that, in order to reproduce, a living unit has firstly not simply to 'disintegrate' its nuclear parts, but rather to reduce them to more primitive components from which rebuilding can commence. A parallel to this can be seen to occur in multicellular animals by the way that a developing embryo goes through stages progressively representing juvenile versions, not of its present self, but of its various ancestral forms.

Evolution by means of the fusion of equals has its limitations. These are illustrated by the paramecium. This creature, in the light of observations made above, can be regarded as a progressively integrated series of individuals contained within one skin. A limit on size can be observed. Another limit is on fusion. This is illustrated by conjugation. This allows nuclear material from two sources to be exchanged and fused, but the two entities themselves do not fuse permanently to form a new whole. A further limit is on immortality. The principle on view here is that, as the paramecium has evolved by a series of fusions and specialisations, so has the consequent complexity made reproduction by simple binary fission a more difficult operation and one that "wears out" the strain, which itself causes it to fail under adversity. Conjugation initiates a new strain with more thorough rejuvenation, as well as selection into mating types and theoretical prospects of further evolution.

The fusion of equals was the way in which earliest evolution worked. The former 'two' became a new whole 'one', yet with duplicated parts that were then free to diverge and specialise within the single being. The evolutionary possibilities of this non-cellular life form would seem to have been virtually exhausted by the paramecium. Life on Earth had to achieve further evolution by other means and these were in the form of co-operating entities called cells, which reproduced by 'sexual' methods.

In multicellular sexual organisms each cell 'represents' two former individuals in the 'fused' condition, but it can ignore the duality and reproduce bilaterally to produce twin replicas of itself by a copying process. Nearly all such body cells have lost the ability to revert to representative earlier forms by binary fission, despite their dual nature; this is the special function of cells that are only produced in the gonads. The sex cells thus produced (gametes) have half the number of chromosomes of the other body cells, within which the chromosomes can be seen as twinned, apparently unnecessarily so.

Non-cellular organisms were evidently limited in the size they could evolve to and this was accompanied by limits on complexity as species. The paramecium represents the free-swimming type of microbe, but other ciliates are found in colonies, clumps of individuals whose actions are independent, but which appear to act in some unison because their aims are exactly the same. In perfected multicellular life the individual cells are in intimate contact with each other, occur in differing categories, and, without being fused, behave in divergent ways to suit the needs of the whole organism, being in turn supported by it for as long as is necessary or possible for the sustenance of the collective identity.

While unicellular and non-cellular microbes can reproduce themselves with immediacy by binary fission, this is impossible in the case of most multicellular beings. In the build-up of such organisms the integrity of the single cell as a concept must be respected; this means that such can divide, but generally do not fuse. One can contemplate each entity as starting off its life as a single cell and then dividing and re-dividing until the complete organism is achieved.

However, this last is not quite right, for it takes no account of sex; all multicellular organisms have this. The implication is that they are all made up of cells of diploid nature. The paired chromosomes suggest an original significant act of fusion among unicellular ancestry, which resulted in such an advantage that any later acts of fusion were inhibited, except for the specialised sex cells.

In early Pre-Cambrian reproduction a possibility was initiated that enabled unicellular entities each to fuse with a neighbour in a way that left the inheritance stock – as represented by the nucleus – separable, while the rest was sufficiently integrated to allow the new combination otherwise to act as a whole. Owing to the inheritance material remaining discrete, the possibility remained for 'de-fusion' when appropriate. This facility became the basis of sex in that, after parting, the resultant haploid cells could recombine with new partners. It is the latent evolutionary advantage of these proceedings which effectively prevented the paired chromosomes of the resultant diploid cells from diverging from each other and specialising, with the important exception of the specialised pair that eventually came to be distinguished on a sexual basis.

To begin with, the sex cells were not differentiated into male and female, but simply formed a means by which material could be exchanged and new strains created. In some ways this resembled the conjugation of the paramecium, but involved the whole unit, rather than just the nuclear material concerned with inheritance. Cellular beings, diploid in nature because of past fusion having caused duplication of chromosomes, could split back to something virtually identical to their former selves and then recombine with different partners. As with the paramecium, the advantage of this was that the genes could be spread by such meiosis between different strains, so that whole populations consisted of interrelated members. In this way the 'integrated species' was created, which provided evolutionary advantages that were not possible for creatures whose reproduction was permanently restricted to simple binary fission (using mitosis exclusively). From the heredity point of view the latter were doomed to consist of continually separating strains that had no means of reintegration, but whose integrity was maintained through the lack of evolutionary opportunities.

The cells that could thus fuse with new partners after meiosis were in essence "identical twins" and can be regarded as being all female. This system was adequate while the cells themselves were independent entities, but once joined together into co-operating groups, the system became inadequate and as specialisation of function of cells progressed the inadequacy increased. The growing complexity of organisms increased the difficulties of the specialised sex cells when it came to getting together. This was of course surmountable by making all such products of meiosis motile. Yet duplication always invites specialisation. The re-fusion differentially of the all-female products of meiosis provided initially the evolutionary prospects that allowed the male and female sexes to emerge.

The advantage of the crossover of material lay entirely in the way in which genes could pervade and reintegrate otherwise divergent strains. Hence the non-hereditary functions of a species could quite readily be provided by just one partner, even if their future development was going to be governed by genes derived from either. From this point of view it makes sense that one partner should specialise in being a well-stocked and durable body of little motility, while the other became short-lived and very motile, pared down to the bare essentials for the transference of genes, but with these drastic reductions compensated for by production in huge numbers.

■ Stem Cells

On this matter of heredity and reproduction an obvious puzzle arises (should one so contemplate) as to how one fertilized cell - given the requisite continued supply of nutrients – can divide again and again to provide the vast array of different cells, tissues and organs, together with the particular appearance, that are characteristic for the species. The answer has to lie within the genes themselves, each of which governs a special aspect of growth or function. While it might be too much to assert that cells literally 'know' where they are, it seems pertinent to postulate that the relevant genes are activated in a cell that finds itself in a certain position relative to its companions. In this way the genes that are present in every cell and seemingly so superfluous with regard to the reproduction of the whole organism, are still active with regard to the reproduction of individual cells, albeit simply by binary fission. Each cell has its full complement of genes, but only suitably specific ones are used, depending on where the cell finds itself. Circumstances hence normally inhibit the activation of the wrong genes.

Yet the above does not apply to all cells. One can otherwise look upon it differently and recognise that a specific kind of cell in a multicellular organism has been controlled throughout its ancestry by a certain group of genes. These were in the original fertilised cell and were predestined to govern the production of generations of cells. This would lead to a specific kind of cell at any stage of its development. However, the system is more flexible than this. There are indeed certain core elements in multicellular beings that are capable of developing into any other type. The ones with these capabilities are known as stem cells and reflect the organism in its primordial undifferentiated cellular condition. However, derived cells cannot revert to being stem cells; evolution does not work in reverse.

Stem cells constitute the basic condition from which all the different tissues and organs of a multicellular organism develop. Apart from this, they persist in the body as a resource that can replenish or carry out repairs on cellular structures that are failing.

Ontogenesis and Phylogenesis

The former term refers to the development of an individual life form during its lifetime, while the latter refers to the development of the species throughout its total history. In the course of the former process, various earlier juvenile stages passed through during the latter may be recognisable.

The evidence suggests that by the end of the Pre-Cambrian representatives of all the phyla of the animal kingdom were present, although there were no vertebrates. It follows that consideration of the ontogenesis of members of the various phyla might provide clues about their phylogenesis and hence of their Pre-Cambrian origins. All animal phyla, save the Protozoa, are comprised exclusively of multicellular animals that produce male and female gametes. Even among the Protozoa some do likewise, such as the 'colonial' Volvox.

CHAPTER 10

The Start and Persistence of Life

■ What Is Living?

Living is the activity of that matter which has life. In this sense matter can be divided into three kinds.

1. Living matter.
2. Dead matter, i.e. that from which life is known to have departed, namely organic matter.
3. Lifeless matter, i.e. that which, as far as can be observed, never lived, namely inorganic matter.

■ What is the Nature of Life?

Although the above third category is always structurally present, most of it has nothing to do with the ability to move by parts of the car of life, but some kinds of it serve as constituent parts of the fuel that is continuously being poured through the system as nourishment, and which occur in the food, drink and air taken in by the higher animals. The rest simply serves as the road on which the car of life is driven.

The problem that presents itself is how some constituents of '3' transformed themselves into '1'. The existence of an intermediate phase during which some were selected and promoted has perforce to be assumed.

The perceived knowledge of what living things look and feel like, are made of and are able to do does not in itself enable us to comprehend the actual and unique nature of life. The study of birth, life and death as they appear to be to the particular observer, isolated as he is in his conscious present time, will never give a satisfactory answer, because life is a product of the past and it is only immensely deep in the prehistoric past that any big secrets will be found to be hiding. Anyone curious about life and death can be likened to an ignorant, albeit intelligent being examining clockwork without a winder, watching the clock perform and then eventually stop; then he wonders why it has ceased to function and how he might get it going again. He may take it to pieces, examine each part minutely and put it perfectly back together again; but to no avail. Without realising that it has to be wound up and that the winder is missing he will wonder on in vain.

What then was the missing 'master winder' that could create life? What on Earth could it have been? Or was it indeed on Earth? It is certainly not to be found there in the present at all, but has been lost at some time in the past. A neat approach is to postulate a single yet perhaps lengthy opportunistic winding up event in the past after which life has always shown a tendency to require some sort of regular small scale rewinding to keep going. The continuity has entailed a still ongoing process involving countless physical and chemical changes. The functioning of the big clock of life eventually was mainly taken over by and still largely comprises a lot of little clocks, the function of which is to wind up replacements virtually identical to themselves before their own allotted individual time runs out. As yet, not only do we mortal folk totally lack the knowledge as to how the big clock could be restarted, should life totally stop, but even restarting the little ones once they stop is quite beyond our capabilities. Life is not the clockwork itself, but its function, and this has to be continuous for life to exist and persist.

It is hence being postulated here that life is a process originally wound up deep in the past by a master winder and it has managed to keep going ever since; but an alternative conception is possible. This can be envisaged as a primitive motor vehicle with a starting handle and a virtually inexhaustible supply of fuel. The vehicle could be in a prepared condition, but, since it was a bad starter, without a driver to initiate activity by turning the handle determinedly and at length there is no chance of it going anywhere.

One can imagine that pre-organic matter on Earth was once such a car, all tanked up and ready to be driven, if only the driver was ready to make a journey. What then was this driving force that first turned the engine over to enable it eventually to fire and then be driven? Let this question be shelved for the time being, except to add now that the machine could not be stopped for refills of fuel; once mobility was achieved, refuelling must take place while in motion. Since it first started it had to be kept going continuously by having a persistent driver and an available fuel supply that was virtually inexhaustible. The original starting handle has somehow been lost since first used to start up the motor, this making any restart after breakdown impossible.

■ Prelife, Infra-life and Proto-life

To be or not to be? That was the question. Whether life was to occur or not can be answered in the positive; but did it suddenly appear or emerge very gradually from a lifeless mineral wilderness? That is the dilemma that confronts us.

If readers who are implacably opposed to the notion of biological evolution have reached this page, they should perhaps cease reading at this point. A choice is necessary. One can neither prove nor disprove that there are any kinds of extraterrestrial living entities in the universe. The same applies to the existence of anything in an ultimate sensitive transcendental form. Acceptance of such notions involves belief that somehow a kind of life force was transmitted to planet Earth and proved able to take root and flourish there. However, can one assume that any such entity would be so interested in Earth that it consciously caused events and set up the conditions that led to life on this spherical speck in space?

Otherwise what type of start could it have been then? In the light of the argument being pursued here one might resist as an analogy the line-up for a cross country race, where the competitors are strung across a field so that all can start simultaneously on receiving the signal to go. Otherwise one can look for a metaphor in a modern marathon, where thousands are confined along a length of a city road. The further back you are the longer the wait before you can get going after the gun. This analogy seems to break down in that the front runners start off fully developed as athletes and can get into their full stride immediately, while those behind are both inferior and hampered until those immediately in front of them have eventually moved off. This smacks of an act of creation before any evolution could take place. Hence this too is an unsuitable example in this context. If one excludes spiritual intervention, the metaphor works best if the beginnings of proto-life are not looked for in an explosive rush of front runners but as a more gentle beginning comprising the infra-life period. Indeed, the start of the evolutionary marathon can be looked at differently again; the runners were not confined to a strait lane, but assembled on a circular track and any small group of them could be the first to start to move gradually outwards and past the others before they knew what was happening. Such a race was a sort of aimless orienteering set off at any point on the track favoured by the starter. Realization that the race was already under way would be transmitted gradually to other competitors, who would then also move off if ready and able.

In reality the actual pre-life runners were not on a track, but grouped on the surface of a sphere. Yet they were not about to embark on a race that only involved covering distance; once they got going the race was one to avoid elimination and ensure survival, even though the runners themselves were quite unaware of such aims. Yet how did the prime movers get going at all? It was no explosive start. The first steps into the race of life were very tentative, the pace of infra-life being much slower even than a tortoise, but with prospects of eventually exceeding the performance of a hare.

But the slow start did not immediately give life in its proto-life form, but in the transitional phase that here has been dubbed 'infra-life'. Infra-life involved pioneering minerals that were gradually being transformed in such a way that in retrospect we can see that they were taking the first preparatory steps on the track leading to life proper. The justification of this infra-life stage is a philosophical matter. All the evidence points back to a time when the earliest life existed in its most primitive form (proto-life). Much earlier the Earth had consisted entirely of minerals (pre-life). Yet among the latter were those comprising elements in the right condition so that, when subjected to a specific sequence of circumstances, they could take the form of ever more complex molecules among which were those that eventually were to be of life-bearing nature. This involved combinations with carbon, but without life as yet being present. (This state was infra-life). Such is an inevitable conclusion based on the certainty that life exists and the evidence that Earth was originally simply a lifeless sphere, albeit already providing certain minerals and about to experience conditions that were collectively going to produce the right opportunities. A sudden transition should not even be contemplated.

In the infra-life phase the relevant materials were gradually transformed and rearranged so that they eventually could perform the five activities of life, namely feeding, respiring, growing, moving and reproducing. Only material reaching this stage can then be said to have embarked on 'proto-life'. Not all stages were reached at the same time. Indeed, primitive forms of the first three can be readily recognized as intrinsic to infra-life. According to the thinking here a forerunner of reproduction as now understood must also have been in existence. The onset of proto-life was only arrived at once self-generated motion was also present, albeit also in primitive form.

Living began with proto-life, but once this was established conditions continued to change; after the onset of proto-life further creation of life was phased out, with all further progress eventually being confined to the preferential treatment known as evolution. Even so, initially progress was very slow and life remained at a stage resembling very primitive bacteria over a vast period of time. Eventually enough environmental change took place that happened to be just right to cause some proto-life entities to develop relatively quickly into ones bearing life in significantly more advanced forms. Once this stage was reached evolution really took hold and drove the primitive living beings into an ever increasing variety of forms. The Earth had become a vastly different place and the conditions that created life comprised a phase that was left behind and with no chance of it ever returning.

Reciprocating Activity before the Cycles of Life

The nourishment taken in by animals is transformed and then ejected through the exhaust system; but the energy in the environment and available for absorption and release by these processes would all be used up in time if it were not for the fact that the basic life activities of plants are to some extent opposite and complementary to those of animals. Plants in general flourish best when the exhaust material of animals is available to them and, while animals absorb oxygen and release the carbon dioxide that plants need, this is complemented by the reverse of the latter process by green plants, at least during the daytime. At night the gaseous exchange of plants is similar to that of animals. If the amount of darkness was unremittingly increased the resulting release of carbon dioxide would eventually be fatal to all life. Sudden total failure of the light would be quickly catastrophic to every life form that has been created on Earth, although some would last longer than others.

Which came first, plants or animals? All animals are ultimately dependent on the plant world for existence, so obviously animal forms could not precede plant forms on this planet. Although most existing plants are to some extent dependent on animals for life, a large part, i.e. the green plants and algae, are capable of directly absorbing selected mineral matter and breaking it down for their own purposes. These plants manage to do this by trapping radiation from the Sun with their leaves and converting the energy so gained to their own purposes. The system by which this is done is known as photosynthesis. However, many plants can only acquire the mineral nutrients they need and hence survive when in the presence of relevant fungi. Otherwise plant life would have reached a primitive plateau if animal life had not appeared alongside it; free carbon dioxide would then only exist temporarily in the daytime ambience after being released at the point of death and desperately absorbed while the Sun shone by all those remaining alive.

Thus is it clear that the life of Earth's biosphere does not have enough energy to keep going on its own. It would quickly run down and the shortfall is provided by the Sun, which has always ensured that the fuel supply is replenished as this essential commodity is used up. Take away the Sun and the car of life would quickly grind to a halt. It then seems reasonable to assume that this external source of energy that now keeps the car going is of the same nature as that which originally started it off. The entity that now provides the essential supercharge is the same as the driver that originally started the motor and subsequently controlled its performance, in other words the Sun (with the original process being photosynthesis in some form). This body has kept the car moving, but has had no control over the steering and has no idea as to where it is heading. Navigation is the job of a co-driver, i.e. evolution, but even this has no idea where it is all leading to and is only proactive by chance, so that life is selectively able to avoid lesser hazards - than (say) the onset of perpetual darkness - as and when they occur.

Earliest life was recognisable in entities whose activities were plant-like in the most primitive sense. Its behaviour was similar to that of most plants now, namely by selectively absorbing and processing matter from the surrounding world. This was accompanied by interaction with gases, namely carbon dioxide (CO_2), absorbed by day and released by night, and oxygen (O_2) released by day and absorbed by night. The habitat of such a proto-life form would be water (H_2O). Yet to justify this description it would exhibit certain needs, namely access to sunlight and ambient gases, the latter at first indirectly. This suggests an ideal location on or near the surface of water where a complex solution could develop towards becoming a 'soup'. The first entities involved were microscopic, namely bacteria in their most primitive form, which, if not extinct are now of rare type, because eventually survival involved evolution into more diverse and complex forms. (Although much too small for us to perceive, bacteria are still around in vast numbers and some of these constitute a persistent threat to our wellbeing.)

Some bacteria still employ photosynthesis and in this sense they are the direct descendants of the relevant earliest forms. It might seem that life was destined not to evolve beyond the bacterial level. Yet forms of living matter eventually did emerge that were organised differently and able to survive by being parasitic on those bacteria that were reliant on photosynthesis. As life evolved to include multicellular entities the light driven kind of living was assumed by subsequent emergent forms of plant.

Whenever the supply of light or air broke down, the way was open for entities in an infra-life form to recycle the de-energised material in something like the manner of a fungus. This situation was very restrictive for evolutionary purposes and the presence of other factors would be necessary to enable progress to be made. Eventually a parallel type emerged that could absorb plant tissue into itself and by its own effort, initially as discarded, but eventually while still constituting part of or the whole of a functioning entity. They were seizing energy created by others as a result of photosynthesis and this was the start of microbes lacking chlorophyll, followed by the fungi and eventually animal entities, both of which last two types thereby finding themselves in a symbiotic relationship with plants that had enormous evolutionary potential, especially for the second of the two, and indeed for the plants themselves.

Animals fuelled their activities by the absorption of oxygen, either as dissolved in water or from the air. Their continual release of CO_2 redressed a hitherto gaseous imbalance in which the amount of this gas in free form was being reduced and would eventually virtually run out. Increasing animal activity ensured that carbon dioxide was being released at a greater rate, while the amount of free oxygen would be diminished. The existence and number of animal entities was completely dependent on the status of the plant ones; more of the latter meant more of the former. Since life forms all contain carbon, it becomes evident that its conversion from a gaseous form in the atmosphere to liquid/solid forms existing and developing in water is an essential component of the activity known as life.

To sum up this section: once there was no life, but at a suitable stage in the Earth's history the rays of the Sun, beating down on the watery surface of the planet for a period each day, gradually started a primitive process that has developed and grown to become life. The prime mover of life is the Sun and its agent on Earth is photoreceptive matter, the latter being the carburettor of the car of life, while the actual fuel comprised light and heat. One can hence well imagine that all these were also the agents that originally joined forces to start off certain lifeless matter along the long trail of transformation by means of the transference of gaseous compounds of carbon into entities found in water and towards the manifold forms of life known today.

Yet these changes, though complex, at first involved lifeless minerals and something else must have developed. The created entities were descended from former 'slave' materials, but these molecules had somehow escaped from the rigid rules that govern normal chemical activity.

■ Molecular Energy

Life is concerned with matter at the **molecular** level, for molecular changes are the ones that govern normal activity on the Earth's surface. The energy used in molecular changes is bound up with the electrical charges carried by the electrons orbiting atomic nuclei. The binding together thus of the involved atoms results from the overlapping of such orbits. However, the initiating energy emanating from the Sun is itself the result of **atomic** nuclear changes in that body; such changes are of much more thorough going and violent nature than even the most extreme due to molecules and the amount of energy released is incomparably huge, as is known only too well from similar atomic changes engineered here on Earth by mankind.

Molecules can be simple or compound, depending on them being composed of one element or more than one. The composition of molecules (fusion) requires the application of energy, while their decomposition (fission) releases it. Alternatively fusion can be said to absorb energy from the environment, while fission occurs when the environment extracts energy from the matter concerned. With life forms these events have evolved to be no longer passive, albeit within certain limits, and control is exercised with a resultant degree of self-preservation of both structures and activities. This highlights the fact that the structural materials of life are to a variable degree fundamentally unstable and it is the state of 'living' that not only enabled them to exist in the first place, but also permits them to remain in existence.

■ The Start of Living

The Earth was originally a rotating spherical entity comprising matter that consisted of nothing but lifeless elements, even if already energised by self-contained electricity and heat. This state can be referred to as 'pre-life'. Before proto-life could begin some of the vital combinations of the atoms present must already have occurred. Among these were some that developed a specific sensitivity to light. Solar activity and Earth rotation combined to allow the circumstances to arise that were to make life eventually possible. So, before life proper there was originally solar-driven activity between relevant molecules that can be called 'pre-life'. This eventually led to 'proto-life' by way of a possible but accidental intermediate stage here dubbed 'infra-life'. Any proposal that the solar powered chemical engine was thus started and progressed to find its way to proto-life is vindicated by the abundant confirmatory evidence now available to human senses and reason.

Prior to infra-life the barren planet Earth basked in the light of the Sun even as it does now, with one half in light and the other in darkness, while the regular rotation gave a combined cycle for these two opposing zones of one day's duration. The appropriate atoms that were to provide the molecular framework for life already existed and were merely waiting for the correct conditions to create their enlivenment in the form of certain compound molecules of ever increasing complexity.

The first essential was for the Earth to have cooled down to a temperature within the range tolerable for life as it is. Furthermore, since liquid is an essential environment for all present-day primitive forms of life to be active, then its presence must have been necessary right from the start and the molecules concerned were either immersed or dissolved in water, or floated at its surface, or even used constituent parts of this very liquid. The ability of light to penetrate water would appear to have been an important factor.

It is not the intention at this point to blunder into the precise details of the chemical changes that could have been the forerunners of living functions. This would complicate the argument at too early a stage. The task is rather to take a broad and simplified view of the pertinent situation that the prevailing conditions must have allowed to come into being. Such an initial simplistic view is absolutely preferable as an aid to easier comprehension.

The route from the beginning of life to the Animal Kingdom is to be followed as a critical path in the pages that follow. This means that plants and fungi will be largely ignored unless they contribute to the laying down of this path. To simplify matters for the time being the process can be claimed to be due to the early existence of different kinds of primitive bacteria.

1. Some had flagella and were motile, while others were simply drifters.
2. Some used photosynthesis, while others did not.
3. Among those that were not motile and used photosynthesis some developed into algae and eventually plants.
4. Among those that were not motile and did not use photosynthesis some developed into the fungi.
5. Among those that were motile and did not use photosynthesis some developed into the animals.

■ The Earth's Rhythm

Let the assumption be made of the existence of certain simple particles that for convenience are labelled 'A' and 'B', which types are 'compatible' in that they have the potential to combine and to develop photoreceptive properties, but essentially when in combination with each other or others. For the sake of the argument particle A absorbed the Sun's energy from light by day, but released it again by night. At an early stage this energy was not sufficient to initiate significant changes, but eventually the stage was reached where the sum of the energy stored and energy being received was so great that daytime fusion was enforced between such simple A particles with any B ones that happened to be adjacent. Such created molecules were at first temporary. Came the night and the incoming energy from the Sun ceased. The stored energy of such 'babies' in this watery 'cradle' of life flowed out again into the environment in order to maintain the internal energy balance of each. Reversion thus occurred towards the original earlier state at night-time. This loss by the hypothetical molecules AB would cause them to split into discrete A and B forms of matter again.

The effects were cumulative. At the risk of very special pleading the following scenario can be envisaged. The relevant matter moved towards living forms because combinations of molecules with life-bearing potential were possible and emerged, and such are still in evidence. However, there was a 'master' form upon which all others were dependent: that was the one with photosensitive properties. The irony is that this master did not need to develop much, but its dependent 'slaves' were in competition for the by-products of photosynthesis. This put a sort of evolutionary pressure on them and survivors became progressively more advanced forms that made their way through infralife. Inevitably their magnificence disguised their descent from 'slaves' who were all ultimately dependant on the humble photoreceptive matter. In addition to being created thus, to survive they must gain access to the energy otherwise produced by photosynthesis, be it first or second hand or even further removed. This drive contributed to the appearance of protolife.

Yet some still served their 'master', i.e. chlorophyll, by providing somewhere for it to dwell (algae and green plants), even after others had broken free (fungi and animals). The latter took what they wanted from the running machine of life as parasites, both harmless and harmful. Many were thus not directly dependent on the source of fuel, but could survive by absorbing examples of those that were, or even one another. The fact is that entities either did or did not behave in one or other of these ways; but those which did not had little chance to evolve.

In the master molecules, AB can be regarded as a permanent partnership. At this point these infra-life molecules were in a position to expand their powers of combination into other areas. They had reached the limiting point where destructive excess energy could only be relieved by a means to lose such, namely by fusion with units of matter of suitable nature (say ABB, ABBB, etc.), but ultimately involving different entities. This is the point where one has to argue that the latter did happen because it has happened.

Changes of position and form are vital characteristics of a universe consisting of matter and energy and the initial process depicted above could not last forever in its primitive form. It is essential to suppose here that at certain locations and at some stage in past time, energy absorbed during the daytime fusion was not all released by night-time fission. This was due to the amount of energy being slowly increased in the photoreceptive parts of the environment due to absorption from the Sun's rays. In this way a bank of energy would eventually be built up in these 'ABplus' molecules which then could endure as apparently lasting combinations, even if their status was still ultimately unstable.

The Sun had still poured its energy onto the scene, creating a surplus every day. In their turn the crowded ranks of the enlarged AB+ molecules would become increasingly energised until a further cycle of changes would be forced upon them. Each day might see AB molecules fused into the larger AB+ types. But as the energy bank built up molecules could continue to grow until a limit was reached, leading to combination with material of a different nature. Hence, fusion would occur with available but dissimilar C, or D, or CD molecules and lead to (say) AB+C, AB+D or AB+CD types. These too would at first revert to a previous state at night-time until the remorseless build-up of energy again led to yet further permanent combinations. Yet this did not constitute life, but did mean entry had been made into the infra-life stage. Virtually all evolution was due to take place in the non-photoreceptive particles, say C, D, etc. In other words the creation of life was never going to occur in the photoreceptive matter itself, but became possible in the matter that was created by itself and due to the ability of chlorophyll to accumulate surplus energy derived from sunlight.

Yet once more the energy bank would grow to allow even such more complex fused molecules to survive the night or revert only partially. This process was becoming ever more greedy for constituents. On balance energy was retained by this operation and as the combinations became more and more complex the ability to retain and store energy was increased. The energy bank would grow to allow even such more complex fused molecules to survive the night. This sort of process was gradual in the extreme, but nevertheless unrelenting; yet just a section of it embarked on a haphazard yet continuous advance towards life-forms. It would persist until at last there would be large molecules including 'protein' in existence that might also be thought of as the simplest of 'cells' and comprising protolife. When over energised such entities would have the ability to absorb others if such were present by selectively over-energising them. Otherwise reversion of themselves would occur. While reversion would lead to increase in numbers (primitive reproduction), absorption can be regarded as the beginnings of growth. In combination these two processes would lead to overcrowding and this was and had always been an important evolutionary force.

In short these hypothetical 'molecules' were gradually entering into the proto-life phase by still developing their incipient 'lifestyles'. Evolution of a kind had been taking place in that those which had resisted reversion were survivors, while those which had reverted might well have become more vulnerable to change by becoming in their turn something a little different. It is among such flexibility of form that the laying down was allowed of the tortuous route followed by such selected pre-life molecules to create during the intermediate phase of infra-life the critical path towards proto-life. None of the survivors were yet alive; they could feed and respire, but it was in their ability to revert and start absorbing again that they were set on the path toward the final necessities for the state of proto-life. Being in water they could be moved, but still lacked motility; nor could they reproduce in a specific controlled manner.

■ Primitive Environmental Factors

Living did not originate in the photoreceptive matter itself, but in the molecular material with which it combined, then transformed and eventually released. But once separated such materials could not have been driven by light, although some of them apparently maintained the capacity for reciprocal or cyclic behaviour. It follows that some other form of energy must have been sustaining the sequences. The only candidates would appear to be heat and electricity. While the effects of light and heat on the Earth's surface were clearly tied to its rotation, heat, unlike light, could be retained by matter. Hence, while the complex molecules of infra-life were being created by light, the carryover of energy was being controlled by fluctuations in temperature, and these effects were not as abrupt as those of light. Light has always been necessary for life to be created and continue. Ambient heat on its own has never been able to sustain life, but temperature did come to define limits within which it could exist.

Any geothermal heat available would originally have been too extreme, too erratic, or even too stable to be useful. The continuous supply of heat necessary in the future biosphere was produced by transforming light into heat. Yet this was never going to create anything on its own. The vital step occurred when certain molecules developed a way of dissipating light energy into the production of something other than heat. Even so, static conditions would not have been suitable to induce sufficient change in the material of pre-life and infra-life. The factors that brought different conditions would include the anomalies of the Earth's relationship with the Sun, namely its elliptical orbit and the tilt of its axis against the plane of this. The movement of the Earth's waters would sweep any infra-life or proto-life entities into different latitudes. All this constantly changed the environment in which the materials found themselves. Survivors were those that chanced to exist in forms that enabled them to withstand or even exploit the range of conditions obtaining at any one time or place.

As throughout time, the Sun still keeps life going by the conversion of light to heat. But such original state of living would never have been possible without the rhythm of the Earth's phases creating activity that was the forerunner of reproduction. Life still relies on the Sun, but rather less on the rhythm of its appearances. Indeed, some life forms now survive without ever being exposed to light at all. This is always an adaptation and such entities survive by using ambient heat. Such specialists are characteristically trapped in their particular niches and cannot survive outside of their select environments.

The new carbon compounds produced by early photosynthesis were not sensitive to light, but to heat. The fluctuations of this were not as rigidly tied to the cyclic movements of the Earth, nor the results of this on the waters of its surface. The compounds were sensitive to heat and hence found themselves subject to more variable circumstances than the rigid light-regime of chlorophyll. This would induce them to develop in a variety of ways. Otherwise, without having to cope with these environmental stimuli, compounds seemingly only capable of mere chemical change could not have provided the basis for protolife. This depended on the fact that one particular element is so flexible that it has a seemingly inexhaustible ability to combine with others in a vast range of complicated compounds. That is carbon.

■ Molecular and Cellular Division

It has been claimed earlier that for stability external energy must relate to internal. Let a grossly simplified fused molecule now be considered, such as those discussed above. At times the external energy will have dropped to too low a level compared with its internal one and it must split to release some of this. Thus it divides and any CC could become C and C; but further application of solar energy would create the form CC again. The initial fission is similar to cell division at a very simple level. A single cell spends its life collecting energy until at last it must relieve the resultant stress, so it divides to release any excess. The existence of the earlier forms ensures that when internal energy is in excess and causes too great a differential with the external energy level, reversion occurs to avoid destruction. However, life could only be presaged in such molecules that could resist such reversion by delaying this process while the destruction of less successful competitors made it initially unnecessary. The eventual absorption of suitable remnants of these would lead to a more complex molecular form. This trend was eventually to lead to individual cells, but life-forms were going to remain relatively minute until evolution could organise some of them into tissues.

However, cells are not tied precisely to any cycle caused by the Earth's rotation, nor do they normally release enough energy to make the chance of fusing again possible. Indeed, any such new apparent primitive 'half-cells' that retained the faculty to re-fuse could not become components of proto-life if they did so. It is rather the case that the true new half-cells of proto-life go on absorbing energy again and their component molecules continue to grow in number, until at last they independently attained full-cell status again. But so much excess energy would continue to be accumulated that eventual release must once more come from further division of the whole if destruction was to be avoided. Eventually the release of energy by binary division would itself become adequate for survival and make internal division of all molecules unnecessary and ultimately the business of specialists, as are chromosomes. This became a feature of life. Such developments began to make materials available for new purposes, while earlier forms of activity gradually changed or became redundant.

■ Molecular and Cellular Predation

Once more let two hypothetical separate molecules of previously fused origin be considered, namely at night in a group and approaching fission. One is ready before the other and thus releases energy into its environment, which will tend to delay further the other's fission. In this way the fission of some molecules helps to preserve the fusion of others. This is indeed one way in which primitive fused molecules could come to survive the night. To any observer it might look as though the survivors were destroying the others. In a way they were in the process of selectively 'eating' them if by dawn some recombination has taken place between such unequally sized molecules and it even looks as though the larger still-fused molecules have actually preyed on some of the smaller reverted ones, taken what they needed and rejected the rest.

Cellular predation may indeed act in a similar way, but at this later evolutionary stage the action might appear to be contrived rather than automatic. A growing cell is hungry for energy. An amoeba will absorb another creature into its cytoplasm and then creates conditions for the prey so that it finds itself in an under-energised environment and must break down to restore the balance. This is perhaps a method used by bacteria to break down organic matter. Seen in this way, they absorb so much energy around the matter concerned by means of such energy hungry activities as growth after fission, namely the results of reproduction of themselves. Any nearby organic matter if relatively over-energised is then bound to disintegrate and the new bacteria are there to pick up the pieces. Such a process may eventually be accomplished or controlled by a specialist chemical catalytic presence such as provided by enzymes. Thus have the energy so released and the resultant fragments been immediately seized upon and dissipated by the living scavengers so that the cycle will repeat itself in the minimum of time until all such surplus usable energy has been drawn from the supply and only a residue of under-energised and unusable matter remains. One might note that with bacteria (non-animal cells) the process is still largely chemical, but with amoebas (animal cells) a physical element has arisen in that the food is ingested before being chemically digested.

Where ingestion is concerned, the energy released by the disintegration of absorbed 'food' is dissipated by the eater by means of performing fusion with the suitable resulting compounds, namely by absorbing them into its own internal parts. Unsuitable residues are first rejected then ejected.

Fission and Division

In living matter at the cellular stage 'fission' normally has become just a case of dividing into two on each occasion. In a cell this 'division' goes right through its components, from the cytoplasm to the nucleus, the chromosomes and the genes. This externally simpler kind of split is a controlled evolutionary process that has somehow developed from the greater disintegration and subsequent re-assembly in different form that otherwise accompanies fission resulting from purely chemical reactions. The control that a cell can exert over the process is by means as yet not apparent. Yet without this faculty life could not have developed further from infra-life.

It may be supposed that a cell is designed to split when it has accumulated sufficient energy. If it did not split characteristically, then further intake of energy would just destroy it. This is perhaps how viruses destroy tissues. When one puts material into a cell, this can be seen under magnification to quiver and then disintegrate. Viruses are obviously insatiably greedy for energy and their aim is to rob the cell of its energy content by using it for viral reproductive processes. The rapid destruction of the cell suggests a sudden increase in energy as a result of the virus's activity. After the defunct cell's destruction the virus can be seen to have reproduced, something it could not have achieved on its own, without having access to the cell's reproductive system.

The Living Cell

This is a unit of life that strives to keep its identity. To do this it endeavours to take in just enough energy to hold it in the characteristic fused state, because a constant and regulated turnover is required for this. An excess of energy taken in would quickly destroy the cell, so it dissipates it by various activities, but chiefly in the absorption of suitable matter, i.e. by growth. Eventually however the limit of size is reached and then the surplus energy must normally be expended by division to give two half-sized copies that can then grow in their turn. In advanced cells the division starts in the nucleus, where the genes arrange themselves along the chromosomes like strings of beads. Then the chromosomes split longitudinally, with each half moving to its own side of the nucleus. At this stage the nucleus itself divides to give the impression of a cell with two nuclei, until the cytoplasm finally nips together between these nuclei so that two young half-sized cells are formed. This is a development that has transcended the earlier system of molecular division found in the simpler forms found in infra-life.

■ Molecular Colonies

The molecules of early living matter, as in other homogeneous substances, were found contiguous to each other. As has already been suggested they have fused to their existing form by absorbing incoming energy. If the energy level continues to rise they must fuse again to make new forms. However, the process may go too far and insufficient external energy may cause fission to occur to restore the balance. Molecules at the spearhead of evolution will not have acquired the ability to make new forms by fusion (this only represents life as growth). The chosen ones will need the ability to split characteristically to survive until suitable readjustment occurs to suit changing conditions and offers a better chance of resisting normal fission and fusion as a means of survival. This random acquisition of characters resulting from the changing of external conditions governing choice of fission or fusion could be one basis for mutation. Eventual preferential selection of entities showing suitable examples of such functions would result from a drive to deal with changes in the energy balance and represents incipient survival of the fittest, the very stuff of evolution.

■ Cellular Colonies

That which has been stated above regarding molecules can also be applied to cells. Mutation of constituent molecules leads to cellular mutation. The drive for molecules to mutate comes from changes external to the cell demanding counteracting internal changes. Cells can exist as individuals, either alone or forming loose colonies, or may be bonded together in tissues. It is in the latter case that they are in the best position to make communal use of the energy available to them. For example, in this close association it will be easy for them to snap up energy released from one of their fellows before it can escape into other areas. In a tissue the conditions are at their best for the individual cells to withdraw from or deposit at the common bank as required and the latter is kept 'healthy' by the vigorous energy turnover. An additional advantage of a tissue is that members may be screened from the attention of hostile forces by the presence of their neighbours. There is hence an evolutionary pressure for cells to congregate in tissues. However, cells not deeply located within tissues are most exposed and hence eventually more liable to commence divergence, i.e. evolution towards the formation of diverse organic structures.

■ Death

This happens when life ceases. A living molecule or cell soon ceases to exist in the same form when it is critically over energised or under energised in relationship to its surroundings. The life-energy it contains may be immediately transferred to resultant new forms, where total death has hardly occurred; or it may simply serve to energise the environment, when it can be considered to be quite dead. Hence total death is the loss of both identity and continuity of units of matter previously having life.

The life and death of cells can easily be comprehended in these terms, but what is the cause of the inevitability of death among multicellular organisms? During growth there is a great intake of energy and this results in the division of the cells so that their number increases all the time. The energy released from division goes toward more growth and other bodily activities. When adulthood is reached the process is modified in that normal growth ceases and, if the intake of energy is not reduced or other compensations made, unhealthy oversize will result. The important thing for a cell is to keep growing till it divides. An adult animal theoretically has a fixed optimum number of cells, in which case it is clear that theoretically for every cell that divides another one must die somewhere in the entity, if numbers are to be kept constant.

Hence, as an approximation, it would seem that for every average cellular lifetime half of a body's cells are theoretically sacrificed to maintain its characteristic size. The energy released by these dying cells must reduce the energy intake requirement to some degree. This would not seem to matter, except that it reduces the energy turnover, and this is an extremely important feature of living matter – the amount of energy passing through, rather than the amount simply locked in its tissues.

The energy released by dying cells reduces energy demand and tends to delay the fission of those still living and this lengthened period between divisions is a feature of ageing, since the delay in cell development reduces even further the energy intake required, which again means a lower turnover and the effect is cumulative. It is the energy turnover that drives the animal machine and as the energy supply fades, so the various functions run down until eventually a vital one fails because of energy starvation. Of course this is failure due to old age and is distinct from premature failures due to genetic defects and incidental damage.

CHAPTER 11

Life and the Lead-up to It

■ Infra-life and Photosynthesis

In contemplating the start of life, consideration of the most obvious observable token should preferably be reserved till last, namely the ability to initiate motion (motility). Likewise one should bear in mind that the road to life started among entities that were so small as to be incomprehensibly remote from our perception, except for those researchers with relevantly scientific minds and special equipment. The minute entities are known as molecules. Even so infra-life came into being among such microscopic items, although they had the potential eventually to be transformed into both relatively large and hugely variable living beings. Given the basic tenet regarding infra-life, such materials had to be already present, rightly placed and prepared to be transmuted by the complex yet selective chemical reactions into new physical structures that collectively would eventually lead to proto-life. The essential difference between infra-life and proto-life is that, while the former was structurally prepared passively to facilitate survival, the latter had in addition acquired the ability to react to circumstances and behave in ways that served the purposes required, which further enhanced the prospects of survival. This process involved incipient evolution. At its simplest level, proto-life came into being when entities reached a stage at which they had some degree of control over the occurrence of fusion and fission in order to increase the chances of survival of themselves as individuals and incidentally their type.

The molecular basis for life was and still is to be found in chlorophyll, even if then in a more primitive milieu: it is the prime agent for harnessing the energy of sunlight to turn lifeless chemicals into living vegetable matter. It now occurs as small organelles found in the cytoplasm of green plant cells and consists of 'plastids'. Such structures (chloroplasts), containing the green pigment (chlorophyll), enable the absorption of light by plants and the convergence of this to energy (photosynthesis) to take place.

Chlorophyll is a complex kind of molecular compound (chelate) in which a large molecule (ligand) is bonded to a single central metal ion at two or more points, in this case four. The relevant metal is magnesium and the flat ring of the molecule (porphyrin) consists of atoms of more than one element (heterocyclic), typically carbon, hydrogen and oxygen, with the bonding being achieved by way of four nitrogen atoms. (Other porphyrins are haem - which is centred on iron and features in the red blood cells of vertebrates - and certain vitamins.) Simply stated, the basic result of photosynthesis is to transmute carbon dioxide (CO_2) and water (H_2O) into carbohydrates (say CH_2O) and oxygen (O_2). This is achieved by negatively charged subatomic particles (electrons) being transferred from water to carbon dioxide. The absorption from light raises the energy of electrons in water to a higher state, which enables them to be transferred first to the chlorophyll and then to adjacent carbon dioxide. While energy from light is available, there is a continuous reaction involving loss of electrons (oxidation) and their gain (reduction). It is this activity that enabled the microscopic ancestors of plants to construct cell-like structures using the resultant build-up of carbohydrates produced by the magnesium-centred chlorophyll molecules.

In the light of above it can be claimed that life started with a metal, namely magnesium that could combine with nitrogen in such a manner that the results were sensitive to sunlight in a peculiar way. It would seem, then, that magnesium is central to the first process leading to life. The energizing effect of absorbed light was to attract further materials that were present in the environment and this eventually led to the creation of chlorophyll chelates. The activity was promoted by the electric charge of the electrons. Constituents of the porphyrin surrounding the central ion could be released as carbohydrate compounds when the energy supply turned off after the Sun set. After sunrise the ion was again in a state of preparation to receive a new supply of untransformed constituents.

As an element magnesium (Mg) is not found in free form in nature. Magnesium compounds are typically white crystals, with most being soluble in water, whereby a suitable ion of Mg is provided. The solution is alkaline and can form the basis for remedies. Magnesium is the eighth most abundant element and the third most plentiful one dissolved in seawater. This shows how it could have been abundantly present and available to be the first step in the initiation of chlorophyll.

It can be noted that the prime relationship of magnesium in these matters is with nitrogen, not carbon. By its association with the metal magnesium the gas nitrogen is an initiator of life, while the solid carbon is a sustainer.

It may seem that chlorophyll is the very essence of life, even if appearing to act as its servant. Chlorophyll is indeed the agent of control, but, while its form remained primitive, the carbohydrate slaves it could produce were the entities that had the potential to develop into other forms of extremely complex nature. These original 'slaves' were due to develop to such an extent that it is hard to realize that they were and still are ultimately dependent on the humble chlorophyll. The start of life was entirely in response to early bullying of other molecules by something now apparently so insignificant.

In biological terms enzymes are substances that act as catalysts to promote compounds, each by a specific biochemical reaction. In this sense chlorophyll represents the most primitive form of such in that it enables the production of carbohydrates from carbon dioxide and water. Since carbonic molecules would always be cloaked with water, even at the surface of this, it might seem that originally the materials in question were totally comprised of solutions. In this case gaseous carbon dioxide was not initially involved, but rather it took part as a solution, a weak acid (i.e. carbonic - H_2CO_3), which is formed when carbon dioxide is dissolved. At such simplest stage of carbohydrate production the basis for the release of oxygen is already evident ($H_2CO_3 > CH_2O + O_2$). The inference is that the synthesis using free carbon dioxide came later.

Although carbonic acid is formed when carbon dioxide is dissolved in water (which occurs during rainfall), it can only persist in the absence of free water. In other words where it is not saturated it will decompose. This may simply involve reversion, i.e. $H_2CO_3 > CO_2 + H_2O$. Otherwise it can be transformed when an ion of hydrogen is detached to leave bicarbonate, i.e. $H_2CO_3 > H^+ + HCO_3$. This is one of the initial processes of infra-life, which became ever more complicated, especially when the elements phosphorus (P), sodium (Na) and Calcium (Ca) also became involved.

The effect that sunlight can have by way of chlorophyll has been represented by the formula $6H_2O + 6CO_2 > C_6H_{12}O_6 + 6O_2$, whereby dissolved carbon dioxide is consumed and dissolved oxygen produced, leaving the simple sugar called glucose.

■ The Nature of Infra-life and Its Limits

It was essential that two of the five features that comprise life were present early on in the infra-life stage. Feeding and growth did occur, in that accretion of additional matter took place when carbohydrates were built up by chlorophyll. The basic source for the magnesium, nitrogen, carbon, hydrogen and oxygen was water, but carbon and oxygen were destined eventually to take part in exchanges involving direct contact with air in the form of carbon dioxide and free oxygen. This was respiration, which form of direct gaseous exchange also took place in the form of absorption in the light of day and the reverse in darkness. It is to be noted that the eventual colonization of land by living matter could not have taken place without this step having been taken. The transition from infra-life to proto-life not only involved the ability to respire, but also the capability to reproduce and the power to self-move.

Any infra-life movement was neither deliberate nor purposeful. The carbohydrate molecules were not motile and only moved as a result of external forces. These could be due to electricity, diffusion, convection or currents. Since the metal taking part was magnesium, magnetism can initially be discounted until iron got involved. Since life in its earliest form was as entities resembling primitive bacteria, the constituent molecules of infra-life were also repetitive yet free units that had broken away from the carbohydrate mass; at this stage any such that happened to remain bonded together were on the most primitive stage of the route leading to proto-life. The molecules of infra-life were presumably one kind among many. The natural forces working on them may have been favourable for some but led to others either remaining passive or being destroyed. However, since reproduction was still at a very primitive and simple level, evolution of any degree of rapidity was not possible. Yet it will have been pertinent that those entities heading for proto-life were survivors and would hence exist in ever greater proportion of a population consisting of the relevant melange of molecules.

Advanced reproduction (rather than replication) was not possible until those molecules staying on the critical path combined with other materials namely amino acids: they were then on the way to include proteins in their structures, which were to become ever more important to them and be essential for them not only to be considered 'alive', but attain greater size.

Reproduction can take place to produce equals, but when it occurs in the form of 'sex' the entities taking part are unequal. This added the topping on the cake of life and allowed evolution to proceed at a greatly elevated rate, but it did not take place until the substance known as DNA made an appearance. These additional materials were not produced as a direct result of photosynthesis, but were synthesized by the relevant molecular clusters themselves using ingredients available in their surroundings.

■ Motility

Movement within entities, rather than that imposed by external forces, is perhaps the hardest step to understand among those leading to proto-life. While the actual cause is obscure it can be seen to lie within the possibilities arising from the chemical reactions available to the relevant carbon compounds and the electrical forces intrinsic to them. Earliest motility would produce movements entirely governed by chance, but with the advanced structures becoming 'cellular' in nature a gradual evolutionary drift towards 'purpose' would develop. Even particles that could be passively moved in a way that selectively increased chances of sustaining themselves would be more likely to survive and hence come to constitute an ever greater proportion of populations. In conjunction with this, chance beneficial protuberances would eventually lead to specific organs of movement on free-swimming cells, such as long whip-like appendages called flagella, and later masses of shorter 'bristles' on the surface showing coordinated movement and known as cilia. The latter only appear on animal cells, being used to move fluids for the purposes of movement, respiration or ingestion of particles. They are believed to be derived from flagella, these originally only used for motility. Yet the flagella found on some bacteria operate in a different way and may be of discrete origin.

In short it might be claimed that motility was attained by an early evolutionary process. The energy within carbon molecules caused them to move relative to others of their kind, or towards a source of light, or to react to other circumstances that happened to be beneficial or harmful. The acquisition of motility would give unstable entities a greater chance to survive. Eventually kinds of movement would evolve that were generally advantageous and chance irregularities on the outer surface may also have proved helpful and hence be enlarged and improved. Members of populations that chanced not to develop in this way would find themselves at a survival disadvantage and gradually constitute a smaller proportion before eventually being entirely removed from the road leading to proto-life.

■ Starch, Cellulose and Sugar

These are the carbohydrates commonly found in life forms. Animals are able to perform the functions of life because they obtain these materials by consuming plants or other animals, or both. Starch is formed in plants, from where its storage faculty is obtained by animals, either directly or indirectly. Cellulose is derived from glucose, a form of sugar: it is the main constituent of cell walls and fibres in plants. Sugar is more important to animals because of its ability to release energy quickly and enables rapid movement. It also became the material from which life is derived because of its essential role in reproduction. The type of sugar associated with life is known as ribose.

CHAPTER 12

The Minutiae of Reproduction and Growth

■ Reproduction, Growth and Evolution

Animals, plants and other forms of life can involve reproduction by both sexual and asexual means, but the latter has to be the primary condition. There is also activity such as the conjugation of paramecia that might be referred to as pseudo-sexual. It has already been claimed above that reproduction is the prime activity of living matter. The actual living is the state of such matter between such acts. Reproduction itself was a development from out of an emergent chemical ability of carbohydrates to replicate themselves, albeit sometimes imperfectly. To ensure that the optimum size and condition of any relevant unit be maintained, a primitive form of feeding leading to growth had to occur in the intervals between acts of reproduction. Reproduction has come to be stimulated when certain specifically defined limits on development and size are reached. It was these activities that were to develop into the condition that is known as life. In other words:

photosynthesis + replication > reproduction
reproduction + controlled growth + other activity > living

However, circumstances of a hostile nature continually kept arising and developing, so the forms involved had to respond with further new structures and activities to ensure survival. Because of this, in order to maintain itself life has always been susceptible to the adoption of adaptive change, albeit involving selection from random occurrences. The survival concept behind this is known as heredity. Hence:

living + heritable change > successful defence > evolution

An early evolutionary process occurred when growth eventually involved the synthesis of proteins from various amino acids. The following is an exploration of the activities involved as far as acquired knowledge will allow.

■ Living

The important consideration about living is not only to be found in the details of how it occurs, but in realization of the fact that certain combinations of matter, although unstable, have nevertheless proved themselves to have had survival potential. Combinations of inorganic matter can only occur when the conditions are right, yet such compounds are generally relatively durable, requiring very specific conditions to change them and very harsh ambient circumstance to destroy them, especially in the case of solids. Organic compounds are much more unstable and vulnerable, with the range of conditions to change or destroy them being much greater and hence with them being more in need of a relatively constant and reliable environment in which to survive. They needed to embody a flexible complexity to be able to create slight modifications that would not hinder their survival in prevailing conditions, but must just be right to enable survival whenever a not excessively hostile change occurred in the ambience or environment. Even so, durability of any single entity remained limited; it was survival of the type that was vital.

Yet the primitive ability to survive was quite meaningless. It still is. This applies to beings much more complex than a bacterium. A cow has no idea as to why it does things that enable it to survive, such as the eating of grass. An antelope does not fully comprehend why it must run from a lion, yet it somehow just 'knows' that it should, this 'knowledge' being an instinct, but being reinforced by experience;

there is a compulsion to escape. Only the reflexive nature of the human intellect fully allows such questions to arise and be posed, and attempts made to find answers.

Changes that occurred could be perpetuated by means of the reproductive system. Important ones happened in the system itself and eventually also in the development of the ability to move autonomously. Any right physical activity to improve chances of survival would become the norm and to any theoretical observer would be the prime indication that life was present.

In this respect plants can be seen as not as 'alive' as animals. Yet plants can be said to be essentially alive, because they can die. Like all life forms they are able to persist in an unstable condition with internal movements of constituents possible. However, in their multicellular forms they have generally not developed autonomous movement. With them discernible action has been overwhelmingly restricted to reaction to sunlight and time by leaves and flowers. They are indeed still close to the infra-life stage in this respect, and also in that they still greatly favour the synthesis of carbohydrates rather than proteins. Despite all this, as a later development some have acquired the ability to fling seeds abroad and even spring traps on unwary insects.

A difficulty in the search for the remote origins of life is that compounds were produced by earlier forms that came to be ever more complex in both structure and function so that they eventually could become dominant. Hence that which appears to be in control should not be assumed to be the earlier feature. A better judgement of primacy is by deciding which of two entities specifically depends on the other.

■ The Bonded Bricks of Life

The basis for all reproduction is replication. Evidence of it occurs in entities of ever increasing size, i.e. bases, triplets, genes, chromosomes, cells, etc. The binary principle might appear to be an essential factor in microscopic life. Yet there is evidence of a tertiary principle being at work at a very early stage. One can observe this in the composition and behaviour of RNA (ribonucleic acid) and DNA (deoxyribonucleic acid).

Among that which follows will be found further indications of the way that cells developed, but the smaller entities that preceded cells had found a different way to become adjoined: they did not form into tissue-like clumps, but rather arranged themselves end-to-end to form 'chains'. These hint at the nature of infra-life.

DNA molecules include multiple copies of a structural unit called a nucleotide, which comes in 4 forms based on either adenine (A), or thymine (T), or guanine (G), or cytosine (C), each of which is bonded to a phosphate, this being one or other out of a choice of three groups. Without this bonding the unit involved is known as a nucleoside (note the spelling difference), an organic compound consisting of a purine or pyrimidine base linked to a sugar, e.g. adenosine, which is such a one consisting of adenine combined with ribose. Thus is a nucleotide a compound consisting of a nucleoside linked to a phosphate group. So adenine (i.e. A) is the particular nitrogenous base involved in the nucleoside adenosine, just one of the four constituents of nucleic acids. A significant compound is adenosine triphosphate (ATP) which by its breakdown in the body (to adenosine diphosphate) provides energy for physiological processes such as muscular contraction

As a contribution to growth, by means of 'transcription' a chain of RNA is produced by one of the dual helical chains of DNA in the nucleus of a cell. The result is known as messenger RNA and it is able to pass from the nucleus into the cytoplasm where are found compatible minute particles of RNA known as ribosomes. Contact with the messenger RNA and selective use of the triplet code causes these to synthesise polymers in chains known as polypeptides and consisting of large numbers of amino-acid residues, these being the constituents of protein molecules.

Genes are chains of DNA arranged along a chromosome. Each of these comprises the four bases A, T, G & C specifically arranged to provide a bonding code that is the basis for the production of messenger RNA. In this A should bond with T and G with C to avoid aberration. The coding is achieved by the DNA bases being arranged in variable groups of three (triplets), e.g. CAA, GGT, etc. The RNA result of such is GUU, CCA, etc. A vital difference between DNA and RNA is that the latter has U (uracil) instead of T (thymine). One might here note that A not only should not bond laterally with G or C, but neither with another A. Such hydrogen bonding across the helices is how the genetic code works. In contrast the arrangement along the chromosomes is without such selectivity. Here the system works through the mentioned arrangement of the bases in triplets.

The code works by replication always occurring between bases of different type, i.e. a purine produces a pyrimidine and *vice versa*. Thus with DNA is A a purine base and T a pyrimidine. RNA is present when another pyrimidine base is used instead of T, namely U.

Should a triplet lose a base it does not become a duo, but borrows from the next triplet along. A shift then occurs along the gene so that the triplet principal is not lost. However this then changes the whole of the coding 'downstream' of the original loss. The effect is reminiscent of playing the solo card game patience (solitaire), where the cards are dealt face up in triplets and the uppermost one can be used in the game. However, when one is able to play such a one from the pack, then the whole sequence of available cards is altered from this point onwards at the next round of play.

Within genes the longitudinal bonding between triplets does not vary; it is the abiding triplet principle that precludes this. Yet at intervals the bonding is weaker, and these places define the genes: any breaks should occur at these points, which is how the genes act as units.

Bacteria

The earliest life forms that are recognised are microscopic and described as 'bacteria'. Yet this term now covers a vast range of simple microbes that have one thing in common: they are neither animals nor plants. They exist in various forms (morphologies) - compact, rod-like, curved, coiled, spiral, etc. They can also be segregated by whether or not they reproduce 'sexually', use photosynthesis, or are motile (those without flagella or other means to initiate movement are as much at the mercy of their surroundings as are inorganic particles).

A bacterium does not have a nucleus defined by a membrane, but rather has a mass of DNA forming a single chromosome: this is known as a nucleoid. The cytoplasm does not have defined organelles, but includes various particles, including ribosomes (RNA) and plasmids (items of DNA). The whole is enclosed in a membrane within a capsule (cell wall) upon which protrude select external features enabling contact or motility.

Asexual reproduction of bacteria occurs in a similar way to the cells of more advanced life forms. It might occur three times in an hour and can result in a very rapid increase in numbers. However, evolutionary potential here is minimal.

They have three other ways of reproducing that are described as 'sexual'. Yet it is a simpler form of sexuality (pseudo-sexual) in that it does not involve meiosis, gametes or zygotes; there is simply a transference of DNA from one cell to another. In transformation and transduction the occurrence is haphazard. In the former process genetic change occurs by the absorption of detached DNA from one bacterium by another. In the latter the DNA is transferred as a by-product of viral activity. More understandably sexual in nature is conjugation. Unless they come into contact with others that have one or other faculty for DNA to be transferred to or injected into a partner, bacteria can only reproduce asexually. Yet there are 'father' and 'mother' strains. After conjugation mothers will only produce fathers. This seems to foretoken or reflect in grossly simplified form heterosexual reproduction as found in higher forms of life. Hence with bacteria, mothers are the products of asexual reproduction and this means that any potential at all for evolution lies overwhelmingly with the 'sexually' produced father strain.

Bacteria are generally far too small to be seen with the human eye, are immensely diverse, exist in huge numbers and are extremely widespread. They can be divided into three types that are based on the way they acquire energy, which they can use to absorb carbon either from organic or inorganic matter, or by the fixation of carbon dioxide. The organotrophs can acquire energy from organic compounds, while the lithotrophs do so from inorganic ones. The phototrophs use sunlight for this purpose, which according to the reasoning here makes them the direct descendants of infra-life forms. The other two have really become parasites of the biosphere. They have developed alongside the main progression from infra-life through proto-life to life. Members of the first (organotrophs) depend on the presence of other life forms for energy, while members of the second (lithotrophs) can acquire energy from the wider environment and use it to absorb the carbon they need for growth and reproduction. Examples of such that are able to survive in the tremendous heat of sub-oceanic fumaroles may appear to be surviving examples of primeval forms of life, but are in fact at the end of a branch-line, being unable to escape from the extreme environment in which they find themselves. Ultimately considered, they are an example of reversion.

The phototrophs are the real descendants of the structures that harnessed light to organise themselves into the web of organic compounds that formed the basis for infra-life. In this respect these, as progenitors, led to the green algae and green plants. They could originally survive by combining with carbon dioxide dissolved in water. Eventually a type arose that dispensed with synthesising carbon dioxide themselves because they developed means to obtain that contained in the proto-phototrophs, the latter being destroyed in the process. Hence these proto-organotrophs were the ancestors of the fungi and the animals. Others eventually could gain access to the carbon dioxide available from phototrophs, without destroying them. This became a symbiotic relationship of mutual advantage that led to the proto-lithotrophs and which appears to have been a process without which life would not have progressed beyond its simplest forms. The fungal branch of the proto-organotrophs could become complex organisms, but these were destined never to be at the spearhead of evolution; this was the role reserved for the branch that developed into the animals. Simplistically stated the three main branches of life occurred thus:

phototrophs > plants
lithotrophs > fungi
organotrophs > fungi and animals

■ Replication, Reproduction and Growth

Organic molecules can exist in a vast array of forms. Certain carbohydrates assumed a degree of permanency once they developed the capacity to replicate themselves when under stress as a result of being over energised. This led to reproduction, but it had only been made possible by the way in which pre-life molecules had already become organised.

At the infra-life stage the heterocyclic products of a photoreceptive porphyry would be scraps of carbohydrate. The potential for variation was great, but some varieties exhibited special properties that made them more suitable for survival by way of organisation and complexity. It was already chemically destined that some of these molecules were going to take on certain ways and that these ways were going to be the basis for infra-life, itself being a necessary stage on the route towards protolife. Basic to everything that was occurring were chemical compounds based on carbon with an inherent sensitivity to temperature change and an acquired ability to control the effects of this and adapt to it.

Replication is at the heart of the reproduction methods of cells, such processes being known as mitosis and meiosis. Cellular reproduction has come to be initiated by growth and occurs once optimum size has been attained. However, while a cell grows volumetrically, there is evidence enough to show that the basic carbohydrates involved in the appearance of reproduction and growth (DNA and RNA) primitively employed a linear principle.

Thus have chromosomes come to consist of chains of genes, the numbers of both entities being variable among life forms, but fixed with regard to each species. But the genes themselves employ groupings of related but variable entities along their lengths, with each group consisting of a selection from three of the bases available. The question then arises as to why such triplets were used, rather than other numerical combinations. This would occur by chance. Serendipity worked on what was available and any system that worked best or at all was going to be the one that survived to work again. For some reason twin or quadruple bases did not work or were otherwise wanting and hence did not persist. Replication and reproduction within living matter could not occur as now known and understood until the triplet system had been determined and its virtual monopoly established. With regard to reproduction, the four bases involved are the constituents of DNA.

James Watson has described how proteins are produced within the growth process. The relationships of the various constituents have a numerical basis. Amino acids occur in about 20 different forms. If each DNA nucleotide had only involved any two of the bases then the variety of amino acid would be restricted to 16 (4^2). If on the other hand the DNA nucleotide had included 4 bases, then the theoretical variety of amino acid would be 256 (4^4). This would be a gross extravagance. The triplet became the favoured grouping because it was the minimum possible to produce the required range of amino acids (about 20) from a selection of four bases, i.e. 64 (4^3), although still with ample to spare. This suggests that the triplet arrangement has cause in the needs of growth and RNA, rather than reproduction and DNA. The latter also has to use the triplet system, even if other groupings were available, because it has developed alongside RNA and the two have always needed to be compatible.

The amino acids are produced by ribosomes in the cytoplasm that 'read' the code of specific messenger RNA encoded by the DNA in a chromosome of the nucleus. The messenger RNA passes through any relevant ribosome, whereby the latter's own RNA 'reads' the code of the other to select amino acids and produce them in the form of a polypeptide chain. Such chains are the basic constituents of proteins.

The question arises as to which came first, RNA or DNA? It would appear to be a chicken and egg problem. However, the argument here is that reproduction precedes growth in the origin of life, and this procedure is clearly the business of DNA. RNA is concerned with activities such as growth and movement. The primacy of DNA might seem apparent in that ribosomal RNA is under the control of chromosomal DNA by means of the messenger RNA produced by the latter. There is apparently no comparative method by which the needs arising in the cytoplasm are transmitted back to the DNA. It is a one-way system under the control of the DNA. It is to be noted that chromosomes do not grow; they use selected material to replicate. However, their numbers are manifold but predetermined within the cells of any particular organism, which can be seen to have been arrived at by growth over an immense period due to acts of fission or failed fusion leading to this multiplicity. This represents growth in the species, rather than in the lifetime of the individual. Simple binary fission still occurs with the simplest organisms, while mitosis also relies on fission. Acts of fusion represent a characteristic still retained with meiosis.

So how then does DNA 'know' when reproduction is necessary? Whatever method is being used, a cell acts like a machine that has a limited lifetime. DNA is designed to become aware when the whole cell body is becoming overstressed, either by the threat of excessive growth or exceeding its allotted life span. The DNA is constrained to act in order to continue to service the organism.

This all suggests that DNA preceded RNA. One might surmise that DNA was the product of the infra-life period, some four billion years ago, but that proto-life only became possible when the possibilities of interaction with and control of RNA arose and were finally realised. Yet both carbohydrate combinations can be regarded as chance creations of the chemical possibilities available that happened to have a certain potential for internalising their ability to survive. This secured their persistence. It was during this infra-life period that certain carbohydrates arranged themselves into chains, which through predestined chemical possibilities led to the extraordinary complexities of form and activity of DNA and RNA by way of the formation of bases into triplets and the eventual appearance of genes.

A reminder would seem to be a good idea here. These extremely early developments involved huge periods of time, together with matter with potential for life at a microscopic level. On no account must it be imagined that the transference from infra-life to proto-life was anything like a sudden occurrence. Some research strongly favours the theory that all life is descended from a single source. In this respect it might be claimed that from out of the mass of carbohydrates constituting infra-life there gradually emerged a compatible grouping that came to utilise the potential of DNA and RNA in a manner that can be called proto-life. Such a self-replicating entity would be the first photosensitive bacterium, whose successors are still with us in the form of microscopic algae (cyanobacteria), constituents of the phytoplankton. This was the progenitor of all subsequent life on Earth and such emergence was initiated in water that had long been illuminated and warmed by the Sun for a period every day, the length of which varied within a yearly cycle. This was caused by the Earth's orbit round the Sun and the time taken for this also involved just over 365 revolutions of the former, which events are known as days.

■ Early Evolution

The progress made by evolution during more recent phases has been put down to mistakes made in the reproductive process having occasionally proved to be advantageous. This has led to them being perpetuated and has been caused by random changes in genes during the replication cycle of chromosomal DNA. This adequately explains the process as it occurs among those phyla comprising complex, multicellular organisms. Yet how far is it applicable to simpler entities, especially unicellular (or non-cellular) ones? Among these a different set of rules seems to be at work.

The way that bacteria engage in pseudo-sexual reproduction has been described above, but this, in fact, simply involves the transference of genetic material from one individual to another and then to become part of its genome. However, such transfer can also occur between dissimilar species, where the process is known as horizontal (or lateral) gene transfer (HGT).The inference is that HGT has been of much greater significance earlier in life's story.

Bacteria contain scraps of material that are necessary for them to function. These potential endoparasites may be descended from feeding activities that were transformed through such becoming useful as symbiotic components of cells. Eventually they became essential features of the specific organism. At more advanced level they could become self-contained, as with chloroplasts (containing chlorophyll) and mitochondria (DNA dealing with respiration and energy production). Such as these are known as organelles.

A cell nucleus appears to be a type of organelle that has achieved a dominance of function; it can bully the others. Other entities enclosed in the cytoplasm seem to represent a phase prior to organelles. Here one might indicate ribosomes and plasmids. The latter consist of dual chains of DNA genes that can be transferred to organisms that need not be related. They are usually found as closed rings. Such transference is common among bacteria, but becomes less frequent when more complex creatures are considered. This facility for HGT among such primitive organisms suggests that descent did not have the characteristics of a branching tree, but rather was reticulate. With such a network being the only way that proto-life could keep going, it is clear that the concept 'species' is hardly applicable at this stage. Indeed the ability of the network to distribute genes all over itself suggests that there was a drive towards there being only one form of life. This indicates that some other process was at work that could allow evolution to take place and override the retrograde tendencies of the network. The answer seems to lie in that from the start HGT did not necessarily replace or destroy existing genes: these carried on independently. Indeed, the number of genes just came to exist in ever increasing and eventually enormous quantities and in collaboration with the other elements with which they were associated and were dependent on in order to constitute and function as the living wholes they had become.

■ The Nature of Cells

Reproduction of cells occurs using one of two available methods, mitosis and meiosis. For sexual reproduction the latter is necessary. The genetic information of cells is contained in temporary bodies known as chromosomes, the number of which varies depending on the species. Each chromosome comprises 'chains' consisting of an immense number of genes. Generally the chromosomes of sexual beings occur as pairs, all equal except for one such pair where difference occurs; but only in males, which have the so-called X and Y chromosomes. The latter is smaller than its X partner, this reflecting a difference in the gene count. All the genes of every cell are arranged end-to-end along each chromosome as an interlinked double helix of DNA (deoxyribonucleic acid), resembling something like a 'spiral ladder', the 'stringers' of which form the so-called 'backbones' of the chromosome.

During normal cell division (mitosis) the two structural 'stringers' of each DNA 'ladder' come apart, taking half of each 'rung' with them. After this breaking of the central hydrogen bonding each half makes a copy of itself. However, these heterogeneous 'half-rungs' of the split helix differ in the arrangement of the chemical bases involved, but after replication the original order is restored, since each half only makes a copy of its departed mate. Two double helices are thereby produced discretely that, along with the division of the rest of the complement, enable two new daughter cells to be formed. But when sex cells are produced by meiosis the chromosomes are haploid (not paired). These still reproduce by mitosis, but are each created from half the complement of chromosomes that are found in a cell designated as diploid, receiving one complete set of genes from either one of each pair of such chromosomes.

In plants and animals generally comprising diploid cells, the haploid condition is the mark of reproduction of the whole being by sexual means. Where non-sexual reproduction occurs division of haploid cells is theoretically the only means required, with the diploid condition under these circumstances appearing to be a gross extravagance. So, what is the point of the diploid condition? Well, without it having occurred, with the inherent possibilities for exchange of possibly variant genetic material, the achievements of evolution as known to us would not even have been approached, let alone reached.

Since it is evident that sexual beings arose from earlier asexual ones, it follows that the diploid cell is a development from the haploid condition. Thus have acts of fusion occurred in the deep past that could be followed either by fission or by further fusion, and so on until at last, depending upon the circumstances, a stage embodying an alternation between fusion and fission became permanent, as observable in life now at the cellular level. However, the way replication took place meant that occasional mistakes in copying were possible in each instigation of fission followed by re-fusion, mistakes that sometimes proved to be beneficial. Bad mistakes would lead to disaster, but others could provide mutations that proved advantageous, the very gist of evolution. Such mistakes would cause differences that at one stage could well have allowed the appearance of sexual differentiation and maybe even the difference between DNA and RNA. Further beneficial errors among such special entities would enhance the differences that had been initiated and which had so much evolutionary potential. Thus did favourable small errors in replication initiate the gradual development of a process involving increase in both size and complexity: this was unstoppable, at least biologically and if not restricted by environmental or spatial limits.

Both growth and reproduction involve replication. The difference is, that in growth and asexual reproduction cells could get by with single chromosomes, while reproduction that is truly sexual demands paired ones. The human complement of chromosomes is 23 pairs; but the actual total is of no account regarding method. The fact that they can form themselves into distinct entities within a defined nucleus suggests that they all represent original individuals that have been assembled by various acts of fusion involving the nuclear matrix alone over a period of time. DNA is a substance capable of huge variety and the inference can be made that many different and independent strands of it once existed, each with a surrounding accumulation.

At some time in the past, when surrounds (nuclear matrices) fused, very occasionally it proved advantageous for survival that the DNA chains themselves sometimes resisted fusion and stayed as paired but only partially fused entities within the new whole. Thus did chromosomes then appear as pairs linked at one point that were theoretically identical in content, but not mirror images of each other. With further fusion of nuclei the result was slightly different in that the equivalent enclosed entities of DNA avoided fusion and remained as discrete pairs within a single nucleus, namely duplicated chromosomes. Then when such a nuclear body replicated itself these may have done the same while still remaining as pairs, which continued the existing process of growth through mitosis. However, occasionally a chromosome of such a nucleus would perhaps for some reason 'choose' not to behave in this manner. Then any eventual re-fusion of such originally twinned 'haploids' may not have produced quite the same arrangement as the original partners and the new 'diploids' would have the chance to be marginally different. Something like this could explain the origin of the dual nature of sex and would greatly increase evolutionary potential.

This process would reflect the behaviour of molecules. While at the lowest level atoms will combine with their like to form simple molecules, e.g. O_2, many molecules that are already compound will readily combine with others that differ from themselves to form ever more complex compounds. Life's beginning is founded on the fact that this property could lead to immense complexities. The element that particularly lent itself to this process was carbon.

Mitosis causes growth in multicellular organisms by increasing the number of cells. Each cell has its ultimate origin in the so-called stem cells, these being capable of turning into any type of cell. They are in fact predestined to become those cells characteristic of the position they come to occupy within the organism and the function they serve. All the cells carry the same genes, apparently another extravagance. The position they find themselves in with regard to the whole developing organism governs which of their genes are activated. This determines both the nature of their structure and the way they behave. What is evident is that the acts of fusion that earlier resulted in various chromosomes to be assembled within the same cell still allowed continuing acts of fission whereby whole cells replicated themselves, as they still do now.

Where primitive cells of proto-life became overcrowded the fission had not only resulted in increased numbers, but also greater consumption of resources and the occupation of more space. While they remained separate entities such cells could still feed independently. However, when they were so numerous as to be contiguous, with fusion often resulting, one thus created 'cell' could then contain more than one chromosome. Subsequent fission after growth then exhibited the characteristics found in asexual reproduction. But where original fusion for some reason did not occur they could remain contiguous, but come to depend on each other, even though not as yet constituting a tissue during the earlier stages on the road to full life. When thus adjoined in numbers, some method had to be found to maintain the inner-member cells of such gatherings if they were to survive. The answer was co-operation.

Cellular nuclei came to exist as such because they were set in an outer mass of material known as cytoplasm. The evolutionary experiment with multiple nuclei, as found with the paramecium, had limited success because it inhibited the attainment of a massive increase in size of the living individual entity. The fusion of partners here is temporary; no tissue results. Successful combination of cells, first into tissues and then into organs, came to depend on their being but one nucleus per cell that was capable of initiating acts of binary fission that would affect the whole cell.

The change from pre-life to infra-life occurred when certain relatively complex substances came into being that had never existed before, namely carbohydrates. The end of infra-life itself was presaged by the appearance of further new substances. Proto-life gradually developed when carbohydrates began to react with amino acids in a very complex, diverse, yet preordained and predictable manner, an important result of which being the production of proteins.

■ Mortality of Cells

Each cell of a multicellular organism 'knows' its own entity and its place in it. The type and number of cells has to be regulated to maintain the optimum form of the whole. Both internal and external presences can interfere with this. The cause for a cell being healthy, or dying, or having its life extended rests on the relationship between energy supply and energy need. Dearth of available energy ultimately results in premature death, but the process in any case leads to undersize of the whole. The availability and retention of excess energy tends to cause extension to cell life, excessive numbers of cells and oversize of the whole organism. This condition can also become dangerous. Particular sufferers in the latter respect are found among the vertebrates. Apart from being undesirable in themselves, extreme conditions also lead to proneness to attack by other organisms.

Cells resulting from mitosis should be immortal; but are not. Even purely mineral compounds can be weakened and fail in the course of time, as with fatigue of metal and age embrittlement of plastics. The difference with living organic matter is that a degree of control is exercised by a cell to delay the final destruction that its own complexity of structure and activities will eventually bring about.

Both meiosis and mitosis require energy. As any type of cell wears out operationally, these functions are increasingly delayed as energy fades and the cell will eventually die before having carried out such activity. With time the increase of such mitosis failure will inevitably cause the failure of any whole organism, even though it has hitherto survived other threats to its existence, while persistent occurrence of meiosis failure could theoretically lead to widespread sterility and the end of the whole species.

CHAPTER 13

Adults, Young and Remote Evolutionary Links

■ Animal Feeding Methods

All the creatures that make up the animal world can be roughly divided into two groups with regard to method of obtaining food. There are, on the one side, those that seize their food by some device and, on the other, those that create currents to whisk food particles into their body cavities and whereby these can be absorbed into their tissues. They can be loosely termed "coarse and grasp" and "fine and whisk" feeders respectively. The latter are confined to aquatic habitats. Among some of the phyla both methods can be found. The division is old in the animal story.

Other and lesser phyla can usually be found to follow one or other method exclusively. Thus the Bryozoans and Brachiopods are fine-and-whisk, while the Ctenophores (sea gooseberries), arrow worms and proboscis worms are coarse-and-grasp. Rotifers are basically fine-and-whisk, but some can also seize prey with their jaw-like pharynges.

Some creatures change over in the course of their lifetimes. Confirmed coarse-and-grasp types like annelids and arthropods often start off their lives as free-swimming, fine-and-whisk feeding larvae, this having phylogenic implications.

Fine-and whisk feeders rely on making a current in order to get their nutrition. In multicellular forms this is done by being provided with special incurrent and excurrent openings. In the case of the Protozoa the configuration of the single opening is such that fluid can pass in and out at the same time. Some multicellular coarse-and-grasp feeders also have only one body opening (e.g. sea anemones), so that after a meal the wastes must be ejected by the same route as the food came in.

	Phylum	Coarse & Grasp	Fine & Whisk
1	The Protozoa	Amoebas, suctorians, etc	Paramecia and other ciliates
2	The Sponges	None	All
3	The Coelenterates	All (using sting cells)	None
4	The Flatworms	All	None
5	The Roundworms	All	None
6	The Echinoderms	Stars, Sea Urchins & Sea Cucumbers	Sea Lilies
7	The Molluscs	Cephalopods, Gastropods	Pelecypods (Bi-valves)
8	The Annelids	Earthworms, Lugworms, etc.	Feather & Fan Worms
9	The Arthropods	Most of them	Barnacles
10	The Chordates	Vertebrates, Acorn Worms	Sea Squirts, Lancelets

■ A Basic Form of Animal Life

To live by either of the two methods of feeding, certain appendages are possessed by animals to pass food to their tissues – in most cases by way of an opening that serves as a mouth. Coarse-and-grasp protozoa often do not have a mouth, for these may take in or eject matter through their very cell walls. Paramecia and vase-like protozoa such as vorticella, on the other hand, use cilia to whisk food into a definite mouth. Sponges are provided with internal flagella to whisk food in through their numerous incurrent pores and wastes out by way of their collective excurrent openings. Other multicellular fine-and-whisk feeders either have internal cilia, like the molluscan pelecypods (bi-valves) and the chordate tunicates (sea squirts), or ciliated tentacles like the bryozoans and feather worms to waft food particles into their mouth openings. A feature of multicellular fine-and-whisk feeders is a filter for trapping the food. Strangely enough, some of the largest sharks and whales use a filter method of feeding, even though, being motile, they do not need to whisk (and indeed cannot).

How former land animals like whales came to feed in this way is puzzling, but ostensibly provide a good example of convergent evolution, for the organs used in this way by both them and the relevant sharks are apparently analogous - rather than homologous - to those of each other and other filter feeders. However, in the difficult case of the whales this method of feeding may point to them never having been fully terrestrial; they may always have retained an adult form that remained marine, i.e. one that returned to that habit in the later stages of its life. Once the younger terrestrial phase was for some reason abandoned, the relevant feeding and reproductive processes used on land were lost. That some other cetaceans - i.e. among those with teeth - are also called 'whales' may reflect a genuine case of convergent evolution; however, these should perhaps better be regarded as big dolphins.

Coarse-and-grasp feeders may be equipped with tentacles fitted with adhesive, sucking, barbed or poisonous organs; or they may simply be active creatures using the mouth itself for grasping food. Others, such as many vertebrates, are provided with organs of divergent origin that have retained or evolved grasping ability in addition to – or in place of – their other functions. The hands of human beings are a good example of this.

The first 'animals' must have been minute 'predators' that had departed from and had begun to live off a basically bacterial living world. Perhaps after an initial existence as plant-animals (like the green flagellates) they abandoned photosynthesis altogether and let their prey do all this energy-collecting work for them; such 'prey' would originally be like those minute specimens of plant life that float about in water and are now major constituents of the plankton. The best way for primitive animals like these to maintain a steady supply of such nutrition was to create a food-bearing current and to have a system for extracting particles from ingested fluid. The standard method of doing this was, and still is, by means of a body cavity where the food could be trapped, treated and absorbed into the tissues, by means of an incurrent opening for fluid and particles and an excurrent opening for fluid and wastes. This is basically the system used by the Sponges now and its soundness is demonstrated by the way these animals have survived with little alteration throughout the ages. Although some have more formal shapes, all sponges are basically alike. Even though they use a form of the fine-and-whisk method, they are structurally distinct from all other types of animal.

Another self-contained phylum is that of the Coelenterates. The sedentary forms all consist basically of a vertical hollow cylinder with a feeding opening at one end surrounded with tentacles. This opening serves as both mouth and anus. Since they are all coarse-and-grasp, they take definite meals and the lack of a separate anus is not such a great disadvantage to them. The poisonous barbs, which they all bear in their skins to stun prey, clearly distinguish this phylum from others.

Among many other phyla there are animals that more or less resemble the above-mentioned hollow cylindrical form, but since they are all fine-and-whisk a separate anus is greatly beneficial. These are the bryozoans, the sea lilies and the sea squirts, while certain others, like the brachiopods, the rotifers and the bivalve (pelecypod) molluscs - while not bearing much resemblance to the basic idea - must be thought of in the light of divergent evolution. Are they not distorted versions as the result of them being subjected to differing evolutionary pressures?

The oldest fully fossiliferous rocks are from the Cambrian. Yet even at that time all the phyla were already in existence, although the vertebrate chordates were initially absent. Evidence of fossils in the Pre-Cambrian is slight. If the contention is upheld that life on Earth is a unique property, a unit, and that all lines – though often now apparently so far apart – have peeled off a single stem at their own particular times, then there must also not only be awareness that the evidence of living forms in Pre-Cambrian times is grossly inadequate, but also that indication of the common origins of the phyla is at present non-existent. It has either not yet been found, or it has all been destroyed by geological events. As it is, the geological picture suggests the sudden appearance of various phyla at the start of the Cambrian with no clue as to their previous development. The contention here is that all multicellular animal life once existed in the described basically similar form at some time during this obscure earlier period and later evolved from it to become the differing phyla in evidence from the Cambrian. The divergence occurred in different ways, yet in many cases clues were left to indicate former relationships. The problem is too complex to explore in general terms and must be tackled with specific examples.

The Eyes of Cephalopods and Vertebrates

The theory of convergent evolution explains the resemblance of the eyes of these two apparently quite distinct groups of animals as being due to the fact that squids, octopuses, etc., are active predators, like many fish and other vertebrates. Yet a great number of vertebrates are not predators. On the other hand many crustaceans, arachnids, insects and even gastropods certainly are, yet they all have eyes of quite different structure. If convergent evolution worked in one case, why not in others? So did the cephalopods and vertebrates have a common or closely linked ancestry that was somehow predestined to produce such eyes, or was there a common ancestral form already equipped with the same? When the first cephalopods and fish appear in the geological record eyes are already present, thus leaving the second suggestion a distinct possibility.

Let a start be made again from the basic form proposed in this work for earlier multicellular forms of life, namely a cylinder standing on or fixed to a bottom or other surface and with a mouth at the other end surrounded by tentacles. It reproduced sexually to give rise to free-swimming larvae, each of these being destined eventually to settle down on a base in order to grow into a new adult. Then let it be assumed that, as an evolutionary process, the stage of growth for settling down in this way became later and later, until at last the free-swimming larva itself became a quite complex animal in its own right, which, however, still needed a spell as an adult on a base for reproductive purposes. This is similar to the condition now still prevailing with the chordate sea squirts.

The next stage is the creation of the free-swimming adult. If at any particular period, and under certain conditions, life in a certain area became more favourable for the larval free-swimming form, rather than the adult sessile one, then in the course of time mutation and natural selection would see to it that reproduction would take place in what had hitherto been an immature state. Under continuity of such conditions the original adult would face extinction – made redundant by competition from its own larval form, which had acquired the ability to reproduce at this earlier stage of growth and formation.

The contention here is that the cephalopods and vertebrates are both descended from such an adult. The problem is to overcome the difficulty of explaining why the two sets of animals are now ostensibly so dissimilar.

Some resemblances in addition to the eyes have already been indicated in an earlier chapter. In both cases the original adult would have neither eyes nor motility and the two strains must have evolved from some promoted juvenile form which extinguished its own immediate adulthood; but it would not be from the same stage of development in each case, and most probably these stages, as well as the period of their defection from the parental stem, were quite remote in terms of time, place and degree of relationship to the old mature form. The cephalopods certainly broke away first, for they represent a juvenile stage nearer to the former adult.

The cephalopods have all retained one archaic feature from the old adult and that is the ring of tentacles around the mouth. This is usually equated with the 'foot' of other molluscs. This seems a bad comparison, for in no other case is a molluscan foot associated with food gathering. Their organ known as the foot simply acts as a means of contact, anchorage or locomotion in association with the solid supporting matter of the environment; tentacles are certainly not of this nature. To find corresponding organs on the gastropods one must look to the feelers that many of them bear near the mouth – former tentacles that have evolved by conversion to sense organs.

The earliest known cephalopods had torpedo shaped bodies enclosed in shells and most modern forms, especially squids, retain this shape, but have lost the outer shell, which was a relic of dwelling on the bottom. They do however often have internal bony or cartilaginous structures. Yet an ancient divergent development was that of the longitudinally coiled shell – not transversely coiled as with most gastropods – and these became very numerous in the Palaeozoic seas, with large examples existing (e.g. ammonites). But such shells, however light, must have been a drag and are only found today among a few diminutive species.

As far as the chordates are concerned, the most archaic adult form is possessed by the sea squirts (tunicates). They must be comparatively close to the original adult, although more complex in structure and function. If so they have lost the tentacles and rely on cilia to create a food-bearing current. In addition they are equipped with a filter-type food trap in the form of gills. Other features that are unusual in such sessile creatures are the possession of a brain of sorts and a circulatory system, complete with heart. These organs were presumably first evolved in the motile juvenile stage and only transferred later to the sessile adult form.

None of the modern invertebrate chordates have eyes and in this respect they are more primitive than their vertebrate relations. On the question of eyes, the cephalopods and vertebrates are more closely related to each other than either of them is to the invertebrate chordates and any other molluscs, or, for that matter, any other kind of animal.

■ Tentacles

As being proposed here, these constitute the most obvious feature of the hypothetical common adult ancestor of so many phyla. One can expect to find them in present day animals, either as something like their original structure, or in some rudimentary form, or in a greatly modified state; or again they may have vanished without any obvious trace. Perhaps the animals that have tentacles most like the prototype are the bryozoans, which are small and simple enough to be compatible with an early stage of development and have mouths encircled by ciliated tentacles. On the other hand, the suckers and other aids to grasping found on cephalopod tentacles are not found elsewhere and these can be regarded as divergent organs, while the 'fringe' round the mouths of adult tunicates may represent residual evidence of former ciliated tentacles on these chordates.

Among the protozoa, tentacles are represented by flagella or cilia, or by the fused clumps of these that surround the mouths of many of the vase-like types. In this way single celled organisms have foretokened structures that were also to develop among the multicellular ones.

The rotifers, although multicellular, display many features that suggest them to be links between the protozoa and multicellular phyla. The feeding method of rotifers shows the typical protozoan way of using cilia, which create a current to pass food to the mouth. These are borne on lobes, which, if they had the power to wave about, could be regarded as rudimentary tentacles. There are normally two or four of such lobes and in some cases, such as *stephenoceros eichornii*, they are certainly very tentacle-like in shape. As already mentioned, some rotifers have tentacles that can be used for coarse-and-grasp purposes.

Among the coelenterates, the tentacles of polyps often resemble the basic type, except for the inevitable presence of their unique stinging cell system. Medusas and sea anemones still preserve the basic pattern of a mouth surrounded by tentacles, but they have generally tended to evolve towards an elaboration of the system as an answer to evolutionary pressures, rather than favouring a move towards bilateral symmetry, which has been the route taken by the adults of so many other phyla.

Some primitive animals have their supporting base or 'stalk' elongated and this may be buried in the sediment of the bottom. Such is true of certain brachiopods, which have the power to withdraw themselves into their burrows for safety. Other brachiopods only have a short stalk that is used for attachment to rocks and other solid objects, with the animals having no other means of protection other than their bivalve shells. Inside its shells a brachiopod bears a pair of long coiled arms called the lophophore, which is provided with ciliated tentacles, these being used to whisk food towards the mouth.

The feather worms constitute another set of animals that use tubes. They bear some resemblance to the brachiopods, but construct their shells underground in the form of hard tubes and can withdraw their tentacled and ciliated 'heads' into these when danger is perceived.

The free-swimming and predacious arrow worms are quite a different matter. They have bilateral symmetry and are coarse-and-grasp feeders. With them the tentacles are represented by bristles retained on each side of the mouth that aid the catching of prey.

The various round and flat worm species are all coarse-and-grasp feeders and frequently parasitic. Some retain external body cilia used for swimming about, but not for feeding purposes. Any vestiges of tentacles are rare.

The sponges are fine-and-whisk feeders and use internal flagella to create currents. They have no trace of tentacles.

Except for the sea urchins, the echinoderms are provided with some sort of tentacles around the mouth. These animals are equipped with numerous "tube feet" upon which many of them can move about, while the starfishes actually employ the suction of these to increase the grasping power of their tentacles. Among the existing echinoderms, the only ones to retain the general posture and function of the proposed basic sessile type are the stalked crinoids or sea lilies, which direct food towards their up-turned mouths by means of ciliated tentacles. All the other types represent former juvenile forms that have been promoted to adulthood by acquiring the ability to reproduce.

Annelids are coarse-and-grasp feeders and, like the round worms (nematodes) referred to above, are without tentacles. However, some of the predatory species are provided with sensory projections that are known by the words 'feelers' or 'antennae' and indicating tentacular origins. Adult arthropods too are normally adorned in this manner, e.g. lobster, bee and butterfly. Very peculiar among the arthropods are the barnacles, which are marine crustaceans. They are well known for their ability to cling to rocks and other hard surfaces and feed by protruding their fringed appendages to trap small animals and other food particles. Yet their 'tentacles' are actually homologous to the legs of other crustaceans; they have been described as "animals which stand on their heads and kick food into their mouths".

The molluscs have already been considered in some depth. The cephalopods all have grasping tentacles. The gastropods, which include the slugs, snails, winkles and whelks, are coarse-and-grasp, but do not have any method of seizing their food except the mouth alone, simply chewing away at it as it is found. (Head feelers are however frequent.) Chitons feed in a similar way to gastropods, but are of much simpler build and completely lack appendages. The pelecypods – clams, oysters and other bivalves – are all fine-and-whisk. Currents are produced by internal cilia so that tentacles are neither found nor needed.

Among the invertebrate chordates, the acorn worms can be described as coarse-and-grasp. They burrow in mud or sand to feed somewhat like earthworms by passing such substances through themselves and extracting food in the process; they obviously have no need for tentacles. Neither do the tunicates, which, like clams, are fine-and-whisk feeders that use internal cilia. The lancelet, however, which is a small fish-like creature, does have a mouth fringed with tentacles. Although it can swim about, the lancelet spends most of its time buried in mud with only its head protruding. Like the tunicates it is a confirmed fine-and-whisk feeder and has internal gills to this end. The retention of mouth tentacles perhaps suggests that the internal cilia have never become sufficiently efficient to cause their redundancy. But the tentacles now seem mainly to serve a sensory function and in addition are able to exclude the entry of over-sized particles.

The vertebrates are coarse-and-grasp feeders exclusively, but none of them retains grasping tentacles. However, some species of fish do possess sensory appendages at the mouth known as barbels and again these can be regarded as vestiges of earlier organs of rather different function. Modern amphibians and reptiles are devoid of such features, but the special sensory hairs or 'whiskers' adorning the mouths of mammals could well represent the result of evolution on sensory tentacles as an adaptation to living on land by animals developing an overall covering of hairs of different purpose. Whiskers are obviously themselves not the remnants of tentacles, but could be their natural successors, being the usurpers of their position near the mouth, if not exactly all their functions.

■ Gills

Many animals have organs that can transfer oxygen from an ingested liquid stream to the circulatory system and, at the same time, relieve the latter of its burden of carbon dioxide. In many cases such 'gills' are also used to trap food brought in with the current, although none are employed solely in this way. Other animals are quite devoid of gills, either because their ancestors never had them, or because they have been lost by evolutionary processes.

Among many of the simpler and – one might venture to say – more primitive animals, all signs of gills and circulatory system are absent. This is true of the protozoa and all those phyla which feed in that protozoan manor of absorbing food, pre-digested or otherwise, namely in amoeboid style from the liquid environment. Among those retaining primitive traits of this kind are sponges, coelenterates, flatworms, roundworms, bryozoans, brachiopods and certain other minor creatures.

Numerous examples of gills are found being used for gaseous exchange (respiration) among the arthropods, molluscs and vertebrates. Gills which are used for both respiration and as food filters occur among the pelecypod molluscs and the invertebrate chordates, all such being fine-and-whisk feeders.

It seems reasonable to suppose that the additional area provided for gaseous exchange by gills was the original reason for their appearance and development and that this was related to a greater degree of activity by the animals acquiring them. The food filter aspect would then be a further advance whereby Nature, as she has so often done, found a new use for an organ originally developed for a quite different purpose. Since it is the rather sluggish pelecypods and chordates that use this method, one can feel rather sure that respiratory gills were evolved by some free-swimming and active juvenile stage before eventually being incorporated into the less lively adults and converted to food collectors.

■ Tentacles, Gills and Other Appendages

Since segmented creatures such as the fan worms and feather worms (polychaete annelids) have gills that serve as tentacles, or tentacles that serve as gills (whichever way one wants to regard it), it is tempting to conclude that tentacles and gills are generally speaking homologous and hence of common ancestry. If this is so, one would think to find clues in modern animals which indicate milestones in the earlier development of gills and tentacles as found in some former and more primitive ancestor and leading to the often dual purpose appendages of the present The basic form would be a non-segmental sessile animal, similar to a polyp, but instead of stinging tentacles it would be equipped with a ring of fringed ones for fine-and-whisk feeding. They would not be specifically gills at this early stage and gaseous exchange took place simply through the external skin (ectoderm) of the animal. However, the tentacles would have a larger surface area compared to their bulk than would the other external body parts and hence respiration would be easier, and thus more efficient, by way of them. When an organ performs a task better than others do, it tends in time to specialise in it and thus gain a monopoly. In this way tentacles specialising in respiration became the forerunners of gills. The picture presented is something similar to a bryozoan, only at this stage with a single body opening to act as both mouth and anus.

A further step towards elaboration was the ability to elongate the body by means of forming into segments and an indication as to how this came about is found among the coelenterates. The jellyfish Aurelia, for example, starts off by growing into a sessile polyp which eventually shows traits similar to segmentation when it changes and comes to resemble a "pile of saucers" as a stage in a reproductive process. Later each 'saucer' peels off in turn from the top of the 'pile' as a tiny free-swimming jellyfish or medusa. Let it then be supposed that the hypothetical basic type could also reproduce asexually in this manner, except that at some stage the expected individuals failed to detach themselves, but that instead the whole remained together as a floating "pile of saucers". This could then either be regarded as a string colony or a segmented individual, depending upon the degree indicated at any time of collective organisation and co-ordination.

One agent acting to prevent individuals from detaching themselves from such a colony would be incipient polymorphism. When considered as still sessile, only the top member of such a string colony would be in a position to use its tentacles efficiently for feeding. Yet if only partial detachment occurred and produced slots between the piled up units, which gradually increased in size, then it is possible that the lower members could use such side-openings to supplement the feeding of the whole colony. This would be in preparation for the time when each in their turn would reach the top to assume the role of chief 'breadwinner'. The point is that such slots between constituent entities, and the tentacles associated with them, suggest a mechanism that could easily be turned into use as gills. Of course, the foregoing might presuppose a common alimentary canal or gut for the colonial animal concerned, even though an anus does not as yet exist. It is clearly connected with fine-and-whisk feeding; the coarse-and-grasp methods used by the coelenterates would effectively preclude them from taking such a step.

Except that the tentacle-bearing polychaete annelids are each provided with an anus, the animal portrayed above resembles certain of them. These are sedentary and have tentacles in concentric circles round the mouth, each circle representing a former segment, the innermost being the first and so on. These segmented worms live in hard tubes built by them and into which they can withdraw when sensing danger. Another kind of polychaete lives in a U-shaped burrow. In this case the tentacles of the head segments have degenerated and it uses those of its body to cause a current of water to pass through the burrow. It feeds on small organisms whisked in by this means, while getting a fresh supply of oxygen at the same time.

An important facet of segmental evolution was increased motility and this can be represented by the free-swimming polychaete *nereis*. Most free-swimmers either grasp food directly with the mouth, or have organs that trap the food and then pass it into the mouth. Where the grasping is done by recognisable tentacles around the mouth, this can be recognised as a primary function of straight descent from early forms and is a characteristic of the division of the molluscs called cephalopods. Where grasping functions are performed by paired organs such as pincers (crabs), claws (cats), talons (birds) or hands (primates), they can be regarded as secondary modifications of members occurring on segments more remote from the mouth and formerly of different function. Even so, such members as found on crustaceans and other arthropods and used for feeding may have evolved directly from the tentacles of ancestors, all record of which is lost. The worm *nereis* consists of numerous segments, all virtually identical, except at head and tail. At the head tentacles are represented by an array of sensory projections (feelers), which arrangement incorporates several fused segments. There are also feeler-like appendages at the tail. All the other segments have lateral projections called parapods that assist in swimming, burrowing and – by means of the blood vessels they contain – respiration.

Thus the *nereis* virtually has a row of rather unspecialised external gills along each side of its bilateral body. With more advanced animals, such as arthropods, it is easy to see that the segments have become more specialised and that the former homologues of the parapods have developed into legs or gills, or other projections or organs, depending largely upon the position along the body. A link between the annelids and the insects is provided by such lengthy terrestrial segmented arthropods as centipedes and millipedes. Being arthropods, they are clearly nearer to the insects, but do relate to the annelids in having a great number of undifferentiated segments. Another creature placed between the annelids and insects is the rare and peculiar *peripatus,* although here the evidence is somewhat disguised from the casual eye by the inherent segmentation not being externally apparent.

■ A Plastic View of Life

The foregoing seems to imply that life not only can exist in virtually any form one cares to imagine, but also has the ability to change into any such form as required. This must to some extent be true, but there are many limiting factors. Living forms can be small or large, simple or complex, squat or elongated, primitive or advanced, uniform or diverse, single or multiple; or at any conceivable intervening stage. However, changes in living forms can never be violent. A horse cannot suddenly become a cow. Cows could eventually change into something else, but the way that evolution works indeed prevents them ever becoming horses, or pigs, or sheep, etc. To do so would require a return to some common ancestral form. But with evolution, once change has happened there is no going back. The drive of evolution is always forward with the track behind always being hidden among a plethora of failed and lost opportunities.

Life continues through the ability of its forms to reproduce themselves in their own image. Yet permanent change can and does occur in species when for one reason or another the reproduction is faulty. A more rapid and predictable series of changes takes place during the lifetime of the individual, when it gradually changes from egg to adult.

The random changes or mutations that affect the species must be helpful towards survival if they are to spread through it and become permanent features. This usually means that one favourable mutation can occur among many, and - for those thus blessed – it may eventually counteract quite fortuitously some specific hostile development in the environment.

Yet it would be wrong to put evolution entirely down to innovation, to mutation that explores new ground. In the course of its evolution an adult may obtain features which formerly were characteristic of its own juvenile form. In the same way a juvenile may obtain features that used to be characteristic of the adult. This is where plasticity, as opposed to mere growth or uni-directional change, can be seen as a characteristic of life. The juvenile can inherit something from the adult and *vice versa* and the two features might even slip past each other over the ages as they shift their relative positions in the life-span that is normal to the species. In a similar way, as has also been mentioned earlier, if the juvenile inherits on a permanent basis the reproductive functions of the species it then becomes the adult and the erstwhile adult, having no essential reason left to exist, dies out. This leaves the new adult and all its younger stages right back to the egg to continue to follow their own conjoined but specific evolutionary paths. If one considers life as a unit, however, every physical aspect of it as known today is connected by a line of descent to that remote point in the past when life itself was being born and showing none of that diversity we know today; and none of the complexity either, except for that of its molecular bearers.

The reader will appreciate that this work is no handbook on biology. Such biological examples as occur are merely exemplary. The intention is mainly to examine and criticise certain principles relating to evolution.

The point being approached is just such a criticism of the methods in use in order better to understand the living past. The criss-crossing of paths of evolution are so immensely complex that it is impossible to take in all the details of the whole at one go. It is like looking at a lot of people and trying to study each one simultaneously. We can contemplate only one or two persons at once of the many, or one group of the several, or simply one crowd of indistinct individuals.

Efforts to simplify evolution in order to ease comprehension are made by representing the concept graphically. The system usually employed has two basic weaknesses. Firstly the representation on a plane (e.g. sheet of paper) is limiting, because evolution is volumetric and hence requires a three-dimensional model to illustrate it at all adequately. Secondly the common use of a branching tree as a symbol for the manner in which evolution has occurred is misleading in that it rivets the dividing points of the phyla to abrupt occasions, whereas evolutionary change is not linked rigidly to points in time at all, except that once it is done it cannot be undone (or at least only by incredible and virtually impossible coincidence).

In Fig 2 certain phyla are shown more in a plant-like way to give an impression of evolutionary change. For convenience evolution has been divided into five stages of increasing complexity of form - unicellular, radial, elongated, segmental and chordate. To get a truer picture of its significance one needs to imagine the blades of the 'plant' of evolution as growing in a clump, rather than spread across a plane. In this way one can see that all the phyla were earlier positively linked at the base, yet already beginning to show the differential pattern that was to grow more distinct as time passed and actual separation was apparent. Full separation was achieved not only by divergence of the surviving and hence successful groups, but also by the extinction of intermediate forms that proved to be less well equipped for survival.

Fig. 2 suggests that the divisions of the more complex forms of life – echinoderms, chordates, molluscs, flatworms, roundworms, annelids and arthropods – occurred at the radial stage. However, one must take into consideration that the diagram deals only with the adult stage and does not – and indeed cannot – take any account of juvenile stages of growth. Thus the break between chordates and molluscs, while occurring when both were radial adults, ignores that their younger stages must already have evolved into motile and, perhaps, even segmental creatures.

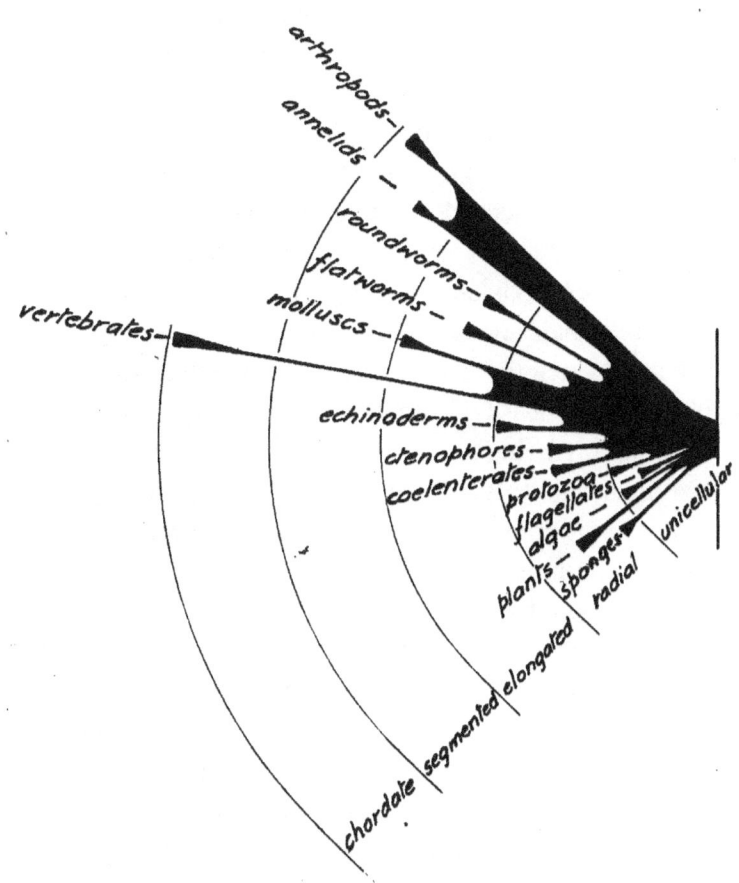

Fig 2. Diagram of Evolutionary Relationships

Thus are there drawbacks with this diagrammatic representation of evolution. The best effect would be obtained by means of a model in which the phyla radiated spherically from a central mass. The diagram is also incomplete because certain other phyla should also be seen as sprouting from the common base, but have been omitted for convenience because of graphical difficulties. Among the more important of these absentees can be mentioned the bryozoans, the brachiopods and the rotifers. The term 'bilateral' does not feature on the diagram, but it will have made its appearance somewhere in the 'elongated' zone. Its occurrence is clearly linked with increased motility and this stamps it as a juvenile trait that has come of age in some of the more advanced phyla.

The diagram hence should be much more complicated. The simplistic presentation just serves to illustrate the principle. The various limbs should indeed be much subdivided so that the effect would actually resemble a sort of sea urchin whose spines were of variable length.

However, even viewed as it is, one might well postulate that the more advanced phyla, i.e. the chordates, molluscs, annelids and arthropods, once constituted what was virtually a single distinct phylum. This means that the deep divisions of today were non-existent, or at least indefinite, among their ancestors. To this hypothetical phylum may be added for convenience the echinoderms. Although at this stage these modern phyla were not yet separated, to a large degree the constituent species were already being prearranged into the relationship groupings they were going to show after division into their discrete phyla occurred.

When new phyla appear, it is because extinction bands have formed in the old one. The most chordate-like molluscs are the cephalopods. If one imagines their forebears as being neighbours of the ancestors of the chordates in the original phylum, the existence of chordate-like features among them could perhaps be better described as a persistence. Of course one could argue conversely that it was the chordates that took some mollusc-like characters with them at the breakaway. The chordates could in fact better be described at this stage as a minor progressive group that broke away from the molluscs (and other evolutionary neighbours) and later came to outshine them completely in performance by means of more thorough elongated, segmented and bilateral adulthood. On the other hand this breakaway group carried with it certain conservative creatures, some of which even clung to static radial adulthood – the tunicates.

This is all a way of restating that all forms of life at some time had their ancestry in a common group and that the ancestors of the cephalopods and the fishes were once close enough together within such a primitive group for them to bequeath certain resemblances onto their otherwise divergent successors. Thus with the eyes of cephalopods and vertebrates. Rather than claiming that such resemblance results from convergent evolution among creatures whose motility and predatory habits give them something in common, the reason could be lack of divergence for exactly the same selective causes. Certainly one ought to avoid the picture of the break between molluscs and chordates being clear cut and sudden, even in geological terms. The links would first go through a long period of strain and weakening before severance could be regarded as complete and final. Even then it would be impossible to judge exactly when it had occurred, while in surviving zones - adjacent to the ruptured links - inherited similarities could be maintained independently if favoured by natural selection. This illustrates another weakness in the classification system, which for convenience needs to simplify and make clear cut that which in Nature is not necessarily so.

To restate the case, there is a tendency to regard all molluscs as being equally remote from the chordates (and hence vertebrates), when in reality, it might seem, the molluscs were once forming into groupings such as cephalopods, pelecypods and gastropods of which one – the proto-chordates – may have developed so quickly and along such different lines that it came to earn the label "separate phylum". The classifications are inadequate to cover every shade of difference, so they cannot accurately represent the whole and true system, which is far more complex, less rigid and hence harder to understand.

One can relate the chordates with the echinoderms in a similar manner, except that, whereas the cephalopods preserve the remnants of an earlier radial adult in their structures, at which level they can be linked to the chordates, the echinoderms relate to the latter at a point much nearer the egg, in that the bilateral larva of the starfish has a clear resemblance to that of the acorn worm (*balanoglossus*). This correspondence does not necessarily mean that the starfish was once a fully formed bilateral animal, as has been claimed elsewhere, but suggests rather a retained contact with some other phylum at the larval level, i.e. the larval stage of the primitive chordates. Even among the chordates a bilateral larva does not necessarily lead to a bilateral adult, as evidenced by the tunicates.

In the race to become free-swimming masters of the oceans, even the most advanced echinoderms were outpaced by both the cephalopod molluscs and most of the chordates. The competition was so fierce that the adult forms of the echinoderms were virtually non-starters and must be satisfied with either standing on or grubbing about on the seabed. The proposed generalised early sessile radial adult is now represented in this phylum by the crinoids or sea lilies, which are tentacle-bearing fine-and-whisk feeders attached to the bottom of the sea by a stalk. Paradoxically, it is from these most primitive echinoderms that the only free-swimming forms arise, when in some species the adults detach themselves to become radial feather stars.

■ Segmentation and De-segmentation

Certain kinds of animals not only have the ability to form themselves into repetitive units, but such mode of growth is actually normal to them. This process is known as 'segmentation' and, that such strings are thought of as individual animals rather than colonies, is an understandable assessment of their status. So, where segments are in lifelong association and co-ordinated by a continuous nervous system, this appears to constitute a single whole animal, which notion is strengthened by it having grown from a single egg. Yet, among the coelenterates, polyps can add further copies of themselves in various ways, but because such additions are not co-ordinated internally one tends to regard them as colonies of individuals. Those created in this way may be due to detachment from the original from which they sprouted in order to live separate lives and this process is known as reproduction by asexual budding. However, one must keep in mind that copying is itself reproduction, whether subsequent detachment takes place or not. One could indeed imagine that any segmental animal is simply the result of asexual reproduction in which the individuals so formed have failed to separate; or rather that at some time in the history of the species some selective pressure has made it favourable for them to remain welded together on a lifelong basis.

This poses certain problems regarding the more advanced types of segmental animals. Apart from the linked nervous system, there is the puzzle of the through-gut, with its mouth at one end and its anus at the other. This might induce one to rather absurdly explain the segments as individuals by imagining that each has its mouth applied to the anus of the one preceding it. However, a great deal of the water of life has passed under the bridge of evolution since the through-gut first made its appearance and the retention of segmentation today in any form must be regarded as an archaic feature and, indeed, it is to be noted that the gut itself does not ever show clear signs of it.

The most consistently segmented creatures are the annelid worms, where the segments are – outwardly at least – usually good copies of each other, except at head and tail. To these one can add the centipedes and millipedes. The advantage of such apparently excessive length and segmentation is to be found in tunnelling, or in crawling through tunnel-like openings. The retention of outward signs of segmentation does not in many cases have an obvious reason and perhaps merely results from it being to no disadvantage, although it does allow flexibility and the ribbing involved must impart extra structural strength to the cuticle (or exoskeleton).

One advantage of segments has been the tendency for them to specialise. Thus in the hermaphrodite earthworms, male and female organs are housed in their own specific segments. In most arthropods specialisation has led to the complete fusion of some segments and also the division of the length of the body into clear-cut zones determined by function. Thus does the insect head consist of six fused segments, the thorax of three bearing the legs and wings and the abdomen of a variable number according to species. At the joints between head, thorax and abdomen external segmentation has been accentuated. An earlier, more evenly repetitive segmentation is suggested by many insect larvae, such as grubs and caterpillars.

With them having no exoskeleton, molluscs show few signs of segmentation, although from the shell pattern of the chitons one can presume that the number of former segments has never been very great in comparison to arthropods.

Among the chordates and vertebrates it is again outward signs of former segmentation that are scarce, although the gill slits of acorn worms and sharks may be mentioned as evidence. However, internal bone and muscle structures suggest strongly a segmented past. The evidence of this has been largely disguised in later forms by complete external fusion of segments on the one hand and specialisation on the other, leading to distortion of both shape and proportion. Specialisation has indeed led to organs originally developed for one function being modified in the course of time to suit a quite different one. An example of this is provided by the first gill arches of the early jawless fishes (now generally extinct), which by evolutionary change in the distant past were brought into service as the lower jaws of the later forms that were eventually largely to displace them and become a characteristic feature of nearly all the vertebrates.

CHAPTER 14

The Vertebrates

■ Introduction

The Vertebrates comprise the sub-phylum that includes man among its members and forms the bulk of the phylum *Chordata*. The vertebrates arrived late on the scene, being the one group whose recorded appearance is later than the Pre-Cambrian, although the chordate stock they are derived from is clearly from some earlier period, even though obscure in the record; such primitive invertebrate creatures have indeed been discussed earlier. The remote predecessor of the vertebrates can hence be thought of as a kind of sea-squirt (tunicate); yet the direct ancestor would not be the sessile adult, but rather a bilaterally symmetrical larval form that had assumed adulthood and was to evolve quite independently from any final sessile stage.

■ Fish

Appearance of the Chordates The earliest vertebrates can be recognised as fish, but were without jaws and paired side fins. Logic suggests that before bones developed the fishes all went through a stage when the skeleton consisted of cartilage, as indeed is still the case with the jawless lampreys of today, together with the sharks, rays and skates (all of which do have jaws). Bones appear to have later developed along the

171

line of the cartilaginous back stiffening, which is itself a development of the notochord of the chordates. The bones of the body and tail of later vertebrates are clearly of a longitudinally segmented nature, which state reflects that also inherent to the musculature.

As with the chordate lancelet, the tails of vertebrates protrude beyond the anus. Such tails have clearly developed as segments that in time have been converted to perform lateral movement along the body. This has resulted in the exhaust opening having apparently been transferred to segments further forward by some obscure means. It is to be noted that the non-swimming chordates called acorn worms do not have such overhanging tails; with them the anus is at the rear extremity, just as with so many other kinds of invertebrates.

The overhanging tail is unique to the vertebrates and some of their chordate cousins. It is a condition of evolutionary importance and demands an explanation. The solution that presents itself is that it represents a development of the stalk by which the redundant sessile adult used to be attached to some primeval foundation surface. This would resemble the 'stalks' found on crinoids. One might hence think of it as some sort of sea squirt attached to the bottom by an extension of its own tissues. The anus would then represent the vestiges of a bud whose other features have apparently all been atrophied through their functions becoming redundant. In that case, all that was left was an opening to provide an exit for the through gut.

But the foregoing, while being helpful as a thought, is misleading through being too simplistic. The vertebrates are extremely complex organisms and there is a gulf between this complexity and the simplicity just described. The gap can to some extent be filled by considering the coelom and the evolutionary developments that occurred after its appearance.

One aspect to be considered is the connection between the common dual coelom and bilateralism. With vertebrates, organs that appear and grow within or in association with the coelom occur in pairs, such as the kidneys and other aspects of the urinary system. Against this one can observe that organs that grow outwards from the wall of the alimentary canal are single, such as the liver and pancreas.

The through-gut of the vertebrates has been created by two individuals remaining in tandem instead of completing vegetative reproduction of the 'budding' kind. This has occurred because of the efficiency of using the two 'mouths' selectively, one evolving solely to ingest food and the other specifically to eject wastes. This represents the creation of a being with radial symmetry consisting of two longitudinal segments. In addition to this simple arrangement a fully developed coelom has evolved in the cases of more advanced phyla, such as the molluscs, echinoderms and the chordates, while members of more primitive phyla can have a more rudimentary cavity, known as a pseudocoelom, while others have no such cavity at all, such as the flatworms.

The coelom is coupled to the reproductive method of the animal and its significance depends on the ontogenetic stage involved. Thus do the coeloms of molluscs constitute small parts of the whole, while those of chordates and vertebrates are relatively large. Since molluscs are evolved from prototypes representing a stage nearer to the earlier sessile adult, it can be compared with the fact that the chordates originate from prototypes at a much earlier stage in their ontology.

Early in their story, some of the molluscs (i.e. the gastropods), lost one coelom and with it many of the organs associated with this dual constituent. Yet much of the bilateral arrangement remained, especially at the rostral end, where is found the head. But in another of these coelomate phyla the changes went much further: with the echinoderms, including the likes of starfish and sea urchins, virtually the whole of one side of an original bilateral creature has been lost: the effect of this can now be seen at an early larval stage of these creatures. During the course of such evolution, the remaining side (with its single coelom and fivefold segmental features) curved around the body containing the alimentary canal to completely surround it and form a radially symmetric creature. Thus were the echinoderms created with their characteristic five radial segments, or a multiple of this number where further segmental division has occurred. The arms of a starfish are pseudo-tentacles in that they represent the 'parapodia' from one side of an unknown adult. True tentacles occur with the sea cucumbers and some have even retained such at the anus, albeit of different design and function from those at the mouth.

The 'tail' as found on all the vertebrates - now sometimes greatly reduced and hidden from sight, as in hominids – can be claimed as homologous to the 'stalk' as found on crinoids (sea lilies). This statement can be substantiated in the following way.

The vertebrate dual coelom contains fluid enclosed in a mesoderm and has evolved from twin outgrowths from the endoderm of the through-gut. Since it is generally associated with the presence of the latter and the workings of the paired organs of a bilateral body, it does not enter into the tail of vertebrates. Hence the vertebrate tail seems to have no apparent internal function relative to the body to which it is attached, so why does it exist and how has it come into being? The answer apparently lies in the former existence of a function that no longer exists. Because of this, the tail became a residual feature that was converted by evolution to other use.

The three 'advanced' phyla being considered here are the molluscs, echinoderms and the chordates/vertebrates. They all have means of attachment to supporting matter, such as rocks or sea bed. Some types have completely lost this facility, e.g. the cephalopod molluscs. But such 'feet' might otherwise be thought to represent evolutionary developments from the 'stalks' of more primitive forms. However they lack segmentation and any stiffening structure. It was segmentation and the provision of longitudinal features that allowed the building up of a stalk, as occurred with the crinoids (sea lilies). This would seem to be analogous with a former function of the tail of vertebrates.

Recalling the "sea urchin" analogy used earlier to illustrate volumetric relationships between the various phyla during their earliest phases, with regard to 'stalks', the vertebrates are related to that division of the echinoderms known as crinoids. Unlike the feet of molluscs such as bivalves, gastropods, etc. crinoid stalks are segmented and contain within them a row of 'discs' consisting of bonelike material. Other members of this phylum have no trace left of any presumed former stalk structures. A difference from the vertebrates is to be seen in the crinoid stalks in that they maintain the fivefold radial symmetry of their kind, with its unilateral cause. The crinoids hence parted from the stalked chordates before they lost their bilateral status.

The primitive sessile adult form of the chordates is now represented by the tunicates (sea squirts), which exist in a host of different species that inhabit the sea bed from the shallows to the depths and in a great range of latitudes. They attach themselves to suitable matter, once their swimming larvae are ready to settle down. The sea squirt larva resembles a tadpole in that it has distinct head and tail sections. It has sticky organs on the head with which to attach itself after which the tail is absorbed and the head metamorphoses into the adult form. This would seem to negate the claim that the chordate tail was once used as a stalk that supported the adult body. This then requires a postulate in that the original stalk of the bilateral adult evolved into a swimming organ in the larval forms. The development of this function made it unsuitable for attachment as an adult and it was gradually replaced by a different method. However the use of the tail as an anchor can still be seen in it being stuck in a hole in the sea bed in the case of the amphioxus with its head protruding for feeding purposes. This method is also still retained by some fish, such as blennies and sand eels.

Appearance of the Fish The earliest fish in the fossil record are the jawless 'ostracoderms' of the Ordovician Period (c450 million years ago), but whose hay-day was in the following Silurian. The feature that draws attention to them as fossils is the bone or bone-like armour that protected the front end – head, shoulders and gills. The skull of modern vertebrates is different from all their other bones in that it encloses a space and hence may provisionally be thought of as descending directly from the protective armour of the ostracoderms, or rather as sharing some kind of common origin. Yet it rather shows a common trend among the early swimmers in the oceans of long ago and that was to protect themselves with such body armour. In the end it was the ability to swim well that proved to be better protection and external armour came to be discarded from most phyla, except for the molluscs, especially those species of which that did not swim about in the adult form.

Vertebrates now set out on life with cartilaginous skeletons, with ossification being an early part of the growing process. Some cartilage is retained in the adult, especially at the ends of bones where there are movable joints. One might argue that the vertebrae evolved round the original cartilage, but using some biological mechanism to produce structures made of bone that had presumably already been pioneered by evolutionary processes elsewhere.

Without jaws the ostracoderms could hardly be full predators and at first were apparently bottom grubbers in fresh water locations, who needed their armour as protection against predation by contemporary invertebrates, particularly by the giant water scorpions ('eurypterids'). It has been proposed that jaws eventually developed from an anterior pair of gill bars.

The appearance of jaws and teeth, together with paired side fins, indicates a tendency to forsake the bottom in favour of a free-swimming existence. Early 'sharks' developed in fresh water. Like their modern descendants they were not 'bony' but cartilaginous, did not have a swim bladder and lacked various other features of form and function that are thought of as typical of fish. They later moved out to exploit the growing resources of animal food provided by the sea and eventually died out in fresh water. While the sharks proper hunted more widely, including in open water at any level, most rays and skates specialised in feeding off invertebrates on the ocean floor.

The possession of skulls by the early fishes may indicate that they arose from a stage of hypothetical larval form that was closer to the adult than that from which arose the surviving chordates. The cartilaginous skull (or "brain case") might hence be thought of as linked to the hard or tough outsides so typical of sessile or slow moving creatures, especially the 'tunics' of the tunicates. In this way our very remote ancestors can be looked upon as being tadpole-like when young and becoming adults by standing on their tail-ends and growing tough (and later hard) cases round their heads. As already discussed above this arrangement was largely discontinued among the chordates/vertebrates. The combination of heavy head supported by a stalk was not suitable for species whose larval forms were increasingly following a hunting or otherwise active lifestyle.

Eventually evolution occurred within a group of fish species wherein the cartilaginous skeleton was replaced by bone. One type acquired side fins that had spiny inserts that radiated more or less directly from the fish's side. The bulk of modern fishes are of this type ('teleosts') and they developed in the seas. They tend to lay huge quantities of eggs and to rely on good eyesight - although some have later evolved away from this – while the primitive lung is not used as such any more, but has been converted into the swim bladder to assist in rising and descending in the water. Such ray-finned fishes arose in the Devonian Period. Yet primitive species of fish with functioning lungs still exist in Africa, these enabling dry periods to be survived.

Vertebrate lungs exist in pairs, but the single connection to the throat suggests that the pairing is secondary. In the bony fish the swim bladder is dorsal and, being defunct as a lung, has migrated to this position for stability reasons. Apart from those primitive ray-finned fishes just mentioned, other fish with lungs are to be found in some tropical locations where prolonged droughts have favoured the retention of features that enable oxygen to be absorbed directly from the air. These lung-fish are not ray-finned, but are related to a group considered to be more primitive, at least as fishes. With these the paired fins do not radiate directly from the sides of the body, but from small leg-like extensions known as 'lobes'; these are hence called 'lobe-fins'. This type is well known from the fossil record, but was thought to be extinct, except for the lungfish, until the first coelacanth was fished out of the Indian Ocean in 1939.

In most fish the olfactory function is served by simple pits on the face. However, on the lungfish and among the extinct lobe-finned fish, the pits are joined by canals to the mouth, thus allowing such fish to breathe air from the surface without opening the mouth. This may be an evolved condition and indicate that the lack of this facility among teleosts is linked to the abandonment of the breathing facility provided by lungs. This again raises the question as to how the lungs arose in the first place. One might remark that the nostrils suggest that the sense of smell was very important among the lobe-fins and that originally a particular swallowing mechanism was used by them to pull water in through the nostrils. It incidentally became a secondary means of absorbing oxygen, a process that was accelerated by the necessity to survive drought conditions, leading to long periods either lying under mud or attempts at overland travel in search of water.

There were indeed probably three reasons that induced certain lobe-finned fish to leave the water and spend more and more time ashore: drought; a need to exploit new sources of food; to escape aquatic predation. It would appear to have been fresh water fish that first deserted the water for land. They were facilitated in this by already possessing four leg-like lobe-fins, lungs, nostrils leading to the lungs and a bony skeleton. The evidence seems to suggest that, while the lobe-fins were failing as fish, they incidentally had evolved features that offered an escape route from the problems of competing with the more successful ray-fins and coping with the drying out of their fresh water habitat. That escape route led to land and the free air above it.

The above indicates and illuminates the principle that it is species under great survival pressure in their original habitat (and hence can be thought of as becoming more and more unsuccessful) that have the greatest potential for making significant adaptive changes suitable for a new one.

Successful species may be thought of as those that have little cause to change, whereas the unsuccessful are needful to make comparatively violent changes of both physical composition and life-style if they are to avoid becoming extinct. Even so, the means to do this is always to be found in adapting existing obsolescent features to new uses. Change sometimes could make unsuccessful creatures successful, whereafter they might well eventually once more reach a stage where they stagnate and again go into a dangerous decline.

■ Amphibians

Modern amphibians comprise the newts, salamanders, frogs, toads and caecilians, the vast majority of which are dependent on fresh water for a considerable part of their life cycle. The first two form a group of elongated creatures with tails, while the next two are relatively shorter and tail-less. The caecilians are elongated creatures that have neither tails nor legs. They generally skulk underground, although species are also known that seem even to spend their adulthood in water.

The newts and salamanders may be thought to resemble the early amphibian form, like a fish that had taken to the land, but they are now in many ways specialised. The skeleton is of light weight, while the known common amphibians of the remote past had heavy protection of the head, a trait perhaps inherited from their unsuccessful forebears among the fishes. The skin too has now lost the primitive ability to bear scales, but is used to absorb oxygen, thus tending to usurp the lungs. The skin is damp and must always remain so; a modern amphibian hence avoids exposure to the Sun, except where there is a need to regulate the temperature upwards.

Such remarks also apply to frogs and toads. They have the additional specialisations of being tail-less and able to hop by using the long back legs. Toads differ externally from frogs by having a warty skin and shorter back legs; they are much less inclined to hop.

The peculiar gait of frogs is perhaps responsible for their survival in comparatively large numbers of individuals and species. Perhaps it is the tremendous acceleration of the hop that has enabled them to withstand predation. It is a perfectly satisfactory mode of locomotion too, being far superior to the awkward crawl of a newt. While even the common toad has its own weapon against predation in its ability to make itself unpalatable by means of secretions in the skin, certain kinds of tree frogs in South America take this to a point of being deadly poisonous. The crawl of newts and salamanders is directly descended from the way that lobe-finned fish could 'walk' out of the water, while those ancestral to frogs favoured a paired movement that led to hopping, using the rear legs. Avoidance of the ground by leaping across from one item of vegetation to another has some appeal. In this respect the life style of some primates may indicate a parallel development, especially with regard to the long-legged build and mode of progress through the forest of the lemurs of Madagascar. This would further suggest that the tree frogs found in tropical rain forests are far from being a late development.

One feels bound to consider the reasons for amphibians having left the water, even if this leads to further speculation. Did the same reason apply overall? One compulsion was a climatic environment wherein the existence of water tended to become seasonal. The lobe-finned fishes had to adapt to such circumstances or die out there. Several options were available. One was to leave; escape to the sea became possible, the route taken by the coelacanths. Another was to survive in the mud while breathing air, the route taken by the lungfish. Yet another was to adapt for survival on land, the route taken by amphibians.

However, adaptation for terrestrial life may have offered further advantages, at least at first. When other prey was in short supply the lobe-fins may have preyed on each other. The smaller ones may have used land as an escape route from this, but along which their predators eventually followed them. In addition the land was already populated by invertebrates, in particular arthropods, and the pursuit of these may have contributed to the trend onto land. The modern amphibians still feed in this way, but it is the larger predatory types that have survived in the fossil record.

The amphibians are all still tied to water, needing to lay their eggs in it. The resulting tadpoles of the common frog are at first omnivorous, followed by a final carnivorous stage, perhaps illustrating the earlier feeding sequence of this species, if not also much of the phylum. The adults have mouthparts that are specialised for catching insects, a particular organ being the long sticky tongue.

■ Reptiles

Amphibians are tied to water, in that each year they must return to it in order to lay their eggs. The next evolutionary stage was the production of eggs that could withstand the vastly different conditions on land. The pressure that brought such a change about was perhaps the onset of a climate where the drought was unrelenting. To survive this the shelled egg was necessary; but one is led to postulate that a form of this must already have been existing among certain types of amphibian, perhaps as an adaptation against predation. The kind of creature that first produced such eggs on dry land was a proto-reptile. The vast bulk of modern reptiles is comprised of lizards, snakes, turtles and crocodiles.

It is natural to assume that the reptiles arose from the ranks of the amphibians, but such reptilian ancestry may not have dwelt particularly long in the amphibian phase. They could conceivably have arisen independently in time and space from all the amphibians we are now aware of as having existed.

One development among the reptiles was the appearance of the plant eaters. Crocodiles and snakes are all carnivorous, but turtles and lizards include both meat and plant eaters in their ranks. It seems rather unlikely that any amphibious ancestor was originally primarily herbivorous. One might postulate that, with the return to a more consistently damp climate, the vegetation would become luxuriant. Some of the amphibians or proto-reptiles may have developed more catholic tastes and from omnivorous tendencies eventually turned herbivorous as their digestion systems adapted. Such plant eaters can be found among the land tortoises, the sea turtles and certain lizards.

The turtles seem to fall into two categories: sea turtles and land tortoises that are descended from stock that turned herbivorous (even if some still take small animals) and the fresh water turtles that have apparently never abandoned meat eating and still feed on fish.

The lizards still retain the basic shape of the primitive amphibians and superficially resemble the newts in having long bodies and tails and short straddling legs. Like the newts, frogs, etc., they are also basically insect eaters, although many have grown to larger size and have graduated to taking birds and small mammals. However, only the chameleons share the specific amphibian method of feeding on insets by using a long sticky tongue. Other lizards are herbivorous, but some species suggest a historically different preference by eating insects when young and only changing to a vegetable diet when adult, with ontogeny thus throwing light on phylogeny.

The crocodiles are all frequenters of shores, either of rivers, or lakes, or even the sea. All are carnivorous, with some species feeding off fish, while others catch animals, often when these are drinking at the water's edge. Thus do they all obtain food in or from water, yet go ashore to lay their eggs. They habitually bask in the sunshine on the bank when required to for control of body temperature. The crocodiles are a conservative group and have changed but little over many millions of years, a clear sign that, once their physical make-up and life-style had reached a certain stage, it was very successful and they have since not come under any really serious adaptive pressure.

In contrast the snakes are much modified reptiles. In their present form they exhibit many specialised features and as a group are comparative newcomers. This is illustrated by many of them living on mammals, although others take amphibians, lizards and, indeed, other snakes. The obvious specialities of snakes are lack of legs, poisonous bite in many species and a varying ability to wrap the body round things, this facility being used by some to squeeze and asphyxiate prey. Coiled snakes can also straighten rapidly (strike). Some snakes have internal vestiges of legs. Rather than laying eggs, some species give birth to live young. The loss of legs would seem originally have been an adaptation for getting through very narrow openings or for burrowing, all in pursuit of prey. In this they mirror the form taken by some rare species of amphibian and also some lizards (e.g. slow worms), while with lizards like skinks the legs are tiny and, in evolutionary terms, apparently on the verge of going the same way.

■ Dinosaurs

These were reptile-like animals that became the dominant fauna during the Mesozoic Era, but died out completely at the end of the Cretaceous Period. Their fossilised remains are well known, this largely being due to some of them reaching gigantic size. Many resembled lizards in general terms and their eggs have also survived in fossilised form. Unlike the modern lizards they formed part of a complete fauna, with contemporary species not only living on land and in water, but even flying in the air. They included herbivores and carnivores within their ranks. While not all were huge, the smallest would appear to have been larger than the majority of reptile and mammal species living today.

Modern reptiles and amphibians have splayed legs, which normally are only made to take the weight of the body when actually walking. Dinosaurs, on the other hand, had legs that supported the body from underneath, as with birds and mammals and standing on them was habitual. The view is also held that dinosaurs, unlike true reptiles, were warm blooded.

The spectacular career of the dinosaurs makes a fascinating subject, especially in view of the relative suddenness and finality of their disappearance from the face of the Earth. As well as those on the land, relatives in the sea (ichthyosaurs, plesiosaurs and mosasaurs and in the air (pterosaurs), all vanished from the scene. It is almost as though the habitat did not matter; they were all particularly vulnerable to crucial unfavourable circumstances arising in their environment.

They had gone from 'success' to 'success' and produced ever larger forms. Once global circumstances turned against them, it would seem that the larger forms were unable to survive, but with their disappearance being so upsetting to the ecological balance that extinction also was passed down to the smaller species, whose evolution was not up to the new requirements. The fabric of dinosaur society was eventually shattered. One might speculate that the basic cause was a climatic change that radically reduced a food supply based on luxuriant vegetation. The reason would be cosmic and it has been indicated that a huge asteroid that struck the Earth gave rise to a temporary but intolerable rise in temperature. (Of course some suspect that a plesiosaur or two are alive and well and still living in Loch Ness, Scotland).

However, these theories do not explain the extinction of many types of dinosaur that took place millions of years earlier, neither is it easy to understand how fish and other marine dwellers and the ancestors of the birds and mammals managed to survive the eventual eradication of the dinosaurs.

■ Birds

These are well known for their diversity of colour, form and behaviour. Yet more important than these are the features that they all have in common:

they are all bipedal:

they all walk on feet that originally had four toes:

the forelegs are modified into wings, so that all birds can either fly, or are apparently descended from ancestors that could:

all birds have a covering consisting of feathers, a feature not found in any other surviving group:

they all lay eggs in prepared places called nests, control the temperature of these by sitting on them (except for rare exceptions) and also protect them in this manner:

the young are tended after hatching:

the mouths of all birds are provided with two horny extensions constituting the beak.

All this indicates descent from one common ancestor, or a very closely related group of ancestors.

While reptiles in general lack many of the above considerations, some do lay eggs in nests and crocodiles defend and tend these, and even take action to control temperature. However, the most likely line of descent seems to be from certain smaller dinosaurs, which survived the extinction of the rest of their kind because their evolutionary make-up had made them suitable for adaptation to changed circumstances and allowed them later to emerge prolifically as birds. It would seem that they survived, not only because they could be airborne, but could fly well and this facilitated escape from environmental threats.

▪ Mammals

Fundamentally these are more diversified than the birds, but:

they are all covered with hair (or were once so):

all are warm blooded, although this is achieved in a way physically distinct from that of birds:

all were originally quadrupeds with five-toed feet, although some have later tended to adopt a bipedal gait, while others have a reduced number of toes:

the young are not born in eggs, but as exposed animals in virtually all cases and are nourished by 'milk' produced from modified sweat glands on the mother's breast.

There are nonetheless certain fairly clear-cut divergences and exceptions.

With the placental mammals the unborn young are nourished by means of a connection to the womb wall and delivered in a relatively advanced state. In some cases they can walk very soon after birth. At the other extreme they can be blind and helpless, but are usually delivered into some kind of rudimentary nest where they are tended by a parent.

With the marsupials the young are always delivered at a very early stage of growth and make their way to the teats, which are found within a pouch wherein the young remain until grown enough to emerge. Later they may simply cling to their mother's fur while being carried about. Marsupials do not use, nor do they need, a nest.

There are two kinds of mammals that lay eggs. These constitute the 'monotremes' and comprise the two species of echidna and the duck-billed platypus, all found in Australasia.

■ Aspects of the Vertebrates

Eggs and Young The laying of eggs is the primitive method of reproduction among multicellular animals. More progressive ones are viviparous, i.e. they give birth to live young, even if in many species these are both undeveloped and helpless. Once a species has become viviparous regression to egg-laying would seem to be impossible.

Fish and amphibians lay their eggs in water and such cannot survive on land; nor can the young that hatch from them. Eggs laid on land need a tough or hard outer shell to contain the liquid and protect the young within. The fluid itself replaces the water environment enjoyed by the forerunners of such eggs with shells.

It is relatively easy to find explanations as to why, long ago, some amphibians came to lay their eggs on land and initiated the first reptiles, but the question as to how this was physically and habitually accomplished is a vexed and unanswered one and is really more vital than understanding how the adults managed to come ashore. There is obviously a requirement here to have recourse to some evolutionary process; yet even assuming that eggs did evolve that were suitable for land, how did mothers evolve in parallel to possess the behavioural trait necessary to lay them there? Again, how did the young evolve so that, once hatched, they were suited to an existence on land and the breathing of air?

At the end of the Palaeozoic Era the amphibians - they being fresh water dwellers - were faced by an ever-increasing aridity. Their habitat was drying out and it would seem they could not adapt to the marine life. This left only the land as an escape route. The conquest of land by their eggs would seem to have been a much more difficult one than that achieved by their legs. However, the method has been the same – natural selection. It is a relatively simple process to understand in the case of legs; slight chance improvements were selected for, with the possessors of such being better equipped to survive and pass on their genes to their offspring. But with eggs the process is more complex; the eggs must evolve towards a shelled type, the mother must evolve towards a land-laying behaviour and the hatching young must evolve to a form tolerant of land conditions.

There is no way that amphibian eggs could evolve shells purely as an adaptation for survival on land. The development of shells must have come first. One might postulate that there was a tendency towards a tougher exterior as a guard against predation. The best of these may have had a survival advantage when the water in which they lay evaporated away. With increasing aridity and continuing predation pressure the trend to harden the shell would persist.

The typical amphibian young hatches out into fresh water; it is a 'tadpole' at first with external gills, a condition reflecting its pre-fish ancestry. Small boys like to keep tadpoles in various receptacles and each notices the eventual appearance of legs and the shrinking of the tail. This is a fairly rapid phase. Once the final form of the gills goes, the little frogs must be let out of the water or drown. With the dearth of water in the Permian Period one can well surmise that it became selectively advantageous for the gilled tadpole stage to be shortened. Eventually this phase would be dispensed with altogether and the young would scramble ashore immediately after hatching. Such eggs were then virtually ready for laying on shore.

The amphibian of today habitually lays its eggs in water, but often in little more than puddles, while even in larger ponds they tend to be deposited near the shore line. The amphibians of the Permian, which were destined to become reptiles, must have laid their eggs close to shore. Yet, even so, why did some change to laying on land? The end product of the process of reduction of the gilled period, as dealt with in the last paragraph, may have been observed by the parent and have resulted in a gradual loss of compulsion to lay in water, leading to an eventual preference for laying on land.

There is an element of secondary or 'behavioural' evolution here, as opposed to the purely physical or primary form. The laying of eggs on land then became instinctive. Its beginnings had perhaps an observational component, reflecting an element of intelligence and some incipient free will in the adult. Yet in purely evolutionary terms one might surmise that eggs became pre-adapted for survival on land and that sometimes this occurred incidentally or accidentally, with the driving force being persistent aridity. Once the aquatic phase became no longer necessary, it then faded out through natural selection. Perhaps it was even safer to lay eggs on the shore, rather than in the water near it.

The amphibians can be put into three groups:

those that failed fully to adapt to land and survived (frogs toads, newts, etc.):

those that failed and became extinct:

those that adapted and became proto-reptiles.

The last were more like newts than frogs, for no tail-less hopping reptile is known. One further observation here is that the proto-reptiles, having suppressed the gilled stage, were already equipped with lungs while still in the egg, this being a situation inherited from their lobe-finned ancestry.

It would seem that once eggs had adapted for land there was no way they could revert to laying in water. Yet many land vertebrates have taken to the water again, but not always the fresh water of their ancestry; many took to the sea. All the known fresh water reptiles (e.g. most crocodiles) and some of the marine ones (e.g. sea turtles) must return to land in order to lay their eggs. Yet others evolved into forms so specialised for swimming that coming ashore at all became impossible. This has occurred with some surviving species of mammal, such as whales, but also applies to the extinct reptilian ichthyosaurs. Like the whales, these creatures could only survive in such an environment by being viviparous and young have been observed both in and near a fossil adult of such. It seems clear that these creatures had become viviparous before they abandoned the land in order to return permanently to the water.

The contemporaneous plesiosaurs of the Jurassic look as though they could have crawled ashore to lay eggs as turtles do now. However, if animals of the Loch Ness Monster type did exist in lakes and are a kind of fresh water plesiosaur, then they must also be viviparous and completely aquatic, for no eggs have ever been found.

Some of the early land vertebrates would be helpless when they first hatched from their eggs and, in the case of those born live, certainly so. The birds came into the one category and the mammals into the other. Both types need nursing to some degree at life's start, with the nest being universal with birds and common with mammals. In order to try to understand how mammals first arose as sucklers it seems necessary to assume that their ancestors were all nesters. The nest was a place where the young were born and nursed through the early helpless stage.

The question of the origins of suckling presents its own problems, as does the split between the placental and marsupial mammals. Unlike nestling birds, young mammals have soft mouthparts; teeth are hence absent. The predecessors of the mammals must have fed their young entirely on regurgitated food. There must also have been a stage when the young left the nest along with the mother by clinging to her fur. All the time their mouths were searching for food and in and out of the nest the only substance on the mother's body that they could absorb - apart from food remnants - was sweat. Then, by an evolutionary process, some sweat glands would gradually come to specialise in producing something more nutritious – 'milk' – and this eventually came entirely to replace regurgitation as the means of feeding the newborn.

Those mammals that persisted in carrying off their young in their fur became the marsupials. This behaviour became their special trait, in that gradually the period in the nest was reduced, eventually to be dispensed with altogether. The problem then remains to us as to how a newborn marsupial manages to find its way from the vent to the teat.

A female kangaroo in labour reclines backwards and the minute offspring then clambers up her belly to the pouch, wherein are the teats. The young is little more than a foetus, a strip of protoplasm with a mouth. One can hardly attribute its behaviour to instinct alone. The mother presumably reclines backwards so that the tiny one does not risk falling off on its first journey. Yet something must attract it to the pouch and teats. As a postulate one might propose that, since mammals have a pronounced sense of smell, the young are drawn towards the mammary glands by a scent they produce. Young placental mammals must be aided in their search for the first meal in the same way, even if some (e.g elephants) initially appear to require a little maternal assistance. Human babies would seem to be the only ones that need to have it all done for them by the mother, the primate sense of smell in any case being relatively poor.

The marsupial pouch would arise because of the young hanging underneath, with its weight forming a pucker. Selection for a condition that could produce a better pucker was essentially the evolutionary start of a bag proper.

Marsupial young are little more than hatchlings from primitive eggs, such as those of fish or amphibians. The placental mammal gives birth to young more characteristic of those that hatch from eggs laid by land animals. The function of the egg as a container is represented by the amniotic sack, which retains the prenatal young in a liquid environment. Until birth, the feeding mouth and the lungs are not put to use because food and oxygen requirements are obtained directly from the mother. Through all the stages of egg, embryo and foetus, the young mammal never loses contact with the wall of the womb by means of the placenta, the final phase of which 'afterbirth' leaves its mark on us as the navel.

Monotremes, such as the platypus, suggest the primitive egg-laying phase through which all mammals have passed and which has been suppressed by evolutionary pressures.

Legs and Feet Vertebrate limbs first made their appearance among the fishes, with land legs originating as the lateral swimming organs of the lobe-finned kind. At points where limbs are attached, the skeleton is specifically strengthened and one can note this in the human frame at the shoulders and the pelvis. The skeleton anchors the musculature, which itself reflects the segmentation of chordates and earlier invertebrate predecessors. The heavier bone-work at the limb attachments suggests a fusion of segments here, the segmentation itself still being indicated by the toes of the four feet. The number of bones along the limbs is variable, for these started as cartilage-supported protrusions on fishes' sides, which eventually gave rise to bony rays. After this development joints of cartilage were retained to maintain limb flexibility and this was accomplished in the way already pioneered on a segmental basis by the vertebrae. The teleost fishes developed their fins with a multitude of slim, flexible, spiny rays. It was the lobe-fins that evolved the jointed bony structures in their limbs and the fusion of the innermost of some of these led to the lobes.

All living land vertebrates have feet suggesting an original maximum of five toes per foot and this applies to their extinct relatives. Indeed, it is the case with all vertebrates, except for many fishes and the marine ichthyosaurs, i.e. from the amphibians onwards. Some ichthyosaurs have flippers that conceal rows of 'toes' in excess of five. This would seem to divide them off from the main stem of vertebrate development. Apart from teleost fishes, all other vertebrates may come from a distinct stock with feet bearing five toes each; even so the reduction in some cases to four or less still took place at a very early stage indeed.

Modern amphibians have five toes back and four toes front and at least some of their extinct cousins shared this arrangement. However, the missing outer toes on their front feet are present in vestigial form on their skeletons.

Crocodiles have a reverse arrangement in that it is the rear feet that have only four toes. On the basis that on feet used for running on harder surfaces the number of toes tends to reduce, one can postulate that with the remote ancestors of the present amphibians the front legs were the main means of locomotion, even though this is not the case now. With the crocodiles the reason is more obvious and the hind legs are noticeably longer than the front ones. It would seem that like the dinosaurs they had started on the development that led to a bipedal gait, but like some of them reverted long ago to a completely quadrupedal state. This reversal would appear to have cause in the abandonment of a terrestrial habitat at an early collective stage by the type for a more aquatic one. One might say that an aquatic lifestyle does not go with bipedalism, although this statement does not apply to the birds.

Modern lizards retain five toes on all feet. Usually their legs are all roughly the same length and if anything they tend to raise themselves on their front legs to elevate their heads when stationary. Yet one or two species take this further and run on their back legs alone. Because of this, their back legs have evolved to suit. One might postulate that the ancestors of the crocodiles and the dinosaurs went through a similar phase. The crocodiles were among the first to abandon this process, the reason being their return to water, for the speedy rush on land was no longer so vital.

The land dwelling dinosaurs all went bipedal. They then divided into two kinds. The reptile-hipped dinosaurs encompassed both carnivores and herbivores, while the bird-hipped ones were exclusively herbivorous. The former quadrupedal existence of the dinosaurs appears to have reduced the number of toes all round to four, suggesting perhaps that they had earlier been quite fast movers. It was this early need for fast movement that led eventually to the bipedal gait.

These became the Triassic dinosaurs, living in dry conditions. In the following Jurassic the climate became wetter and the vegetation increasingly luxuriant. Gigantism took over. The herbivores grew so huge that they are thought to have spent a great deal of their time in water, as for example Diplodocus and Brontosaurus. Others were not so huge and remained on land, defended by armour plates and club-like tails, as with Stegosaurus. Even so, they still betrayed their former bipedal phase through their back legs being longer than the front ones. A feature of the dinosaur back legs is that they became tucked under the body and arranged for powerful movement in line with it. This feature was retained after they reverted to the quadrupedal stance. That the under-body arrangement of the rear legs is mainly due to the bipedal phase is indicated by the front legs always retaining a degree of the splayed stance of reptiles.

A major exception to the rule is betrayed by the Brachiosaurus, for long the largest known dinosaur, whose front legs are the longer. The build of this animal seems designed to allow it to walk in very deep water.

With the carnivorous dinosaurs the bipedal stance was retained, but there was a tendency to increase in size at the very top of their range and for the forelegs to shrink towards disuse.

The bird-hipped herbivores retained species among them that remained bipedal, even if partially aquatic, such as the duck-billed dinosaurs. Others became quadrupedal, while surviving on land with the assistance of armour, such as the Stegosaurus of the Jurassic mentioned above and the Ankylosaurus and horned dinosaurs of the following Cretaceous.

The legs of fresh water turtles are generally like those of land tortoises, suggesting a relatively recent entry into the water and they retain a good deal of mobility on land. Those of their herbivorous marine relatives have evolved into flippers that consist basically of five fused toes on each foot. In this they resemble the prehistoric plesiosaurs. (It can be noted that baby turtles, when they hatch, leave the nest and rush towards the sea, use an alternating gait. When they eventually return to become mothers they have lost this and use their flippers in laborious unison, as though still swimming. The way of the hatchlings is instinctive and harks back to the time when these animals were basically land dwellers.) The remains of plesiosaurs do indeed suggest some shared ancestry with the turtles and one might even think of them as carnivorous turtles that did not need heavy armour because they were fierce and speedy and evolved long necks as an aid to catching fish and to compensate for the poor manoeuvrability of the flattened shape they had inherited from the time when they dwelled ashore.

The extinct flying reptiles (pterosaurs) had their forelimbs modified as wings. Their rear limbs were degenerate and five-toed. There is nothing to indicate an origin among the dinosaurs with their bipedal gait. Indeed, the pterosaur wing is supported by a four-toed limb. Since the outer toe is lacking, by this reckoning one might even suspect that they shared a very distant ancestry with the modern amphibians.

The most active and persistently bipedal of the dinosaurs were the carnivores. With them the inner toe of the four-toed foot moved round to the back and this is the same arrangement as found with many birds. Among aquatic birds such as gulls and ducks, the feet are webbed across the three front toes. However, in the group that includes the gannets, pelicans and cormorants the webbing includes all four toes. This suggests that these took to the water before the inner toe was irreversibly pointing backward and, it would seem, earlier. With larger birds that are habitual runners on the ground, the back toe has tended to move up the leg and become redundant (at least for support), as with cranes. With flightless birds like the ostrich even further reduction in the number of toes is in evidence.

Birds of prey have sharp claws and strong feet, which are used to catch and kill their quarry. Some, such as vultures, have become completely reliant on carrion, resulting in the evolution of less powerful feet.

The standard arrangement of feet is retained by most birds, but some have progressed to having two toes pointing backwards, including such diverse groups as owls, woodpeckers and parrots. It must be presumed that the original reason for the single backward turned toe was to help with deficient balance and achieving stability when using the bipedal stance. Only incidentally did this provide potential 'opposability', which allowed the ancestors of the birds to move up into the trees and perch.

At first the degenerate front legs would serve an auxiliary climbing purpose, (for the main grasping organs would be the hind feet with their opposable back toes). They may also have come to serve an auxiliary feeding function in support of the mouth. But eventually the ancestor of the birds found a new use for them; they became wings. All birds have, or have had wings. From their arboreal stage they have branched out into every conceivable climate and habitat, sometimes with resultant loss of flight. In the case of the penguins the wings have been converted into flippers with which they can 'fly' through the water. The various auks of the Northern Hemisphere also 'fly' under the surface using their wings; however, they can still fly through the air.

Mammals have legs that indicate that their ancestors did not originally take to a bipedal gait. The upright stance and hopping of kangaroos can be seen as a later development among these herbivores. There are "hoppers" elsewhere, and one might observe that hares may also have started independently along the same road, except that lack of a suitable tail held them back. The bipedal hopping of kangaroos may mean that they first commenced with the hare's habit of sitting up to view the terrain for danger and it has already been mentioned in the case of the frog that the acceleration of the hop is an excellent means of escaping the onrush of a predator.

The first mammals were quadrupeds with five-toed feet, with each toe bearing a claw. The latter has been an important aspect of mammalian evolution. Some animals of the Mesozoic Era showed 'mammalian' features (the mammal-like reptiles), but these apparently died out and it was at the start of the Tertiary that the true mammals came into their own. The early mammals can be imagined as being small, skulking and partly or completely arboreal, using their claws to scamper about up in the trees.

From such beginnings they were able to spread out into all habitats, with their feet evolving to suit. On the ground the feet could become digging tools or, where fast running was the case, body support arose by raising on the toes, which eventually became the norm. Reduction in the number of toes in ground contact occurred and the claws were transformed into hooves. Certain of these ungulates came to walk on two toes per foot, the "cloven hooves" of cows, sheep, antelopes, deer, etc., while others reduced bearing by the foot to one toe only, as with horses.

Where animals reverted to water there was a tendency for feet to become webbed. This has occurred with quite diverse types of animal, such as otters, beavers, seals, etc., and even the platypus (a marsupial). Seals' feet have indeed become flippers. The true seals have legs that are of little use on land, while the fur seals and sea lions retain limbs that also serve rather better for walking.

The whales and sea cows are disparate, but each group has evolved so that the front limbs became flippers and the back ones have vanished. This all suggests that they became aquatic much earlier than did seals; while the sea lions were later still into the sea.

With many carnivores the claws have become modified. Dogs are persistent runners on hard ground, so their claws are blunt, yet still useful for digging. The most specialised are those of the cats, which are protected and kept sharp by retraction, only the cheetah being different in having more dog-like feet. Normal cat claws are used both as weapons and for overpowering prey.

The feet of hares are modified for fast running, with a hopping gait on hard ground; the claws are hence blunt.

The common ancestor of the mammals can be thought of as rat-like in appearance and indeed many groups still have members that tend to resemble this general outline. One can mention the mongoose, the weasel, the shrew, etc. Rats' feet are relatively weak, but those of arboreal rodents, like squirrels, retain stronger claws for clinging to bark. However, the feet of many climbing rodents have developed additional grasping functions; the dormouse can grasp twigs and the harvest mouse the stalks of corn. This full grasping is usually done by animals that normally climb in vegetation too small to run along and are relatively slow movers. However, the swift squirrel can be seen still to retain grasp in the front paws when observed eating a nut or nibbling a cone.

Life in the trees caused limbs to evolve in different ways. Many animals concentrated on the rapid running facility provided by clawed feet. One can see it in the squirrels and those arboreal predators, the martens, whose ground dwelling weasel relatives have feet far less clawed. Their larger relative, the badger, does have powerful claws, but these are primarily used for digging.

One group of arboreal animals called tree shrews resembles the envisaged rat-like basic early mammal shape, including ground dwelling insectivores like shrews, hedgehogs, etc., but are of special interest in being considered as being near to the stock from which the primates evolved. They are diurnal omnivores, who descend from the trees to forage on the forest floor, mainly for vegetable matter.

Slow climbers tend to be graspers and the somewhat detached inner toe of the mammalian foot came to form an ever more efficient opposable member. All species of the primates have, or have had, this feature. This development led to a change of primate claws into flat nails, although some species have retained a claw on one toe for some special purpose, e.g. grooming.

The lemurs are among the more primitive primates and are mostly relatively slow movers in the treetops, especially so in the case of the 'slow' loris. Perhaps all the primates went through a slower phase in the relative safety of the slenderer branches of the canopy, before quickening again when circumstances demanded it. However, monkeys can move about at great speed and it is hard to imagine that as such they were ever as slow as many of the lemurs.

The grasping qualities of the feet of primates developed even further in the case of the forelimbs, so they became arms and hands.

Monkeys and apes use their hands to examine objects more closely and to manipulate food. Chimpanzees pick up objects and use them as primitive tools. All this constitutes the puny beginnings of human manipulative power, which culminates in the amazing dexterity of the concert pianist and other artistry. Yet our hands are still good for climbing, as demonstrated by gymnasts and various circus acts; but our feet are not so. The human foot would seem to result from the onset of permanent descent from the trees.

In the Tertiary Era numerous animals grazed on the plains. Man (as do chimpanzees now with monkeys) would occasionally augment his diet in the trees by killing any smaller animals he could catch. From such a tendency he changed into a communal hunter who eventually descended to exploit the bounty of the plains, yet without losing his earlier practice of being a forager. This last tended with time to become the prerogative of the female of the species. This style of life was accompanied by cultural changes in which the development of hands played a large part.

Apes can be bipedal when required, yet are reluctant to adopt such a posture. Only the smallest kind of ape, the gibbon, habitually walks on two legs when on the ground without "knuckle touching", although it rarely needs to descend thus. The anthropoid apes all retain the opposable toe on the foot. This is a hindrance for a bipedal ape and in man evolution has led to the inner 'big' toe being located close to its companions and opposability has been completely lost.

Many land dwellers, especially those fleet of foot, have legs that are restricted to fore-and-aft movement. However, life in the trees encouraged the development of movement and strength in all directions and especially so with the forelegs. In the case of man the legs have some lateral performance, (even if kicking a ball sideways cannot be done very well). Multidirectional exertion of force is better developed with the arms. The structure that allows this is the shoulder. Most animals do not have this abrupt widening of the body behind the head and neck. There is little sign of it in monkeys and one must look to the anthropoid apes for comparison, although even there the feature is not quite as drastic as with man. The shoulder was originally developed as a result of swinging through the trees using the arms, but the feature has also enabled the arms to use a large range of movement, without which the hands could not have developed as they have.

Other animals have also tended to leave the trees. The baboons moved onto the plains in a similar way to man, but their ancestors were monkeys, not apes, and their limbs still reflect this. Other mammals found their way beyond the trees, but in their case by acquiring mastery of the air: they became the bats. Their forelegs evolved into wings, while their rear legs became useless for normal locomotion; their feet are five-toed and clawed and serve as hooks whereby bats can hang upside down.

Tails The tail of a fish is a bone- or cartilage-stiffened extension of the body protruding rearward of the anus. It can be swung laterally, thus propelling the creature through the water. All higher vertebrates retain tails, or vestiges of the same, even if their uses vary widely and external evidence of them is missing in some species.

Newts and salamanders have retained tails in all phases, but frogs and toads lose theirs on reaching the end of the fully aquatic tadpole stage. Members of the former group still use their tails when swimming, while the latter employ their back legs in a way similar to a human breast stroke swimmer.

All the reptiles have retained their tails, although with the turtles the member is simply a stump. The extinct ichthyosaurs became completely aquatic and their tails reassumed a fish-like appearance. Other aquatic reptiles came to swim by using lateral tail movement, e.g. the extinct mosasaurs, the crocodiles and the marine iguanas. On land, lizards are found that often seem to have tails of unnecessary length. Very often this has a sort of protective role in that a predator can be left with the detachable tail in its mouth while the lizard escapes.

The tails of dinosaurs were often enormous and, because of the creature's weight being taken on the rear legs even after the bipedal gait was abandoned, the tail served as a counterbalance for the rest of the body. In the case of the reptile-hipped dinosaurs, some, like the stegosaurus, developed the tail as a club-like weapon of defence.

The tails of birds, originally reptilian, had their functions usurped by the feathers they bore, so they evolved into stubs, but with a wide range of movement. With aquatic birds the tail takes no significant part in propulsion. But it is an important balancing organ when perching and, with its capacity to fan out, acts as a brake or auxiliary steering device when airborne. It is also used for signalling, especially during territorial disputes between mature males. In species that have developed magnificent plumage the tail often forms an important part of the display, as with the peacock.

The tails of mammals have evolved into diverse forms. Among some rather primitive kinds they retain a reptilian quality, as with the great anteater and the pangolin. However, where reversion to an aquatic life-style has occurred the tail is never swung sideways to gain propulsion, as is the case with fishes, amphibians and reptiles, but may be flattened horizontally and used together with vertical undulation of the body to propel the animal. In fresh water one can well observe this lateral development in the case of the beaver.

The whales, dolphins and sea cows have undergone parallel evolution with each other in that the tail has developed horizontal flukes to make it a very efficient organ of propulsion. But with the seals and sea lions the tail is a mere stump, its function being usurped by the rear limbs. They confer the advantage of being very flexible in all directions, thus giving an added degree of manoeuvrability. This would seem to indicate that, when the seals became aquatic, the tail was already of diminished size, suggesting a creature of badger- or bear-like shape. The whales and dolphins on the other hand must have had a tail of sufficient size, strength and flexibility to compete successfully with the rear legs in the functions of propulsion and steering and finally to supplant them completely so that the land could be totally abandoned.

Many mammals use their tails as signalling devices to provide indication of mood, primarily for the benefit of their own species. This can be observed in the tail-wagging of dogs. Tails may be provided with tufts, fringes or patterns to emphasise the signal. Even the humble scut of a rabbit has a white underside that flashes as it runs. Tufts of hair also enable many animals to employ their long tails as flywhisks, as with domestic cattle and lions.

The tail provides a vital organ of balance for swift-running arboreal creatures, such as squirrels. Monkeys too need their tails for this purpose, even though the effectiveness here is often visibly reduced. The baboons, being largely ground level types, have a reduced tail, which however apparently retains a signalling function.

In the New World small primitive monkeys like marmosets have bushy squirrel-like tails. Among the more advanced New World monkeys many have tails that are prehensile and act as a fifth limb, a feature not found among their Old World relatives. It may be significant that in the relevant areas of America are found unrelated creatures with prehensile tails, such as the kinkajou. Here the development among the monkeys may even have resulted in part from mimicry, a like process being denied the Old World monkeys due to lack of precedence.

In the Old World, primates with grasping hands developed a way of climbing that rendered the tail redundant. The slower lemurs lost theirs, as did all the apes. The anthropoid apes developed the arms for swinging through the branches, a manner of progress in which the tail took no part, so that externally it vanished completely. Yet the loss may have been initiated by a primitive slow-climbing phase. Slow progress is still a feature of the Orang Utan of south-east Asia, although the gibbons found in this same zone are the most agile 'swingers'. They are the smallest of the apes. The American spider monkeys proceed in a similar manner to these, but have retained long tails; this has to be ascribed to the advantage conferred by their prehensile function.

Reduction in tail size is common among large and heavily built ground dwellers, such as bears, badgers, elephants, hippopotami, etc. In such cases retention depends on uses such as signalling or fly whisking, for these can override or counteract the trend to reduction when the balancing function is not required. In bears the tail can be preserved as a flap to protect the anus from the cold, as is clearly the case with the polar bear.

Wings Some vertebrates have the forelegs adapted as flying organs. Flight has been developed independently among another group of animals, the arthropods. In this case it is confined to the insects, although even some primitive members of this stock appear never to have developed wings, as with the silverfish.

To digress from the vertebrate theme for a while, insect wings are clearly of disparate origin from those of vertebrates: the way they function is quite different. Their design primarily facilitates hovering, so that backward flight is usually possible for them. Gliding never takes place and the rate of wingbeat is usually extremely rapid, but with the butterflies forming an obvious exception here.

There is a fundamental problem; insects must have wings to fly, so where did wings come from? How did they evolve wings from no wings? It is impossible for such wings to start from nothing, so one is led logically to the conclusion that they evolved from organs already existing for a different purpose.

Insects are essentially small creatures and one might postulate that, like certain small spiders, there were among their ancestry some that could be windblown. In any case such tiny creatures can fall from great heights and survive.

Wings grow out of the front two segments of the three forming the thorax of an insect. The process that could have led to their formation is that known as metamorphosis. In the final stage of this the pupa bursts open and the perfect insect emerges. Observation of an emergent dragon fly shows that the first split occurs at the wing attachments. The possibility thus arises that some early insects evolved moveable projections on the back specifically in order to open the pupal case. Assuming that such insects were arboreal, it seems conceivable that in a falling or windblown situation the projections, being moveable, could initially be used, however slightly, to influence the manner of descent and, as a result of continuing evolution, grow to successfully resist gravity as proto-wings.

Insect wings must be parapodial in origin, as are the legs found on the same segments. Their presence is due to a degree of residual radial symmetry that allowed a multiplicity of appendages per longitudinal segment.

Flying vertebrates are found in two groups – the birds and the bats. All birds have wings or the vestiges of such, but bats only form a section of the mammals. Apart from fliers there are also vertebrate gliders. The so-called 'flying' fish can maintain themselves over the water for considerable distances, but they are clearly distinct from other winged vertebrates, which can be considered as all being arboreal in origin. Gliding is accomplished by stretching out a membrane and there are species of 'flying' frog, lizard, snake, opossum, lemur and squirrel. In the case of the mentioned mammals 'flying' is enabled by means of skin stretching between fore and rear legs. Of them the opossums are marsupials.

Bats must have started off in a similar way, by launching themselves from one tree to glide to another on outstretched skin surfaces. However, in their case the area of skin has been extended to include the fingers and also the spaces between the back legs and the tail. In this they have gone further than the extinct flying 'reptiles', which extended the membranous wing on a gross elongation of the fourth digit. The first three digits remained as claws and presumably retained some use as climbing and roosting organs.

Smaller bats are generally insectivorous and nocturnal/crepuscular. They can fly at night using their echo-sounding equipment and locate and catch flying insects by the same means. A few species specialise as nectar feeders, blood suckers, or even as fish eaters. The group of generally larger species that are herbivorous are known as fruit bats. Bats are placental mammals and it seems probable that they all went through a gliding phase similar to that of the flying squirrels. However, in their case the spread out forepaws also added marginally but significantly to the gliding area so that any small fillet between the 'fingers' developed to become a complete web and flight was in prospect. However, present gliders, even after long glides, can always hit the chosen target tree, so some form of steering is apparent; but there is no sign of flight and, any case, there is hardly a vacancy for any new true flyers.

Bats are nocturnal animals and it would seem that they exploited a gap that had not been filled adequately by the birds; diurnal bats did not arise, for there were no vacancies for them in the house of light, but they could do things in the dark beyond the capabilities of any bird. Historically, hardly any birds have developed a nocturnal life-style, but it may once have been a universal condition among the earliest animals recognisable as mammals.

The bats may be related to the primitive primates. One can well imagine a bat as being descended from some sort of nocturnal lemur, but this would be an unknown one through it being long extinct in its original guise.

Many birds can glide, but none do this without being able to fly, albeit sometimes laboriously. This indicates that the ancestral bird could already fly by the time it had acquired flight feathers.

A bird not only needs wings to fly now, it also needs suitable feathers and the problem arises as to how birds first evolved flight feathers to enable them to fly. The appearance of such only seems possible by postulating their evolution from something else. Feathers are derived from scales and the primary purpose for such modification could not be to allow flight. The most probable reason was to provide heat insulation. As airborne creatures birds must have been earlier gliders, using skin stretched between their limbs, as is the case with modern gliders and bats, as well as the extinct pterosaurs. But the birds are not related directly to any of them, for their legs and other features betray them as having descended from some sort of dinosaur.

The ancestor of the birds would be a kind of small dinosaur that had its body insulated by a covering of feathers, these being scales marvellously broken down into intricately branching and air trapping structures. Up in the trees, such a creature would progress mainly by using its hind limbs, the forelimbs being only auxiliary to locomotion. Similarly to bats, the early birds would start off as gliders, but only using skin stretched between forelegs and the flanks. However, the skin had a covering of feathers to counteract excessive cooling of the blood and which quite incidentally gave a marginal increase in support from the air they were passing through.

By evolutionary processes the amount of support derived in this way would increase as the feathers responded by growing larger and stiffer, so that they gradually became the main organs of flight and eventually took over completely; the skin flaps became redundant and disappeared. Similarly with the tail: once gliding and flying were embarked upon, the feathers on this once longer organ would also be encouraged to evolve to longer and stiffer forms and become the marvellous control organs that now grace flying birds.

Skin This forms the outer covering of the animal body, but is penetrated by the essential orifices and bears various other features, either embedded in or growing out of it. The skin of larger mammals is generally known as 'hide'.

Sharks have hard processes embedded in the skin known as denticles, but the bony fish generally have tough, overlapping flattened outgrowths called scales, which provide an overall protective and flexible covering. The lobe-finned fish took scales ashore with them as part of their amphibious tendencies, although modern amphibians now all have skins devoid of such. The reptiles and dinosaurs inherited scales and the latter passed them on to their bird descendants in the form of feathers.

Mammals have evolved separately and are covered with hair, which, when dense, is referred to as 'fur'. The physical origins of hair are hard to discern, but, like feathers on birds, the original function was presumably the insulation of creatures with warm blood. Hair on mammals is now put to various additional uses, such as camouflage, adornment, fly whisking, etc.

Mouths. Being the inlet end of the alimentary canal, the mouth of a vertebrate is homologous to orifices of similar function on many invertebrates. The earliest fishes were both jawless and toothless. Jaws and teeth developed later and remained as features among all descendants, although subsequent loss of teeth is a common phenomenon. Modern amphibians have no teeth, but extinct ones were thus furnished and some are even named after their dentition.

The great variety of dental systems found among vertebrates is imposing, but perhaps more intriguing is why teeth disappeared in many cases. In the cases of frogs and newts and the reptilian chameleons, loss of teeth seems to be associated with the swallowing of invertebrates whole; the mouth has evolved to be big enough and the prey small enough to be engulfed once seized. In the case of turtles the function of teeth has been usurped by a horny beak-like surround to the mouth. One cannot exactly see this as being an improvement on teeth, but one might postulate that the ancestral turtles went through a "soft feeding" phase and progressively lost their teeth as being unnecessary for this. Eventually they returned to "harder feeding" again, with their mouths becoming beak-like in order to cope, since the loss of teeth was irreversible.

The birds too are all without teeth. In the present every bird has a mouth extension consisting of a horny bill, even though such are found in a vast array of shapes and sizes to suit a huge variety of feeding habits. The origin of the bill of birds may seem to be similar in cause to that of turtles; there was hence among the primitive toothed birds some which came to specialise in soft food, leading to the loss of teeth. Subsequent evolution among this particular group, due to a widening diet, led to the development of the bill, a horny extension of the mouth. However, the suspicion is strong that the early ancestors of surviving bird species were insectivores and their life-style resembled that of amphibians like tree frogs. The hunt for insects, or their larvae, led to investigation of flowers and fruit; this led to omnivorous species some of which were eventually to become fully herbivorous. Others became able to include ever-bigger prey in their hunting and became carnivorous. One might note that many smaller species still include insects as a major part of their intake, especially as nestlings and fledglings. The extremely vocal nature of nearly all birds makes it unlikely that they were derived from a stock whose life-style development ever resembled that of the turtles.

Teeth themselves seem to owe their origin to material produced originally to form the denticles as found in the skin of cartilaginous fish. Again, the ultimate origin of these may be shared by the spines found in the skin of starfish and sea urchins. Vertebrate teeth can in some species grow to exceptional size when used for specialised purposes. Such are known as tusks.

Inside the vertebrate mouth is the organ known as the tongue: it is the seat of the sense of taste and this was probably its original function. It has developed in a variety of ways among the various kinds of vertebrates. Many reptiles have protrusive tongues with which they can 'taste' their surroundings as they advance into them. The frogs, newts and chameleons have extensible tongues with sensitive sticky ends with which they can seize prey. The tongues of birds are generally undeveloped; birds in general cannot lick. A notable exception is provided by the humming birds, which have a long extensible tongue that can reach the depths of flowers to obtain their nectar. Woodpeckers are similarly provided with long tongues, which they use to extract their prey, often the grubs of insects, from within tree trunks.

The tongues of mammals have developed to provide a number of functions, including being the seat of taste. By protruding the tongue surfaces can be licked, either to clean areas around the mouth or other body parts, or to taste surfaces and pick up part of them, as do cattle at a salt-lick. Many mammals drink by lapping the surface of water, rather than by sucking. A prehensile tongue, like that of a giraffe, may be used to strip vegetation. Other tongues are long and sticky and used to extract ants and termites from their recesses.

The tongue is also often used to fine-tune the sounds produced by the larynx. Among mankind this has become the function mostly associated with it, for it has been developed to an amazing degree: the interplay between the tongue and the various areas of the roof of the mouth has played a major part in the development of 'language', with this very word being derived from the French for 'tongue', with the latter term still being found in use as a synonym for the former. The human tongue is also used as an expressive device, mainly to imply derision, both visually and aurally, and it has even been brought into use as an organ of sexual significance. Dogs, being devoid of sweat glands, use their long tongues for cooling down, by protruding the organ and panting over its surface.

The edges of the mouth are called 'lips' and among mammals they are capable of a degree of independent movement. This is especially true of the primates: man again has the greatest capabilities here. Like the tongue, lips are used in language to modify sounds produced by the larynx (vowels) and to produce sounds of their own (certain consonants). Man also uses the lips to suck up fluids (the tongue not being used for this) and in the kiss they can have social and sexual import.

Below the lower lip of humans is the chin. No other mammal has this protrusive feature and it was also lacking in the early hominids. The chin has nothing to do with the workings of the lower jaw, but is the anchor point for the muscles that operate the lower lip. This has evolved entirely as a result of this lip's important role in the forming of the labial and labio-dental consonants, as well as the vowel sounds. One can appreciate the chin's function simply by placing a finger or two on it. Then slowly say "wasps, bees and flies" with some deliberation.

Noses These are organs that house the nostrils, which with man are two channels stretching from the outside above the mouth to a point behind it. Primitive nostrils, as found in fish, are simply pits that are sensitive to chemicals in the water. These are really 'tasters'. With land animals, the nostrils, being thoroughfares, are the breathing organs remaining, should the mouth not be available. The original breakthrough was apparently made at the departure from the fish stage, perhaps through occasional malformations of the pits proving the benefits that were possible.

The chemical sensitivity known as taste requires the oral surfaces involved to be wet. Smell is confined to air breathers. However, for it to work the mucus membrane has to be kept moist.

Mammals generally retain a very powerful sense of smell, but many have suffered a partial loss for various reasons. The primates have been deprived badly in this way. This can be associated with increased reliance on good eyesight and has cause in the adoption of an arboreal existence. While taste and smell have been vital to man's earlier and later ancestors respectively, their temporary loss, as during a cold, is no more than an inconvenience now.

The nose on most other mammals is just a blob on the end of the snout, although with whales the nostrils have migrated to the top of the head, where together they provide the "blow hole". With man the snout has virtually been absorbed into the face and it is the nose itself that juts out, supported on its cartilaginous rod.

There is considerable racial difference to be found among noses. The Negroid nose is large, with wide nostrils and is adapted for inhaling plenty of air in a hot climate. The true Mongoloid nose, on the other hand, is small and set deep in the cheeks, this being more suitable for a cold climate. However, the Caucasoid nose tends to be large yet narrow. This seems to be an intermediate form, with an original large nose being narrowed to cope with a worsening climate. While it tends to preheat the air better than a Mongoloid nose, it might seem less well adapted as this to the cold because of the greater risk of the end getting frost-bitten. The Caucasoid nose is not particularly well adapted to severe conditions.

Nose shapes are also always subject to another factor, evolutionary modification by sexual selection, just as this can also affect all external features. Genes representing a character approved by the other sex stand a better chance of being transmitted.

Ears The ears of fish are comparatively simple and their style is perpetuated by the fluid filled inner ears of the higher vertebrates. More sophisticated equipment was needed for hearing in air, which led to the inner and outer ears of animals such as man, these being separated by the membrane of the ear drum. This itself is the primary receiver of sound waves and also serves as a barrier against extraneous matter. Within this shield the middle ear's essential members have been created from two bones that originally constituted part of the reptilian-style jaw. The outer ear may be a plain hole, or be graced by some form of external flap. The latter is now what we generally refer to when we talk about an 'ear'.

Fishes, amphibians, reptiles and birds do not have these external flaps, but such are typical of mammals. However, some have largely or completely lost them again, particularly those that have returned to the water, such as whales and the true seals. The loss of the flaps does not seem to impair hearing at all, although one might presume that their presence does improve sound reception to some small degree for land dwellers, while the two ear flaps working together would seem to improve direction finding with respect to any sound, which in any case is an important faculty of the twinned ears. Among the higher mammals the flaps often have independent movement (as with cows) and may be adorned by colour patches on the tips (as with tigers) for communication purposes, or adopt different postures (as with horses) to give indication of mood.

Generally among the primates the ear-flaps are situated on the sides of the head, have become flattened against the latter and the insides have developed ridges. The reasons for these changes are not clear, for they seem to offer no improvement to hearing and can hardly be held to look attractive in themselves, at least to human eyes. Perhaps the ridges constitute a device to compensate for loss of protrusion and to increase rigidity, while appearance was originally of less importance to humans through their ears tending to be masked by hair.

The human ear-flaps differ from those found on apes and monkeys in having a lobe at the bottom. Perhaps this was an evolutionary development to assuage the sense of ugliness imparted by the rest of the ear. Because of the presence of hair, it was the part of the ear most likely to be seen and took a shape and texture considered to be more attractive. It certainly still receives much attention, being either a target for kisses or a point from which to suspend ornaments. Very often the lobe is deliberately pierced for this purpose and in some cultures the hole has assumed such significance as to be gradually enlarged so that the lobe dangles down towards the shoulders.

While ears may be thought of as the features we hear with, the original fish ears were probably primarily organs of balance, as indeed are man's inner ears now. The vital nature of this function, normally taken for granted, is brought to prominence should damage occur to the inner ear and serious loss of balance from this cause can occur in later life.

Eyes The eyes of vertebrates are all of similar construction and function, while the resemblance to those of cephalopod molluscs has already been commented upon above. Nocturnal animals frequently have enlarged eyes to make the most of available light and this can be observed among the night-active primates. The iris can contract to exclude light, either by means of a vertical slit or a circular aperture. Eyes need to be lubricated and kept clean, so are fitted with ducts and lids to achieve this. Some mammals have a further protection in the eyelashes that grow on the edges of the lids.

Birds have a weaker sense of smell and their hearing, while good, is hardly brilliant. Owls provide an exception here. However, the main avian sense is eyesight and with some of them it is phenomenal, especially the raptors. With mammals eyesight is more variable. With those retaining a good sense of smell it is often poor, as with the insectivores. Likewise where hearing is exceptional, as with the bats. Vertebrates that live underground or in caves are liable to complete loss of sight, as many examples show, but this never applies to birds.

The presence of light-sensitive cells in the skin is common to many forms of life, but it is where these have become concentrated together into recognisable organs that they are known as 'eyes'. Structurally they can be of quite different origins in different phyla, as between the vertebrates and the arthropods.

An important function of eyes is sensitivity to colour. Insects have acquired a very wide range of colour vision and this is related to many plants having developed coloured flowers in response to this capability. It seems to follow that, before this, insects could appreciate colour for a different reason. It would appear that, with creatures that were evolving eyesight as a major sense, colour vision was an advantage because it aided definition and widened recognition by providing more contrast.

Birds all have colour vision, but most mammals view the world in monochrome. It is specifically the primates that have acquired colour vision, although with not quite the same range as either the insects or the birds. The lack of colour vision in mammals is reflected in the drab coloration of most of them compared with the splendours of so many birds and insects. Mammals often go in for pattern, rather than colour, and one can observe how they may use shades from black to white, or gradations of grey or brown between, with camouflage often being the basic requirement. The bulk of such animals do not display the blues, reds, greens, etc., of the birds and insects; although there are exceptions.

The achievement of colour vision among the primates presumably relates initially to the search for or pursuit of food that was well distinguishable by colour. Insects are often colourful, and the facility must have been helpful in catching these. However, the advantage would only exist with a life-style that was not nocturnal. Thus can one perceive an evolutionary trend developing among some mammals that were starting to feed more in daytime and were fast moving arboreal creatures who caught arthropods by grasping them by using one forefoot with opposable digits. Once acquired, colour vision was too widely valuable ever to be lost again; it can indeed be seen as an important one of the bases from which sprang the artistic nature of the human being.

Another important feature often found among vertebrates is binocular vision, with both eyes facing forward and concentrating on the subject in hand. This provides an improved judgement of distance, particularly at close quarters, and can be noted as a characteristic of certain carnivorous dinosaurs and birds. Flesh eating mammals also have it, but it proves additionally advantageous for dwellers in the treetops. It was probably the arboreal life-style that really instigated binocular vision among the primates, along with the need to catch insects. This existing capability was later to be a contributory factor in the ability of man's ancestors to adopt the life-style of a hunter, especially when this came to involve the casting of missiles.

Some primitive vertebrates and even reptiles have a so-called pineal 'eye' on the top of the head. The skin covering this organ is somewhat transparent, so that light can reach it. Its name relates to the fact that it is connected to the pineal gland, behind it in the brain. Perhaps this feature can be taken as a token of the vestigial retention of rudimentary light-sensitive radial characters at the head end of animals that have evolved to become elongated and bilaterally symmetrical. The head, with its skull, is the representation of the original radial adult phase of the basic vertebrate form.

Other External Features As already discussed, tentacles are commonly found among invertebrates, but do tentacles proper manifest themselves among the vertebrates? Many fish have outgrowths near the mouth known as barbels. These serve a sensory function and have no grasping powers, but, even so, can be thought of as the degenerate remnants of tentacles. They are absent from the higher animals, but might be thought to be represented by the stiff hairs known as bristles found at the beaks of some birds and the whiskers of mammals. These are certainly not of directly tentacular origin, but may make use of the nerve functions formerly evolved for such.

Many animals have external hard parts for defensive or offensive use. Horns on the head are common among herbivorous mammals.

As defensive weapons these could not just appear out of nothing with the passage of time and their origin must rather lie in their offensive use during sexual or territorial disputes among males. The method of combat by butting heads together favoured those with thicker skulls in the common areas of contact. Both bone and skin responded by developing bosses on the skull and horny coverings over them grown out of the skin. From this have come the horns of cattle and antelopes and the antlers of deer. A feature to note is that the larger species not only tend to have the larger horns, but that theirs are generally also larger in proportion to their own body size. It is in the larger species too that the females also have horns, although usually inferior in size to those of the males. The smaller species may be totally devoid of horns. The size of the horns would seem to be roughly proportional to the mass of the animal and its ability to apply it against an opponent. The evolution of horns would also seem to be related to their status as secondary sexual characteristics and this would apply especially to the deer.

Ungulates that do not butt their heads together (e.g. horses) have not evolved horns. The horns of rhinoceroses are distinct in that they are made of compressed hair and must relate to a time when all these beasts were hairy. Such 'horns' would seem to originate in the use of the top of the snout as a pushing agent, as in disputes.

Many animals have enlarged teeth that usually protrude beyond the mouth and are called tusks. In the case of the elephant it is a pair of incisor teeth that are so modified. These are clearly to be linked with the evolution of the trunk. Without the latter such tusks would seriously hinder or even prevent the animal from feeding. It follows that the evolution of the prehensile nose and the tusks occurred in parallel, with the former presumably being the initiator, albeit originally quite short and with its first advantage simply being flexibility. One can see the arrested rudiments of such trunks on other animals, such as the snouts of tapirs and perhaps even pigs.

Internal Organs The basic bilateral symmetry that is so evident externally among vertebrates is also observable within the body, although some organs are unilateral. It has been claimed earlier that the through-gut was formed by non-separation during asexual reproduction and that, of the two original individuals concerned, the one provided the mouth and the other the anus of the new creature. From this one might expect that organs derived directly from the gut would be single, while those derived from the mesoderm and grown in the coelom would be paired.

This rule applies quite consistently; apparent exceptions can be taken as being due to subsequent development. On the gut one can note such solitary organs as the liver and the pancreas, while in association with the coelom are found paired organs such as the kidneys, ovaries and testes, as well as most of the endocrine and lymph glands. Some of these have become fused into a single unit through occupying a central position, such as the thyroid gland. The reverse is true of the lungs, which were originally a single organ derived from the front end of the gut, but which has been secondarily split into two parts.

By this argument our complete sexual equipment should be duplicated, but the final parts of both male and female organs are single openings. Yet the incipient duality is still there and evident in the form the apertures take, which in each case occurs as a slit aligned longitudinally with the body. The original bilateral symmetry of the mouth is evidenced by a ridge along the centre of the pallet, the pairing of the teeth and the groove down the middle of the tongue. After this reasoning the present mouth opening and cavity go with the nasal channels and are not formed out of the gut.

CHAPTER 15

Review of Organic Origins and Developments before Man

■ The Nature of Life

The concept known as 'life' may be defined by attributing to it the various activities considered in Chapter 2, i.e. feeding, respiring, growing, moving and reproducing. Living matter, as perceived scientifically, is based on molecules containing the element carbon and when life ceases the resultant dead matter consists of carbon compounds that prove too unstable to survive, except under special circumstances. Under ambient conditions, survival of such dead matter is always rigorously curtailed because it is inevitably returned to the general cycle of life by being absorbed into the tissues of organisms still alive, be they bacteria, fungi, maggots, vultures, hyenas, etc.

■ Origins and Constituents

Living matter can be divided into two kinds:

1. In this individuals always die
2. In this they need not die, except by misadventure.

In case 1. living is always perpetuated by a form of sexual reproduction, whereas in case 2. individuals can apparently damage themselves severely by division, but reform immediately into new smaller albeit replicated entities. However, the individual cells of case 1. resemble those of case 2. and it would seem indisputable that life was originally exclusively of case 2. It would hence appear irrefutable that life forms were earlier of much simpler composition than the simplest known forms today, because organic matter is an 'invention' of 'life' that must have been built up from rudimentary constituents.

At one stage the Earth was completely barren and had no complex carbon compounds (and hence no organic matter, so therefore no life); but the ingredients were already present. The original gaseous and liquid spheres of the earlier planet were cooling down and the heavier minerals were forming the core, mantle and crust, while an outer cloak of gases and liquids continued to envelope the solidifying central mass. The carbon was perhaps present in gaseous form only, basically carbon dioxide (CO_2) and methane (CH_4) The other life element present, and which is no longer found naturally uncompounded, was hydrogen (H). As cooling proceeded, free hydrogen was all eventually destined to combine with oxygen (O) to form water (H_2O). The CO_2 and water vapour of the air and the H_2O of the fresh water seas were all prepared for combination into the primitive hydrocarbons that were eventually to lead through amino acids to proto-life, but perhaps initially producing the simpler molecules of carbonic acid. That carbon dioxide was potentially much more evident in pre-life atmospheres is inferable from the amount now being released by the burning of fossil fuels, although the exact combined form of carbon with oxygen then may have been dependent on the ambient conditions prevailing at any time; carbon dioxide is the form suited to the more recent. The gases that are thought to have contributed in a major way to primeval atmospheres are methane (CH_4) and ammonia (NH_3), neither of which can survive long as distinguishable entities in present conditions.

The primeval seas of ancient lifeless Earth must have been the milieu wherein complex carbon compounds were originally built up by inorganic means, for without such compounds the emergence of life was impossible. Such entities do not exist now, for they cannot survive. Any survival is perhaps precluded by the presence of free gaseous elements, especially oxygen, but in any case there is now a huge amount of living matter about, ready to dismember and absorb the relevant constituents of such compounds.

The primitive constituents of carbohydrates and proteins (basically comprising carbon, oxygen, hydrogen and nitrogen) had arisen because conditions had allowed them to form; they were somehow invented and, although they appear highly unlikely now, their primordial existence is indisputable. Their creation may have been facilitated by the atmosphere being much more penetrable in those early times by formative cosmic rays, but it does seem highly probable that any such process was accelerated by events of a cataclysmic nature, such as violent electric storms. Such events would seem to have invented new compounds that proved to be relatively stable under the prevailing conditions.

While such compounds will have been stable up to a point, one can postulate that so much complexity would enable the existence of a variety of related but marginally differing forms. Small changes in form could slightly energise or de-energise their immediate environment and affect their neighbours. Conversely one can argue that external changes of energy affected certain complex molecules and set up chain reactions that produced mutations among neighbours.

The external source of energy that was to have the most decisive effect was the Sun, whose light was switched on and off on a daily basis. These extra-atmospheric regular fluctuations in energy application would create the effect postulated in Chapter 3, whereby such radiation would generally build up the store of energy of the great molecules so that the direct link with the solar cycle was eventually broken for such periodic reversions, while some of these did not recur at all, thus creating a new even more complex range of carbon-based molecular compounds.

Nevertheless, it was the initial reversibility of the changes incurred by these sensitive carbon compounds, produced by quite modest fluctuations in the energy available both within and without themselves, that eventually distinguished them so markedly from the irreversible changes that were the norm in the inorganic world (that is if one ignores reversible changes involving different forms of the same substance, as with the evaporation and freezing of water). The initiating change eventually became protracted and can be identified with growth, while the reversal to the original condition became comparatively quick and can be recognised as primitive reproduction. Even so, the reversal may not always have been exactly to the original form and this might be identified with mutation. At this stage, the repetitive collections of dissimilar groups of organic molecules can be identified as cells and during the growth period came to evolve specific and characteristic activities of an autonomous nature that can be collectively thought of as 'life'.

Life at its lowest level can be recognised in the form of the most primitive cell possible, which acts, grows and reproduces in a very simple manner (reverts). From this stage two clear strands have emerged: super cells of more complex nature that could reproduce themselves by pseudo-sexual methods; agglutinated cellular beings that could reproduce sexually by means of specialised reproductive cells. In both cases the use of sexual methods introduced the certainty of mortality to all parts except for the sex cells themselves, even though some of the simpler multicellular organisms, as well as some of the more complex non-cellular ones, have retained the ability to reproduce by non-sexual means, such as budding and division.

The ability to create life continues to fascinate and preoccupy mankind. The fact that primitive protein molecules can be created in the laboratory is a great achievement, but in itself does not achieve 'life'. The reason can be expressed in several ways. Identifying 'organic' molecules (under an electron microscope) of more and more primitive type really takes one back past the creation of life without noticing the fact. Protein molecules produced in the laboratory are created 'inorganically', while living molecules must be organically produced. Life is an ongoing process that cannot be restarted once stopped. The development of living forms has always been a very slow process, not only when measured against the individual human lifespan, but even against the time that homo can be said to have existed. The same can also be claimed for the very creation of life, whose products could not have withstood the same cataclysmic interventions as can be postulated for the origin of the appropriate carbon compounds themselves. The change from inorganic non-life to life was a phase of many tiny steps and of great duration. If this claim is just, then the chances of creating 'life' as defined above, to include growth, activity, and reproduction, cannot be achieved in the laboratory unless some means can be found to recreate the changing conditions that have obtained over a vast period and to compress them into a time span that is reasonably measurable against the normal human lifetime.

It is tempting to think in terms of life starting off as one particular form of organism (and even as one species) and then, as time passed, for the stock to be continually branching into diverse forms that became irreversible genetic entities. On the other hand one can regard life forms as being divided in some way in a fundamental manner, perhaps first as a fluctuation between one form and another by one type of integrated living matter, and then later as divergent entities that could survive more permanently in their own specific life forms.

The basic cause for the creation of different types of living matter was the reciprocating effect of days and nights, as put forward in Chapter 3. At first molecules combined with other matter in the daylight, while reverting to their original forms at night-time. But there was a tendency to build up the available external supply of energy during the period of daytime plenty, so that night-time fission need not take place and changes became somewhat stabilised and some even permanent. But no matter what changes were to take place in the successive steps towards greater complexity, they would originally always need to be alternated on a daily basis with a corresponding "nocturnal form" until any change became thoroughly stabilised.

The above merely suggests that at any stage there were complex dynamic organic molecules that might or might not revert to an earlier stage at night-time, depending on the degree of stability they had attained. It completely ignores the notion that 'nocturnal' molecules may have been stabilised in a similar way. One can postulate the existence of nocturnal molecules from the assumed 'fact' of diurnal ones. But such nocturnal forms could only persist indefinitely if they were in some way unsuited to activity during the day and this would basically mean that they had lost the ability to synthesise matter by means of light; they were no longer photosensitive.

Light sensitive molecules must have been the originals, for they were the ones that had the ability to absorb inorganic matter by means of photosynthesis. These were the primitive ancestors of algae and must have existed on or near the surface of the water, especially near shorelines. The energy supply was so abundant that overcrowding and wastage were inevitable. The surplus matter would be reabsorbed by awaiting nocturnal molecules, who extended the scope and habitat of life by surviving and resisting change away from the light, either in deeper water or in the shadows, but whose existence depended on the demise of diurnal types.

But all nocturnals did not simply await the end of surplus diurnals, for some, by using energy manipulation, were able to accelerate such demise and then reabsorb the appropriate organic compounds into themselves together with any chemical 'releasers' they had emitted to destroy the 'prey'. The term 'energy' in this context has to be understood as chemical energy, i.e. that used to cause fusion or fission and not as physical or electrical energy. It is clear that the use of chemical energy alone is the way of those living forms we can broadly call 'plants', while the use of physical energy prior to chemical energy in the securing and absorption of nutrients is the way of animals, even the simplest of them. Only a very few more advanced plants have developed physical energy in any way comparable to animals, such as those with triggered insect traps. 'Rapid' movements among plants are generally associated with reproduction and feeding, being means to ensure the distribution of their seeds or survival when normal sustenance is scarce.

■ Pseudo-Predators and Parasites

While primitive photosensitive 'plants' were the prime movers in the origins of living matter, there would seem to have emerged subsequently a variety of life forms that could live either as predators or parasites on the plant world and later still also on each other in the same way. The bacteria are unicellular organisms that are plant-like, although they cannot use photosynthesis, yet exist in myriads of forms and individuals as parasites on the whole living world. Yet many do a useful job in accelerating the decomposition of the debris left behind by other life forms.

The fungi are multicultural organisms without photosynthesis that also accelerate and absorb decay, whereby many of them are essential for the very presence of plant life. Yet, like bacteria, in some forms they attack other living organisms, usually distinctly to their detriment and sometimes bringing about their destruction.

The protozoa are unicellular (or non-cellular) animals that use physical means to trap food. The amoeba engulfs food into its cytoplasm using its pseudopods. Many of the non-cellular protozoa are ciliated, using banks of tiny 'hairs' to whisk food particles into their interiors. Others are aggressive predators that seize and engulf other microbes. Like the plants, unicellular forms of animal could combine into tissues and organs to become multicellular organisms – the Metazoa – but the emergence of such cannot be regarded as a phenomenon of life alone, but also of reproduction, these being the two basic aspects of organic existence.

◼ Primordial Sex and Reproduction

Sexual reproduction is found among both plants and animals. This suggests that the basic principles of sex and reproduction arose among primitive multicellular entities before the clear-cut distinction between plant and animal had arisen. Although sexual reproduction of plants is confined to the green ones, not all such use it and all plants can (or can be made to) reproduce asexually. This suggests that the multicellular animals somehow arose as a form of sexually reproducing green plants, while the single celled or non-cellular protozoa are historically more closely related to the asexually reproducing green plants (which proposition is strengthened by the existence of the photosynthesising microbe Euglena). In the light of a claim made above, one is able to propose that animals actually represent an evolutionary descent from the pseudo-predatory night molecules. These were the ones that depended entirely on the solar energy absorbed by their diurnal counterparts. The green plants photosynthesised during the day, building up carbon compounds from the absorbed carbon dioxide and at the same time releasing oxygen, but during the hours of darkness reverted somewhat and released carbon dioxide. The nocturnal primeval 'animal' forms, being unable to photosynthesise, secured carbon compounds by ingesting plant matter, while releasing carbon dioxide all the time. In order to balance this activity, animals need continually to absorb oxygen.

Once the principle of 'predation' had been established, such plant-eating microbes (or molecules even at an earlier stage) need no longer be confined to nocturnal feeding and could extend their activities into daytime.

The ultimate ancestry of animals, including human, goes back through the sex organs. It is the gonads that are directly descended from primitive bisexual ancestors, while the rest of the body - including all the most wonderful functional embellishments - consists simply of evolutionary additions in order to ensure the continual survival of the gonads and their ability to produce gametes.

Strictly speaking, it is not necessary to have male and female sexes to carry out sexual reproduction; the combination of undifferentiated cells from separate entities of the same species to form a new individual is 'sexual' in essence. This must have been the original 'sex' before the advantages of specialisation of the combining partners asserted themselves. Yet this would arise at a very early stage.

Initially all cells must have been haploid (containing one set of chromosomes and hence genes), yet of a more simple nature. But in order to reproduce, such cells in one sense do become temporarily diploid; the genes construct replicas of themselves that are sufficient to assemble twin chromosome sets in readiness for cell division. At a very early stage of cellular evolution, growth, replication and division must have been less clear-cut activities than now, being both less protected and distinguished by evolutionary accretions than are included in the cells of later times. Such cells were relatively unstable and fluctuations between different forms more easily accomplished than now. Such variation implies reversion as well as progression, the form taken being dependent on the relative energy level obtaining outside of the organism.

The ability to 'revert' only remains at the cellular level. The faculty of reversion to a marginally more primitive state is hard to observe in modern forms, having presumably been evolved out, because its usefulness ceased at the proto-life stage, after which it became harmful. However, it may still contribute to the appearance of mutations. The night-time gaseous exchange of green plants is a reversion activity that is not of sufficient duration to necessitate any accompanying physical change.

While marginal reversion is now a difficult phenomenon to pinpoint, the more drastic forms of the event are primal with regard to life and have been formalised to a point where growth leads to duplication and then reversion by halving. In mitosis the chromosomes divide and migrate to opposite poles in the cell to found two young cells upon division of the original entity. In meiosis the duplicated chromosomes of a diploid cell do not divide, but each entity migrates to opposite poles within the cell to found the nuclei of two new haploid cells.

The pertinent questions would seem to be: "How were diploid cells originally formed, and why?" One should observe a "Which came first?" situation here, as with the popular chicken and the egg poser. Diploid cells are only required for sexual reproduction; so why should they appear at all in the primary non-sexual world? One can perhaps best attribute their emergence to chance. Haploid cells become briefly 'diploid' before they divide. It is not too hard to imagine that under circumstances of high external energy such division could be delayed and the cell arrested in a virtual diploid condition. From this position one can perceive two developments. The cells could evolve so that such excess core material became multiple nuclei. This is the route taken by the non-cellular protozoa and one might also recognise it in the syncytium tissue observable in some other organisms. However, the most significant development (taking a chauvinistic human view) was the initiation of sexual reproduction among multicellular organisms and the evolutionary opportunities that were thus provided. The duplication of nuclear material was restricted to one operation and the power to revert to the haploid condition has always been retained by those parts of the organism that lay in the direct line of its reproductive history. In animals this is represented by the gamete producing gonads. It is clear that initially sexual exchanges would take place between organisms that were not sexually differentiated, but gradually the evolutionary needs for differences between gametes that were pairing up to make new diploid wholes would occur in parallel with the advantages inherent in having the adult partners specializing in many aspects of their physical and reproductive development and behaviour. However, it is evident that gamete differentiation was the prime mover for sexual dimorphism, for the latter is not a significant feature among many organisms, except for the reproductive organs themselves, and this is particularly common among the sexual plants, where however the everyday role-playing as found in animals can never take place.

■ Colonies

Most life-bearing molecules, in organisms big enough to be seen, came to organize themselves into the entities we know as cells. The essential parts of these were to do with reproduction and they were concentrated in the nucleus, which itself was the representative of the whole phylogenetic history of any species, while over the years additional material had been added, both in and around the nucleus. This resulted from the activities of the cell becoming more and more complex. In multicellular beings development of the essential nuclear material occurred in parallel with the ever-increasing primal phylogenetic and ontogenetic respective significance of the gonads and gametes. As life forms became more complex, the gonads came to have the appearance of many such organs with specialised functions, they being specifically reproductive, as against the others being specifically feeding, protective, sensory, locomotor, etc. Such lumping together of organs disguises the primary nature of the gonads.

So specialised groups of cells formed organs and these in turn were gathered into colonies. At this stage the cell is the individual, for groups of organs could only form true individuals by means of the unifying powers of the nervous system. By this assessment no plant can be an individual in the same way as many animals; even a solitary tree can be viewed as a colony, which condition allows many of them to respond vigorously to certain quite severe mutilations, capabilities that are exploited by human activities such as pruning, pollarding and coppicing.

Among primitive animals sponges are truly colonial, for they have no nerves. Many coelenterates are colonial in that they share food by means of a continuous protoplasm, even though each polyp has an individual, if simple, nervous system. Others, like the free-swimming medusas, are primitive individuals, as are the freshwater hydras. Among higher forms of animal, independent nervous systems and individuality become more and more pronounced and such is generally accompanied by a diminishing ability to regenerate damaged parts.

Abandoning for the time being all reference to the fortunes of the plant kingdom, one can perceive the precursors of the highest examples of multicellular animals as being groups of cells descended from the "nocturnal molecule" side of life. At first they would be amorphous gatherings of cells that drifted about in the water. Later, as specialisation progressed, divergent and derived diurnal feeding types of animal life would appear and become numerous, shapes would become more regular (i.e. towards the spherical) and rudimentary swimming would gradually replace movement dependent entirely on drifting. Here can be recognised a volvox-like stage of development, with swimming adults giving rise to motile young by sexual methods. But the waters were not still and were full – at least in the shallows – of smaller life forms. Larger feeders were encouraged to sit on a rock or plant stem and wait for food to blunder into their vicinity, where it could be caught and ingested, either by coarse-and-grasp or fine-and-whisk methods.

◼ Asexual Reproduction and De-segmentation

Considered as a generalization, sedentary adults were of a hollow cylindrical shape, with a mouth at one end surrounded by tentacular feeding organs. However, their sexually produced young were free swimmers and ensured a distancing from their parental gonads before settling down on their own. In some cases evolutionary tendencies caused the young to approach the adult shape more and more before assuming the sedentary life. At the same time the adult has always retained a capability for reproducing asexually by fission or budding. Both of these activities could be arrested if evolutionary advantage might result and this could result in multiple colonial members that were held together for convenience, but open to acquiring specialisation of function. Such asexual reproduction of new 'adults' by fission could occur both radially and longitudinally and, when arrested, resulted in segmentation. All segments carried vestiges of tentacles and these had the potential to evolve into a whole host of organs of varying function across many species, especially once segmentation had been transferred from a radial to a longitudinal form, e.g. parapods, legs, feelers, mandibles, gills, etc. and even eventually wings.

Often the need for clear segmentation was superseded by the advantages of cramming together the appendages of a number of segments into one mass, where they could specialise and co-operate more effectively. Such de-segmentation has been common and for convenience was dubbed 'retro-fusion' on earlier pages.

■ Emergence of the Phyla and Later Creation

In early seas opportunities were especially present for life forms to evolve into more complex and often apparently bizarre forms and consequently to split up into discrete biological groupings now recognizable as 'phyla'. All the phyla were created before the Cambrian, since when no new ones have arisen. However, the gulfs between the present phyla certainly would not occur at the same point in time, nor are they of equal magnitude. Subsequent innovation has been restricted to the formation of the lesser divisions, but such has generally been of a somewhat different nature from that observable in the creation of the phyla. The Pre-Cambrian zoomorphic phyletic changes were initiated by the multiplication of assets, through exploiting the possibilities of arrested fission and retro-fusion, to produce radially and longitudinally agglutinated creatures with multiple parts that were ready for specialisation. The Palaeozoic and later changes, together with the earlier developments that can be regarded as being after the formation of the phyla, occurred without the help of agglutination and on the contrary often involved the sacrifice of segments, while progress was made by the transformation and specialization of those that remained. It was this exploitation of existing assets that was instrumental in the creation of new species in this later world.

The mind is misled now in that the later changes are of a nature that can be fairly adequately represented on a page by a graph or flat 'tree', whereas the original divisions into phyla should only rightly be imagined when considered as volumetric in nature and hence require a three dimensional model for proper comprehension. Using this concept the proto-phyla progressed outwards from a central mass that was tending to be differentiated locally, but not originally branching apart. Eventually extinction zones did appear, which were not the result of species branching off a main stem (as was more the case later), but occurred somewhat arbitrarily between great groups of creatures that only formed continuous breeding populations as long as they remained contiguous. It could thus occur that, because of the rather arbitrary nature of extinction zones, entities sundered from each other by such a one might still carry features in common that did not exist across the whole of their respective unbroken zonal spreads. The phyla creation model is like a globe showing the World itself, except that it is increasing in size with age and this expansion causes ever widening gaps to appear in the surface.

All of these early life forms were of necessity aquatic, but one must assume that the water they inhabited was vast and variable in nature for geographic reasons, thus causing differential evolution. The extinction zones would be created by the irruption of impossible habitats, such as the emergence of new landmasses. However, this argument only applies to those multicellular animals that shared a common cellular background. Some phyla must have been created at an even earlier stage, because of fundamental differences at the cellular level. This would apply to the sponges, coelenterates and some others.

One can indeed imagine several 'pre-phyla' divisions of the animal world, in the protozoa, sponges, coelenterates, ctenophores, rotifers, etc., as well as the ancestors of the other large and progressive group. Members of the last had things in common that the others lacked and they were: distinctive bilateral stages of growth, the through-gut and the coelom. From among these have emerged phyla that contain only small and, from the human point of view, generally unspectacular creatures, while others would seem to represent the main stream of life, both in water and on land. The latter types comprise essentially the echinoderms, annelids, molluscs, arthropods and vertebrate chordates.

■ Usurping Juveniles and Extinct Adults

One can dub the grouping of the advanced or 'super-bilateral' animals as the "modified worm-like phyla". All had a tendency to produce elongated, longitudinally segmented juveniles and most of them eventually assumed maturity in some such motile form through the ability to reproduce having been usurped from the original sedentary adult; the culprit was its own juvenile form. As each individual grew from egg to adult it passed through all the ontogenetic stages characteristic of its species. However, during the earlier Pre-Cambrian formative epochs, species had the ability to select that which was useful from various stages and to incorporate them into other stages, with the whole arrangement being plastic, so that characters could flow up and down the ontogenetic scale to give the best evolutionary results. One such occurrence resulted in the acquisition of reproductive powers by the motile juveniles and this was likely to bring about the extinction of their own sedentary adults, and in many cases did.

A unifying feature among these emergent phyla was the through-gut. The origins of this were proposed in Chapter 11 while discussing the bryozoans, whereby the arrested fission of a young one, produced by asexual branching or division, led to a shared food supply on a permanent basis and with the appropriate changes involved in this also becoming embedded in the reproductive system as an evolutionary process.

■ Some Lesser Phyla

There are some phyla whose members are small, either in size or in numbers, or perhaps both. These can be regarded as being somewhat separate from the main thrust represented by the progressives; they lack one or more of the attributes common to the latter. The comb jellies (ctenophores) can in fact be as completely divorced as the coelenterates, for they lack longitudinal segmentation, bilateral symmetry, an anus, a coelom and directional definition in their swimming.

The rotifera are multi-nuclear microbes that exhibit bilateral symmetry and a through-gut and resort to an anchored position when feeding in a fine-and-whisk style. They also seem to presage certain larger invertebrates in having opposable hard feeding parts (constituting a pharynx) for seizing food coarse-and-grasp style. The sedentary life-style of some extinct adult form would seem to be adoptable at will, together with the retention of the free-swimming propensities of the developing larva. The rotifera give a hint as to how the through-gut started: it was in some small primitive sedentary creature similar to a bryozoan, but originally only with one opening. However, the rotifera are distinct from the progressives in that they are only multicellular when young and as they mature the cell walls disappear so that they become a sort of multi-nuclear protozoan; they then constitute a syncytium. Other primitive phyla involve syncytia in their tissues, e.g. the roundworms.

The flat worms are an advance over the coelenterates in that they have cellular tissue between the ectoderm and the endoderm, thus forming the mesoderm (but unlike more evolved phyla does not outline a fluid filled cavity called the coelom). Most flatworms are simplified because they are adapted to a parasitic existence on higher animals, but free-living ones are generally sharply bilateral with regard to general shape and mesodermal originated organs. Yet their gut is many-branched and blind, a relic of a former sedentary adult form. The bilateral body may first have occurred despite the adult being a fine-and-whisk feeder, but one must assume that the free-moving larval forms used coarse-and-grasp methods exclusively and had no need for a through-gut and that all hope of eventually acquiring one was lost with the extinction of the sessile adult.

The roundworms generally have a through-gut, but can be set apart from the main progressive group by their lack of a coelom. This also applies to the proboscis worms. The flat, round and proboscis worms can all be seen as promoted larval forms that use coarse-and-grasp feeding and with lethargic life styles requiring relatively uncomplicated bodies.

■ The Progressive Phyla

The progressive group led principally to the vertebrates, molluscs and arthropods, but also produced less advanced creatures like the brachiopods, echinoderms, annelids and chordates. The brachiopods developed a life-style like that of the bryozoans, but added certain advantages derived from earlier bilateral symmetry, including a double shell in place of the bryozoan one-piece cup. The echinoderms retained the through-gut, but abandoned bilateralism in general by maintaining a radial symmetry derived from one half only of a former bilateral larval form. By chance this resulted in the number of radial segments usually being five, but less frequently a multiple of this. This ontogeny illustrates how longitudinal segments can be converted to radial ones using the plasticity of life-forms to obtain over an immense period the eventual reversion of a bilateral free-swimming larva into a new type of radially segmented bottom-dwelling adult.

The chordates, in the form of the sea squirts, maintained a bryozoan-like sessile adult status, together with a bilateral larva. The lancelet pursues a similar life-style to this larva, but, like the rotifers, has retained for itself the ability to swim and change feeding station. The acorn worms, on the other hand, are firmly coarse-and-grasp burrowers.

The bivalve molluscs are filter feeders of bilateral form, but many species show some ability to change their feeding stations and even to escape the predatory activities of starfish. The cephalopods became free-swimming and fast predators, but retained the archaic feature represented by their tentacles, but in highly developed form. The gastropods are all direct-by-mouth feeding types, but became univalve through the loss on one side of a number of the bilateral segments, thus restricting themselves to only one set of the associated duplicated organs.

The aquatic arthropods, in the form of the crustaceans, all developed strongly the concept of the bilateral adult. The former adult sedentary life-style is hinted at by the barnacles, but even they have drawn into their adulthood segmental features that are properly at home in a promoted larval form. The barnacle represents a transitional form that was necessary between the free-swimming larva and the earlier sessile adult. The latter was made extinct by the transitional form exceeding it in efficiency and taking over completely the sessile state, while at the same time usurping the reproductive function.

■ Invasion of the Land

In discussing the evolutionary developments that led to the appearance of the various phyla, the inherent weaknesses of man's conceptual power, of his biological category system and of the capability of language are all exposed. One can only deal with one point at a time. Yet in fact all sorts of developments were taking place simultaneously, while overlapping and influencing each other in ways that are largely hidden from us. So in order to get a comprehensible view of the progression of living forms to the point where some were ready to leave the water it is necessary to present a grossly simplified case.

The fine-and-whisk feeders were all precluded from leaving the water. Even so, the early precursor of those progressive creatures that were destined to conquer the land was a bryozoan-like fine-and-whisk feeder, but whose larval form was a bilateral coarse-and-grasp type. Once the adult form had been dispensed with such species were ready to go ashore, pending the development of adaptive features. The annelids did it in the form of earthworms and the gastropods as slugs and snails, but the main impact was through the arthropods, first tentatively by means of the crustaceans, but later in the more complete pre-adaptation of the insects and arachnids.

Initial colonization of the land had to be done by the green plants. One might propose surface-dwelling algae as the most likely to accomplish penetration of a rain-sodden but hitherto quite barren landscape. They would be followed by herbivorous and eventually by carnivorous animals. The very organic debris left behind by these life forms would slowly increase the fertility of the terrestrial habitat where they lived and died.

While various round and flat worms came ashore, they were inherently too primitive to compete in an evolutionary way with the progressives and many of them found survival best accomplished by the parasitic and scavenging ways of life. The land gastropods survived by creeping about on their slimy 'feet' and munching away at anything edible they came across, but they never mastered the problem of desiccation. All such land molluscs stayed basically similar and retained a strong resemblance to their gastropod cousins who stayed aquatic. The land annelids were similarly inhibited and showed the same similarity of form to each other and an inability for evolutionary progress. The crustaceans were already differentiated segmentally along their bodies when some started to come ashore, but were unable to beat the desiccation barrier and it can be noted how woodlice will immediately seek cover when suddenly exposed to the light.

Centipedes and millipedes also like to stay covered. They are also inhibited in that they are descended from creatures that left the water in multi-segmental forms and it is apparently now impossible for them as land creatures to evolve away from this repetitive body-structure.

The ancestral insects, which had such a successful career ahead of them, would already have their de-segmentation and retro-fusion behind them when they left the water. They were probably tiny promoted larval forms, a development from the Crustacea. At first they were wingless: relic species from this stage can be seen in the shade-seeking silverfish and earwigs. Eventually the Earth's surface became drier and some of the insects proved to be pre-adaptive to this change. It is specifically from among these that flight was developed.

The immense success of insects on land, their huge numbers and their diversity of form, can be attributed to their ability to metamorphose. This enabled quite radical changes in form to take place in a minimum of time. Insects already had the ability to moult as an inheritance from their crustacean-like ancestors, but the spearhead species developed the ability to enclose themselves in cocoons and emerge as something quite different in the final adult form. This may originally have been a device to allow aquatic larvae quickly to adapt for adult life on the land, as with the dragon fly now; but when in many species egg laying became the norm, metamorphosis was not only employed to create an instant fully winged adult, but also to ensure that such adult, unlike the larva it sprang from, was not only able to withstand desiccation, but within limits actually to relish the sunshine.

Among the arachnids, the scorpions retain crustacean-like characters, resembling in particular the primitive king crabs. They are all predators on other invertebrates. They also remain primitive in that they tend to shun the light. The spiders on the other hand are hunters that concentrate mainly on catching insects. With their remarkable thread-producing capabilities many have even managed to specialise in capturing flying insects by means of their sticky webs. Such spiders have generally evolved a fair tolerance to light, even though they cannot metamorphose. (Dangerous though they are, spiders in their turn are hunted by certain insects, in particular certain solitary wasps that subdue ground spiders with their stings and pack them into larders as a first food supply for their eventually hatching young.) It seems clear that the ancestors of the arachnids were predators that followed the proto-insects out of the water. The transition to air living would seem to have been eased by the arachnid habit of carrying the eggs on the outside of the abdomen, which with the spiders has been further elaborated in the provision of an egg sack.

Among the insects transition from the water must have been fulfilled in other ways, for it seems likely that they originally tended to abandon their eggs once laid. Perhaps the pertinent insect trait here is the common habit of laying eggs where the right sort of food will immediately be available for the hatchlings. This occurs now, even when the particular food is quite unsuitable for the perfect adult.

The last group to leave the water was from among the vertebrates. The fish had once emerged as a type from among invertebrate chordate beginnings. Although they were presumably derived from an extinct fine-and-whisk adult of sea squirt type, they themselves were promoted bilateral larval forms that, unlike the lancelet, adopted coarse-and-grasp feeding methods, which had converted them to food seekers and eventually to good swimmers. They had a great future, unlike the chordates, whose life-style was to leave them aloof from significant evolutionary advances after the Pre-Cambrian.

The notochord of the ancestry of fishes developed first towards a cartilaginous skeleton and then into a bone one. This produced an animal of some bulk, but of sufficient internal structural strength to be able to support itself out of the water if need be. Certain fish had pre-adaptive features, such as lungs and lobe fins, which led to air breathing and walking becoming possible. From among these emerged amphibious adult forms that were well adapted to a terrestrial existence, but still must lay their eggs in water and went through an aquatic larval phase complete with gills. However, the land surfaces of the World were becoming ever drier and the finding of suitable water for mating and egg-laying will have become increasingly difficult.

Fish evolved from the chordate condition through a variety of primitive forms that included jawless and finless types, with heavily armoured head parts and were only suitable for a life grubbing about on the bottom. The muscular body represented a great deal of former segmentation and eventually became able to produce the protrusive organs that became the various fins. The segmentation was also responsible for the numerous gill slits that in the bony fish were eventually coalesced into the more compact and efficient organs they now possess. But it would seem to be from among the front ones of the bones supporting the gills that a new use was found, in that they became the basis for the moveable jaw.

The leaving of the water by some bony fish to become land animals may have had cause in seeking land plants or the pursuit of invertebrate prey, but, no matter how well adapted thus, they were obliged to return to the water for breeding, for their eggs must be laid therein. However, the eggs may to some extent have been pre-adapted for survival on land and it only required females to evolve and inherit the necessary behaviour in this respect (presumably accidentally acquired) for land living to occupy the whole cycle.

Two stages of land colonisation by amphibians can be illustrated by the different behaviour of existing salamanders and newts with regard to egg laying. While the latter go through a seasonal aquatic stage, wherein they mate and lay eggs, the adults of the former are commonly completely terrestrial and only approach and enter the water once a year and specifically in order to deposit their eggs in it.

The laying of eggs on land may have had an element of compunction about it, in that, with increasing aridity, females may have been constrained to lay their eggs in puddles, then in mud and then in damp or even dry earth, depending upon what was available. Eggs evolving in the right direction might survive and give rise to hatchlings that were going to produce further survivors. With the behaviour eventually becoming habitual, it was sustained even when suitable areas of water again became permanent features of the landscape. In other words, while the eggs had incidentally become suitable for laying on land, the female would frequently lay them there because there was no choice. However, females may have contributed to the development through some sort of observation based behaviour. This being by way of a dim realisation that laying in water was no longer imperative and sometimes impossible, and eventually that it was not necessary.

Those amphibians that changed in this way were ancestral to the reptiles, dinosaurs, birds and mammals. The common reptilian instinct for burying eggs would be associated with an original need to immerse them in mud, for exposure to the hot sunshine would be fatal then as now, The growing aridity was the cause of the eventual elimination of most of the amphibians and all of the larger ones, but created the reptiles and other terrestrial vertebrates, even though many of these took to the water again - the sea in particular – in later times. Their former land-locked existence is illustrated by their need to come back ashore for reproductive purposes. Only certain viviparous land vertebrates were able to adopt a completely aquatic existence again, e.g. the whales and dolphins.

■ Land Vertebrates

The large heavy amphibians were replaced by the land-adapted reptiles and dinosaurs. Only the relatively small frogs, toads, newts, salamanders and caecilians survived in numbers and became widespread round the planet. The reptiles proper were hampered from progress in some directions by their failure to develop an internal system for controlling body temperature. This does however appear to have been achieved among the group known as dinosaurs and certainly by the ancestors of the birds and mammals. These ancestral forms must have been derived from a progressive group among the dinosaurs in the broadest sense, with the birds being produced directly out of the latter, while the mammals appeared by a less direct route. However, some of the distinctions between the various classes may already have been predestined at the amphibian stage. The birds would seem to be descended from a true form of dinosaur, while the mammals appear to have diverged from the 'dinosaur' line at a more primitive stage. The dinosaurs proper pioneered the bipedal stance and gait, but when it came to the crunch seem to have been too adapted to a certain kind of climate and habitat and it was the derived mammals and birds that proved pre-adapted for survival when the going got harder for a time.

The conquest of land areas by certain species was dependent on eggs being progressive and 'legs' being provided. Dinosaurs and birds went through a stage when the rear legs became props that supported the body from directly below and allowed standing as well as walking. The bipedal stance was facilitated by the upper body being balanced by long fleshy tails. Many of the giant dinosaurs reverted to a quadrupedal stance, but with the front legs still betraying the 'outboard' aspect of reptiles and amphibians in general. The birds never reverted thus and developed the front legs differently. They took to the trees, where feathers derived from earlier scales and forming an insulating layer were converted in conjunction with the 'outboard' forelegs into wings. The feathers first augmented the gliding capabilities of membranes attached to forelimbs and body, eventually to replace these completely. The mammals originally did not adopt a bipedal stance and gait, but evolved legs whereby all four supported the body from below. Only much later did mammals of disparate kinds go bipedal. The early mammals proper appear to have taken readily to the trees and some eventually developed membranous wings of much greater extent and efficiency than those found on primitive birds and became bats.

It would seem that several classes of terrestrial vertebrates were destined to produce fliers. The extinct pterosaurs may well have been warm blooded, but could not have been the direct ancestors of the birds, even the smallest of them, if only because of their weak legs not being at all bird-like. These do indeed put them nearer to the bats, but their wings and other features were of much more primitive form than those of the flying mammals.

Mammals

After the demise of the dinosaurs the mammals started to produce larger species, on land and in water. Terrestrial herbivores could be huge, although hardly challenging the biggest dinosaurs with regard to length and bulk. They did however provide a source of food for the larger mammalian predators, especially the so-called sabre-toothed tigers.

Other mammals continued to exploit the arboreal life-style without resorting to flight, although several orders produced gliding types, including some among fairly primitive types, i.e. the marsupial opossums and, among the primates, the lemurs.

◼ Primates

Some mammals' feet developed grasping properties used in climbing and in manipulating food. A rudimentary version of this is observable among the rodents, but it is particularly well developed among the primates, where all four feet can indeed be 'hands', complete with nails rather than claws. The early primates were probably insectivorous, but came to supplement this with plant food. Divergent evolution into a variety of more specialised feeding habits has taken place in time, although there has been a reluctance for any primate species to become out and out devourers of other vertebrates. The most forward in this respect emerged in the genus *homo*.

However, the grasping hands were also useful for collecting and catching food, as well as for transferring it to the mouth. With the higher primates the skill developed in handling food could be applied to other objects, at first to the gripping of sticks incidentally broken off. These could be used for building nests, or as tools or weapons.

It was among such tool-using primates that fully bipedal apes arose who were destined to become self-conscious: we have come to know such as *homo* or 'man'.

◼ Chordates and Vertebrates: Thereby Hangs a Tail

A feature found on vertebrates is a tail; this includes some primates. In the section discussing 'tails' above it was remarked that most vertebrates have visible tails. Even in the cases where these are apparently lacking, they still retain internal vestiges of such, as with frogs and apes. This statement even includes *homo*, where the vestige at the end of the spine is known as the coccyx. As already mentioned, the normal vertebrate tail is an elongated organ that protrudes to the rear of the anus and, apart from the chordates and those fish with cartilaginous skeletons, the stiffening element indicates the segmentation that characterizes the whole of the bony spine. Possession of a tail is a condition that applies to the most primitive vertebrates, i.e. the fish (both existing and extinct), as well as to even more primitive chordates, such as the amphioxus and whose ancestry lies in the Pre-Cambrian.

This overhanging tail seems to have been created as a swimming organ, the prime purpose of which was to drive the chordate concerned through the water. But the question arises as to how it originally came into being. As a useful feature, did it gradually appear by chance or was it created by conversion of an existing organ of different purpose? This is an important question, since the answer can lead to a better understanding of the relationship of the chordates (and hence the derived vertebrates) with other phyla. In order to progress it is necessary to claim that the tail has been developed from an organ that originally had a different purpose. It now remains to determine what that organ was and how the conversion occurred.

In order to approach this problem it is helpful to go right back to the way that the zygote develops into an egg at the very beginning of the life of a multicellular animal. Right from the start of the formation of the egg of such animals there is a noticeable rift. On the one hand are the protostomes, which consist mainly of the molluscs and arthropods, but comprise many others. On the other hand are the deuterostomes, which consist mainly of the chordates, vertebrates and echinoderms, as well as a few others.

The schism starts at the blastula stage, which is the name given to the initial bunch of cells that results from their remaining together after repeated replication. The first feature to show on the surface is an opening to the interior, known as a blastopore. The way that this develops indicates the essential difference between the protostomes and the deuterostomes.

Among the differences between these two groups are the mode of cleavage and the way in which the mouth and anus appear. 'Cleavage' is the resultant distribution of the cells of the blastula after replication.

With the protostomes they follow the process known as spiral cleavage, while with the deuterostomes the process followed is known as radial cleavage. This is of no import with regard to the argument being pursued here, except that it is coupled with the order in which openings appear that are destined to become the mouth and anus among multicellular animals. With the protostomes the first opening (blastopore) eventually becomes the mouth; with the deuterostomes the reverse occurs with the first opening eventually becoming the anus.

The appearance of the blastopore in the blastula is connected with the creation of a cell-lined sac, which is the beginning of the gut. At this stage the embryo is a cup-shaped structure having three layers of cells and known as a gastrula. It is at this juncture that the differences between the protostomes and the deuterostomes start to appear. The inner sac is extended to become a through-gut known as an archenteron, with one opening being the mouth and the other the anus. The schism between the protostomes and the deuterostomes then becomes observable. With the protostomes the original blastopore is now the mouth, but with the deuterostomes it has become the anus.

Before the through-gut is established, twin organs known as coelomic sacs are produced and are situated in the fluid filled space between the outer cells and the archenteron, having been created out of the cells of the latter. In each case they are created in the zone that will eventually have an opening that serves as the mouth. With the deuterostomes this is not the original blastopore, but a new opening that is destined to serve this function.

The origins of the through-gut have been considered on previous pages. In short, it is derived from budding; a bud that did not separate from its parent could provide a through-gut, with each component then contributing an opening that sequentially could take in food or dispose of wastes. Eventually this arrangement evolved into a single animal, with the effect finally being made permanent by it pervading the whole life cycle and especially the reproductive system. In order for this to occur it was necessary for the two openings to be coordinated into a better system, a one that performed better than the earlier one. This was achieved once one came to act solely as an inlet for food (mouth) and the other solely as an outlet for waste (anus).

Thus was created a multicellular animal with two openings. The result was a type of creature with two of everything. This condition was made beneficial by the twinned organs being used to instigate bilateralism. Yet in the case of organs associated with the gut one set of them tended to be superfluous and in the course of time was eliminated. Sometimes full bilateralism has been abandoned, as can be seen with the gastropod molluscs, these being constituents of the protostomes. In the case of the deuterostomes a whole phylum of through-gut animals has evolved in which bilateralism has been virtually abandoned, namely the echinoderms (starfish, etc.).

In considering this specialization of the two relevant body openings, it might seem reasonable to assume that it was the bud that eventually provided the anus; but this is difficult to demonstrate. For the sake of argument, it might be claimed that the protostomes represent the condition in which the original two-way blastopore became the one-way mouth, while the one-way anus was provided by the bud. However, this did not work out in the case of the deuterostomes; somehow the mouth function was taken over by the bud. It all seems to boil down to the early evolution of the coeloms. Sometimes they happened to appear first in the bud and this was crucial to the creation of the deuterostomes. This development may imply some relationship with the evolution of the overhanging tail, as found with the chordates and the vertebrates.

After consideration of these matters, a cluster of three phyla can be observed in which links between them can be observed. It has already been observed that the larva of the starfish (echinoderms) resembles that of the acorn worms (chordates). For various reasons the latter are borderline members of the chordate phylum and are mainly included because of the possession of a rudimentary notochord. Yet after the larval stage, the echinoderms do not go on to share development features with the chordates/vertebrates; but they do come to share a feature with one class of the molluscs. The gastropods all partially lose bilateralism, while all the echinoderms do this completely. Loss of bilateralism does not occur among the chordate/vertebrates, but links with the molluscs have been noted earlier, namely with the cephalopods, which likewise retain bilateralism.

There remains the need to indicate a more substantial connection between the chordate/vertebrates and the echinoderms. This has much to do with the evolutionary story of the tail. If the chordate/vertebrates are to be associated with the echinoderms, to which class of these are they most closely related? This would have to be one that shows a relevant affinity and there is indeed only one candidate. Rather surprisingly the only class that satisfies this stipulation is that of the sea lilies, for some of these have larvae with tail-like features extending from their bodies. On reaching adulthood such a 'tail' is used to attach the creature to some firm surface at the bottom of the sea. Thus might it be asserted that the stalk of a sea lily and the tale of a vertebrate are homologous. In that case the vertebrates are not only related to the echinoderms, but to the most primitive class of these. The main split occurred because the remote ancestors of the chordates/vertebrates resisted any tendency to follow their contemporaries and relatives among the ancestral crinoids in their move away from the bilateral condition. Such a common ancestor is of course elusive, since it existed and died out at some time in the Precambrian Era.

PART 2

The World of Man

CHAPTER 16

The Stages of Awareness

◼ In the Beginning

The premise is that life has evolved from a state that was merely one of mineral lifelessness, even at its higher levels and as known to man the thinker. For the human mind - that most enquiring of entities - confrontation with the most difficult question concerning the understanding of life - namely its origin - is then in prospect. Yet a bio-chemical explanation of the way whereby the gap was bridged between lifeless and living matter has already been endeavoured on previous pages. From this state various stages have been passed through, that somehow led to the condition observable today, whereby it is possible for beings, consisting of mineral matter, to harness energy in order to reach amazing heights of achievement, but in parallel with persistent disappointing lapses that sometimes have led to disastrous failure. Even so, failure itself is often the point from which progress has sprung.

◼ The First Stage – Unawareness

Perhaps no one would argue with the statement that lifeless matter – or even dead matter – suffers from lack of awareness both of itself and its surroundings; yet, in spite of this, it does have aspects in common with living matter. They are both energised to some extent and both are subject to change due to influences in their environments.

The physical and chemical activities that are possible among solid, liquid and gaseous compounds are of the same sort as those taking part in life's processes; the essential difference is one of complexity, coupled with exclusivity. Hence, in addition, there is a control and special co-ordination among life's activities, whereas elsewhere the presence of such apparently purposeful conditions cannot be perceived by our biological senses. Yet even the fact that lifeless matter is ready to react to physical and chemical stimuli at all must be taken as 'awareness' at its rudimentary level and it is from this point that any perceptible degree of awareness must be traced as the indicator that life is present.

Visible dead matter does differ from lifeless matter in that it is more unstable, is constituted of carbon compounds and peculiarly attracts other forms of life. Under normal circumstances the more complex forms still team with life, but not of their own constitution, but including microbes, these being largely bacteria of which some are potentially destructive.

■ The Second Stage – Microbes

As far as can be judged lifeless matter is quite passive towards changes which affect it. Living matter is never completely so. Lifeless matter does not seem to care what happens to it, yet the fundamental activities of all living matter seem to be designed to protect itself and its kind and to ensure survival in so far as it is able to achieve this.

Life is a co-ordination of energy and matter that can react positively towards its environment to serve its own ends. These normally involve no more than the preservation of the well-being of whatever particular form it takes, no matter what the cost to everything else. This may seem a selfish aim, but there can hardly be any moral values attached to the activities of mere matter, even if it has assumed some special complexity and happens to be energised in a peculiar way. The apparent imperative to preserve the state of its own matter is the very gist of life. The truth is that early forms of life either had to be specially geared to survive or be eradicated. The earliest forms will have been strangers in a hostile changeable world, precarious innovations consisting of promoted molecules that were always under the threat of being relegated to the ranks. Presumably the vast majority of these upstarts were constantly being thus demoted, but a few somewhere must always just have had the right formula to scrape by and survive, and sometimes even prosper.

The newest forms, being the spearhead and the most complex molecules in existence on Earth, must always have been vulnerable to the slightest change in the environment. This could result in anything, ranging from either their complete destruction or to reversion to a simpler, more primitive form. But their very lack of stability was also their salvation as a type or group, for this must have made them able to nibble at and exploit the enormous number of combinations of their available atoms that were possible, but which did not yet exist. Basic matter is able to exist in more forms than those in which it is found, although on Earth it only occurs in combinations that are possible under the prevailing conditions. This does not imply that at any one time the possibilities are exhausted. Extraordinary conditions may favour the creation of quite new materials from out of the old ones.

One of the stumbling blocks against understanding life is its complexity. Yet once circumstances had produced those first steps towards instability and diversity, further complexity followed automatically. The number of failures was probably colossal, yet there were always some hardy survivors to continue the trend. Wherever random changes in structure were suitable for quite fortuitous changes in conditions some survival would occur.

Even today survival depends on the continuation of conditions so stable that all environmental changes are so mild that they allow life forms the possibility to adjust. Changes that are too rapid and too violent are always fatal to life. Where combinations proved to be successful for survival these tended to proliferate and become entities that we might recognise as being microbes in the form of the most primitive bacteria.

The Second Stage may be summarised thus:

Unstable and minute living particles came into existence that were able to provide slight but constant and multilateral diversification to ensure the survival of some, no matter what external change may occur, as long as it was not too rapid, violent and widespread. They eventually took the form of a variety of microscopic, repetitive and independent living entities.

■ The Third Stage – Tissues

Just as any number of lifeless molecules may tend to join together with their fellows to form homogeneous compounds, so do the minute repetitive units of living matter group together into tissues. However, with the latter, through them being exceptionally unstable, the differing conditions outside and inside the group would again lead to variety of form and to specific varieties always occupying their allotted positions. The outside ones were the most vulnerable and also the most 'aware' of energy gradient. The kind of process envisaged is comparable to the oxidation of metals, but far more complex and reversible when required and possible.

While tissue formation proceeded, independent microbes still carried on following their own evolutionary courses, and embraced forms ranging from non-motile types of bacteria through to complex types of protozoa.

The Third Stage may be summarised thus:

The state where numerous repetitive units remained contiguous proved to be an advantage. It also led to certain cells being exposed to different conditions from others; this initiated specialisation.

■ The Fourth Stage - Organs

In the simplest of tissues energy is diffused from cell to cell and not transported along special ducts, as is the case with more advanced organisms. The former is a rather cumbersome business, rather like pushing the cart through the houses instead of along the street. If such a body finds itself with too much energy outside, the cells of the exterior must absorb some of it or be changed themselves, perhaps permanently or disastrously. To keep on absorbing in this way they must get rid of the excess and the only way they can do this is to pass it on to neighbouring cells. If they do not succeed in this they will become over energised and destroyed. To cope, chains are set up through the tissue, with layers passing on energy inwards. Yet each layer would be able itself to absorb some of the excess energy, so that the net result would be that the outer layers protected the inner ones from harm. If the case arose that the external energy was reduced, the action would be reversed, but with the outer layer giving up more energy than inner ones. Where the energy change was localised, the energy flow pattern could extend hemispherically into the tissue, to be absorbed by a greater number of cells and hence more easily.

Should the claims on foregoing pages that evolution is a basic trait in life be considered acceptable, then it seems reasonable to expect in such circumstances that the outer cells would come to specialise in protection, while the inner ones, more cosily situated, would be free to develop along various lines. But any claim that each cell already had a fixed position and function in an organism is as yet presumptuous.

The fact that some inner cell has received an impulse from its neighbour will have made it aware that change has taken place somewhere, although without a brain it could not of course interpret the message in those terms. What was needed to approach this condition was even greater co-operation between the cells. Prior to this, with each cell acting as an individual (or indeed as an immediate neighbour), feeding itself and reacting to stimuli to suit its own ends exclusively, the organism as a whole would remain completely uncoordinated. But specialisation and interdependence were the answers to all problems. Healthy cells will always have energy to spare and evolution will in the long run give them additional and specialised functions to dissipate this. The outer cells, as discussed above, became protective as well as receptive, while some other cells tended to specialise in communication, transmission and producers of agents.

Energy, like matter, exists in various forms. One form is electricity and this has the ability to traverse suitable matter and at great speed. Some cells became good handlers of electricity and, either singly or collectively, could receive and release charge, depending on the manner and degree of stimulation. In this way any stimulus felt by one cell could instantly become common knowledge of those needing to know.

Information about advantageous and disadvantageous conditions is useless unless something can be done about it. This made the ability to move essential. Single celled animals have means of moving and one can imagine that with uncoordinated multicellular ones the individual cells might react independently to such effect that they would move the whole entity in a way which would be a consensus of their reactions; but this may not have been to the best advantage of the whole organism. If the communication cell system could send messages both ways, those close to the movement cells may have been able to initiate instructions to override the self-centred reactions of the latter. But there would certainly need to be a sorting house somewhere, for purposes of co-ordination and to make the messages intelligible to the recipients. This development would move the animals well away from the plants. Co-ordination was replacing consensus.

The Fourth Stage may be summarised thus:

Thus far the presence of receptor, communication, transmission, action or movement and production cells have been mentioned, but co-ordination coupled with rapid movement is still missing. This could not occur until it was possible for information to be stored as well as passed.

■ The Fifth Stage – Nerves

In certain organic combinations the communication cells grew to become a continuous network. Electrical impulses could flash from one end of an organism to another, with the disposition and frequency of the impulses imparting details of what was going on, not only to the thus integrated animal, but to every relevant cell in it. These specific messages would bring out the necessary action from the particular cells for which they were intended. Each cell could recognise its own orders and was bound to try to obey, because impulses and reactions to them were governed by the same laws of cause and effect as were valid in the world of purely lifeless matter. This was evolution in action; those cell-types which failed to provide the necessary response put themselves at greater risk of destruction and left the field of the future to survivors, who by inference were more compatible to requirements. Items discarded in this way were lost forever.

Thus, at any instant, a living body could be aware as a unit of the suitability of its position and state, but the picture thus presented to it would be an ever changing one, for it was like a cine-shot of that elusive moving target known as the 'present', with no hope of stopping the projector to examine more closely individual frames; nor could the film ever be rerun for a second viewing. Animals at this stage could be said to benefit from the past, but it was by means of evolution, not memory. One facility to develop was the ability of some cells to change shape. Once co-ordination within organs occurred, deliberate movement of the whole animal became a possibility.

The Fifth Stage can be summarised thus:

5a. Instant common awareness by a living entity that those cells concerned have been activated in the organism by relevant cells constituting a nervous system.

5b. Certain nerve cells tended to band together and form bunches (ganglia) to give controlled action in the area thus served and this led gradually towards co-ordination of the whole animal.

5c. The achievement of motility by multicellular animals.

■ The Sixth Stage – Instinct and Intelligence

In all the preceding stages the various activities attributed to animals have been explicable on a simple "cause and effect" basis. A change applied to such an animal would produce a predictable reaction from it, in the same way as a magnet is bound to attract a nail; there is no choice. At the stage now reached, the whole animal – including the nervous system – is governed by this predictability and this form of organisation is the basis of the kind of response that we call instinct.

However, from this point onwards the nervous system, or rather one section of it, became less predictable and progressed towards an ever greater degree of assessment of situations before taking steps to deal with them and, in parallel, showed a corresponding reduction in blind reaction. While the blind or automatic reaction called instinct arose as the direct result of evolutionary processes, its universality was about to be challenged.

Intelligence – the ability to assess and make decisions – arose differently, or at least only did so indirectly. With instinct, both the ability to react and the reactions themselves are due to evolution, while with intelligence this only applies to the ability. Intelligent beings always retain a degree of instinct, but they will normally decide to adjust or override most of their instinctive reactions after assessment. This implies that differences in opinion within the nervous system were continually in need of being sorted out. A major exception to the application of such intelligence occurs when external circumstances make the time available too short.

This gradual changeover from automatic reaction to deliberation would seem to be the crux of this business of awareness. The development of the nervous system in animals can be followed from being simple nerve cells, through ganglia to the brain, which is the seat of intelligence; but even here all the brain is not involved, for parts of it are always concerned with those automatic internal activities that ensure the continuity of life in a body. Yet intelligence is seated in the brain and is due not so much to mere awareness, but to consciousness, which includes anticipation and memory as well.

The development that made consciousness possible was the perfection of the external senses, such as sight and hearing, and the ability to store the information obtained from these within the brain, i.e. memory. While the development of the senses can readily be explained as an evolutionary process whereby specialisation of sensitive cells became more intense and efficient with the passage of time, memory presents a more complex and difficult phenomenon.

Intelligence is the result of co-ordination between three concepts, namely awareness, memory and anticipation, which give to an animal knowledge of the present, past and future respectively. It depends ultimately upon the ability of the memory to store information and release it again when necessary. Memory is located at a level below both the aware and actively conscious ones, in other words in the vast majority of the mind. The conscious mind can at any one time only deal with a minute portion of the total knowledge and ability of the whole mind. The mind can indeed be likened to a lake in which scraps of information are slowly sinking lower and lower, while the conscious mind is a man in a boat who keeps fishing them up again. However, once they sink too deep they become out of his reach (although still there).

The region where information is still accessible is the subconscious mind and can be equated with one's active memories, whereas the lower depths, beyond one's reach, constitute the region of lost memories – the unconscious mind. The loss of contact with memories does not mean that these have been destroyed, for disturbances might always bring them to light again. Even without this they still affect the personality in ways beyond the realms of conscious knowledge.

The conscious mind is always being bombarded with messages from both inside and outside the body and continually ransacks the memory for precedents to guide it in its decisions. The making of decisions is the very way in which intelligence works, being an acceptance or rejection of solutions thrown up by the storehouse of the memory in answer to problems set it.

It is the intention later on to explore further the method by which the memory works, but for the time being let the method be considered by which information is stored away. Nerves transmit items of information, but once such an impulse has passed through it would appear to be lost. As mentioned above, the memory is situated in that complex gathering of nerves known as the brain, and there seem to have been two broad methods by which it could work. Firstly it might be imagined that brain cells themselves have the ability to take messages in and hold them in the form of energy, which can then only be released by a special mechanism operated by the conscious mind. This may sound plausible enough, but the problem remains as to how cells originally intended to pass messages were adapted so that they could hang on to them. This solution seems to lack the required dynamism.

The second solution can be indicated by suggesting that messages that enter the brain and pass through cells transform these marginally for an extended term, even though the basis can be variable. To obtain information the conscious mind just has to reach down to match current circumstances against such old 'scars'. Considered in this way the memory works by means of precedent, insofar as any match tends to reactivate the whole neural pattern of activity whereby the memorised material was laid down in the first place, but making at the same time adjustments to compare and accommodate any disparities with the new circumstances.

The Sixth Stage can be summarised thus:

6a.   The ability of nervous systems to store messages –memory.

6b.   Awareness of past events and their lessons – experience.

6c.   Responses to situations arising based on experience, but with minimum or no consideration – instinct.

6d.   The ability to foresee the desirability to take action that will be beneficial, or even crucial, before the necessity actually arises – anticipation.

6e.   Co-ordination of information from the past (experience), of the future (anticipation) and of the present (awareness) – i.e. consciousness.

6f.   The ability to react to decisions made as the result of consciousness – intelligence.

6g.   Primitive emotions.

■ The Seventh Stage – Reason

The previous section explored the breakaway achieved away from those reactions limited to stimulation of the nervous system by direct and contemporary physical and chemical influences. Many such reactions are designed to prevent tissue from being damaged and are automatic and immediate in effect. For example, let light be taken as such an external influence. If a slumbering man opens his eyes to be met by a blinding light he immediately shuts them again, not so much because he knows he must, but because he simply cannot keep them open. This is an instinctive and automatic protective reaction. However, if he had opened his eyes and found himself confronted by a lion his eyes would certainly not have shut; his instinctive reaction would have been one of escape from the danger, but the means of any attempt to overcome the danger would be left to a decision of his intelligence. Should he lie still and hope for the best, stand his ground, or run for it? Why not climb an accessible tree? Presumably most unarmed men would decide to climb any suitable tree under the circumstances, if time and agility allowed.

However, intelligence without relevant experience can prove less adequate for a given occasion than is instinct. If the animal had been a leopard, an animal about which he was rather ignorant, it would have been a mistake to climb a tree. This over-simplistic example is meant to illustrate how reasoning works best when the facts of experience are known and observed, while blind assumptions, which a reasoning mind tends to make in areas of ignorance or in moments of panic, can lead to worse reactions than those of pure instinct. In emergencies such wrong reasoning is dangerous. Reason does not like to be rushed.

Indeed, the only circumstances under which reasoning not based entirely on known facts should be justifiable is at the frontier of knowledge. It is just there, if any progress is to be made, that the reasoning mind needs to step forward some way into the no man's land beyond and then look around to see, as far as possible, whether the terrain is as imagined. This is the frontier where imaginative speculation on the nature of the truth lying beyond our ken sometimes pays off, for without inspirational ideas of this kind the advance of knowledge would have been very slow indeed. Yet in a way this can be thought of as reason behaving unintelligently.

Reason, as against intelligence, seems to be able to work in a biological vacuum. The term intelligence, as used above in the sixth stage and here, covers those activities of the brain that are stimulated by outside influences and also physical and chemical changes within the body, such as hunger, pain, and so on. But reason is the activity that can stimulate itself out of past experiences and awareness of a future.

The basis upon which reason is founded is cause and effect, yet not on the mere existence of this, but on an understanding of the concept involved and with constant reference to change and time. An intelligent animal will recognise a cause when it arises and feel the effect, but then superimposed reason might step in with the realisation that, under like circumstances, a certain effect will always follow a specific cause. In other words, a recognisable cause will have an equally recognisable effect. Reason depends on this as an undeniable principle. Yet more advanced reasoning minds have often found that in practice the principle does not seem to work; this is always due to wrongful recognition of the causes, owing to flaws and weaknesses in the powers of observation and memory. Diagnosis based on mistaken understanding of cause naturally leads to erroneous prognosis due to false expectation of effect.

Except for rudimentary reasoning to do with learning resulting from sight, reason as we generally understand it is only in action among humankind. The cause of this is quite simple: reason proper, like life itself, creates its own causes and effects, but it is a faculty that cannot exist in an individual animal. Once behaviour has moved on from the state of learning by example, it is based on the accumulation of ideas, not reactions; and the ability to have ideas could not evolve in the purely physical world. Ideas are collected by the senses, assessed by the intelligence and stored by the memory. Yet a common fund of facts and the ideas developed from them can only grow where the individual members of a group have an adequate means of communicating them to each other. Only the human race has a system that is complex enough to fulfil this, and that is language. It is true that other animals pass sound signals to each other, but these, like their visual signals, are mainly indications of mood and generally useless for the transference of ideas, except at the most primitive level. Only human language is sufficiently complex, flexible and comprehensive to fulfil this purpose adequately.

It would seem that language could not start without reason, but neither could reason start without language. The beginnings of reasoning power must have been like early attempts at human flight; there was great difficulty in getting the cumbersome contraptions off the ground, but it was proved to be possible in the end. The most important step for reason was the recognition of the passage of time and the evolution of sounds to express this idea.

Man's forebears must have hence reached an evolutionary stage where they became aware that life was a path they were all travelling along together and that behind them was the past, with its individual and collective world of memories, while ahead lay the unexplored wastes of the future. Just as yesterday had been today, so will tomorrow become today. Days alternated with nights and combined to make months, while the latter and seasons combined to make years – repetitive events with predictable effects driven by hidden causes that needed to be investigated.

Observation creates experience, which then can lead to reasoning. Watch a deer eating leaves and a wolf eating a deer. These observations can be repeated and become experience. If examples of A are seen to eat those of B frequently enough a rule is established that will always be followed. Another rule established is that B (deer) never eats A (wolf). Reason has established that a pattern of activity that extends far into the past can be expected to continue into the future.

Primitive man became conscious of time because of his observations of change. Indeed, these two concepts are really two ways of looking at the same thing. The Earth is subject to cyclic changes that give us days, months and years, while time is an artificial construct, being merely man's method of understanding and assessing such regular and predictable changes as measured against each other and also against his own individual rate of change, as well as – most important of all – against his own lifetime. In short, if all change stopped, so would time!

Until recently man could not leave Earth and his concept of time and the way it was measured was governed by this. Away from Earth the way that time behaves is not absolutely immutable.

However time is an abstract idea, perhaps the first of such to occur to man. It is abstract because it does not exist except in minds equipped to appreciate the idea. Change exists outside the human mind, but time does not. If change could stop, then so also would time. Even today "change" is thought of as something that will happen "in time", whereas change is really the inevitable and irresistible force behind all life. Time is just an idea that follows in its wake to bring about a degree of order to the vast and complex changes apparent on Earth and elsewhere in the Solar System.

Once one idea had become established in the human mind the production of others was easier and inevitable. It was likewise inevitable that, with his limited powers of observation, primitive man was destined to stray frequently right off the narrow path of truth in the course of his observations, meditations and derived convictions. Even modern man is still struggling to shake off the effects of thousands of years of erroneous reasoning and every now and again new insight helps in the struggle to get some way back towards that continuous but narrow way – the truth.

But let this not be a cause for dismay, for the alternative to erroneous reasoning by forebears was certainly no reasoning at all. Without telescopes and microscopes they had not the slightest chance of understanding the Universe. Even now with these and other advanced observational aids, man should be chastened by his oblivion to the lessons of the past. Modern man should not be too cocksure of contemporary conclusions, nor even in the strongly and widely held convictions based on reasoning in the past, no matter how capable were the brains involved.

Early man soon started to err in his reasoning when, as a result of his contemplation of cause and effect, he developed the concept now described as the association of ideas. This is valuable for reasoning, as it is the means by which reason stimulates itself – one idea touches off the memory of others as part of a chain reaction.

A by-product of this, that eventually loomed very large, was the idea that if things looked the same they were the same. Another one was that matter once forming a whole remained an intrinsic part of the same whole even after being severed from it. These are in fact the basic principles behind the anthropological terms used for the two arms of sympathetic magic, i.e. imitative and contagious. They seem to have their origins in the fear of change and dread of the unknown. Sympathetic magic allowed men to think that an actual change had not really occurred and deluded them into being convinced that they knew about things, when all that was really available to their minds were their own conjectures. ('Magic' is the worst aspect of this, it being a blind alley.) Modern man still retains many attitudes that are vestiges of these traits. For example, the use of euphemisms is normally to disguise the effects of change, or to hide or lessen the impact of an unpleasant prospect, as when saying balding for bald, elderly for old, progress for exploitation, quaint for ramshackle, and so on.

With the growth of abstract ideas emotional pressures developed in step with them and self-deception was often the only way for man to face up to the problems that his own advances in thinking power were creating for him. All along the line man has been faced by a choice; either to decide whether deliberately to seek out the naked truth, whatever discomfort or dismay such a quest may bring, or to create self-deceptive illusions to make life, if not exactly pleasant, then at least more bearable. Those men who have in the past set out deliberately after truth have usually suffered, either within their own minds, or at the hands of others who regarded them as rockers of the smug-boat. Among such pathfinders Socrates, Copernicus, Galileo and Darwin come readily to mind.

The growth of the abstract idea and the complexities involved led later to the concept covered by 'opinion'. The element of doubt that is intrinsic to the word illustrates the uncertainty and lack of definition that is inherent to abstract thoughts or those containing abstract elements. The validity of an opinion normally depends upon how deeply it can be investigated and, should penetration be deep enough, one would frequently find lurking under the surface some emotion-laden misconception. Such myth-based opinions make compromise extremely difficult and while they widely persist the prospect of permanent peace will be impossible for humanity. It is regrettable that well-intentioned folk are as equally prone as the evil-minded to having their reason overridden by their opinion based emotions.

Reason has been and still is the ability to come to conclusions about self and environment. However, senses that were designed to appreciate degrees of light, sound, heat, pressure and chemical suitability were initially quite inadequate for probing the secrets of matter, energy and the Universe. In any case, any such honest and valid probing that might have occurred would give little comfort against that which the mind could invent. The interchange of information between individuals, which was the very essence of reason, soon became the means by which emotional 'truths' were spread rather than real ones. This in its turn led to a quagmire of doubt and diversity of opinion that bogged down any search for the extremely obscure path to any real 'truth'. As time went on it became increasingly difficult to make this passage and it would only be attempted by those with special gifts and ready and willing to risk the perilous journey.

There is still a strong tendency to blot unpleasant facts from the mind, while clinging tenaciously to unsupported ideas because they give comfort. It is tempting to say that the truth is the one thing not to have changed; but this can only be considered to be correct if one recalls that the truth itself is actually the change that really has occurred, is occurring and is going to occur. This truthful change has certainly and undoubtedly taken place and most of it has been without the knowledge and help of the human mind. It is hence not easy for us to fill in the details correctly. Human beings all make the mistake of confusing what they hold to be the truth with the real truth of what is, was and shall be. The two concepts may overlap, but will never agree entirely.

Finally, let it be reiterated that reason has not developed from a conscious search for the truth, but as a result of mankind of necessity grappling with the unknown and the misunderstood. Doubt and uncertainty keep dragging reason out of the comfortable stable of intelligence. It is a fine looking horse; but does it run all that well?

The Seventh Stage can be summarised thus:

7a. The growth of language to spread information, rather than just mood.

7b. Awareness of the endlessness of time – a fleeting present set between the limitlessness of past and future.

7c. The ability to understand the concept of the abstract and the creation and development of such ideas.

7d. The ability to doubt and the spread of opinion as important factors of thinking.

7e. The fight between reason and the compulsive demands of emotions.

7f. Sympathetic magic.

■ The Eighth Stage – Imagination

Apart from interpreting the fairly reliable information reaching it by way of the senses, the mind also exercised itself by pushing outposts into the unknown. Trails of thought were blazed into the unexplored outer wilderness of human experience and knowledge, with no reliable facts to mark the way. Although this created an obvious great danger that the mind would no longer necessarily follow the way of Truth, since only the mind itself was there to check this, the process contributed vastly toward the promotion of creative thinking and induced the appearance of the phenomenon called inspiration, which bears a similar relationship to thought as does mutation to evolution.

The Eighth Stage can be summarised thus:

8a. The creative spirit.
8b. The quest for knowledge for its own sake.
8c. The deliberate harnessing and following up of random thoughts, i.e. inspirations.

◼ Comments on the Inter-stages

The stages of awareness enumerated above must only be regarded as highlights in a gradual process and not as clearly defined divisions. Apart from this, even the sequence cannot be taken as absolute and much overlapping must be allowed for. Perhaps the consideration of theoretical inter-stages is more important from the evolutionary point of view.

Inter-stage 1-2 – Unawareness to Microbes

This would be the first breakthrough on the road to life as we know it. Certain units of matter began to develop on lines that enabled them to modify themselves slightly and haphazardly to give them a chance to combat unpredictable hostile changes in the environment. Although this gave the appearance of matter developing a will to survive, it was not so in fact. Units were reaching such complex and unstable forms that they had to evolve a potential self-preserving mechanism in order to avoid the certainty of destruction. The fact that some did so is one of those strange truths that are always confronting the student of nature, but happened because they were possible. Without this the earliest life could neither have begun nor survived

Evolution is coupled to reproduction and they have always progressed together, hand in hand. This was the pioneering pair at the outset. Certain types of lifeless matter found a peculiar way of tapping the reservoir of energy and movement already inherent to the Solar System. This resulted in the continual formation of even more complex forms, of which some would always be in a position to repeat the process and advance further. As stated earlier, the new energy of any structure thus achieved was used in preserving the identity and continuity of the form it occupied, i.e. in living. In order to achieve this the form itself could be modified slightly during reproduction and retain the changes if these proved crucial to survival, and this was evolution.

Inter-stage 2-3 – Microbes to Tissues

The tissue can be regarded either as a number of similar cells that have banded together and are co-operating, or as the descendants of a single microbe that failed repeatedly to move apart after division. In either case overcrowding would seem to be suggested, or rather the existence of sufficient food and energy to make overcrowding possible. Accepting that evolution is a basic and necessary partner of life, as suggested above, then the advantages of co-operation with fellow cells, rather than struggling to outdo them, must have automatically ensured the eventual formation of tissues.

Inter-stage 3-4 – Tissues to Organs

Co-operation was at its best when accompanied by specialization of function and this in turn pointed toward the ultimate formation of cells into organs, i.e. cells existing co-operatively as a variety of homogeneous groups (i.e. tissues). These thus had the potential and acquired the ability to combine into more complex and diverse multicellular groups of living matter.

Inter-stage 4-5 – Organs to Nerves

The specialisation of some cells in communication and control enabled diverse organs to act with unity of purpose, so that - as whole animals - such could react very quickly to stimuli.

Inter-stage 5-6 – Nerves to Intelligence

The nervous system grew in complexity and an additional function was the ability to act as a storehouse for messages, so that a fund of experience was amassed. Because of this developing capability known as 'memory', assessment of situations began to replace blind reactions.

Inter-stage 6-7 – Intelligence to Reason

The ability of individual animals to communicate with each other by means of various signals grew and culminated in human speech. The latter potentially allowed access by individuals to the whole fund of experience of the species from its living members, and the step forward was eventually consolidated by the development of writing, whereby one could learn from the dead as well as the living and from those far away as well as those nearby. The first important result of reason was a conscious and communal appreciation of time.

Inter-stage 7-8 – Reason to Imagination

The reasoning mind was restless. This often made it impatient and dissatisfied with the messages reaching it by way of the senses. Extrasensory probing into the unknown helped to alleviate the frustration caused by a greater amount of energy being available to reason than was necessary simply to deal with 'facts'. Although it can generate comfort, imagination also has its drawbacks. As well as imaginary truths, there can also be imaginary doubts, so that while wishful thoughts can be latched onto and accepted readily as truth, hard facts presented by the senses can be rejected. In other words the imagination can delude the mind both ways. Especially in the case of deteriorating brains, with over-stimulated imaginations the difference between reality and unreality can become blurred, or even disappear.

CHAPTER 17

The Brain and the Mind

■ History of the Central Nervous System (CNS)

The CNS of the Vertebrates is derived from a simple dorsal nerve cord, such as can be seen among the primitive chordates. This can be observed in the mammalian embryo as originating as a trough that forms externally on the back. The edges of this trough later in life meet and fuse so that a 'tube' is created. Subsequently the nerve cell 'bodies' move in towards the centre of the tube to form the 'grey matter', while the nerve 'fibres' remain more peripheral as the 'white matter'. This ontogenetic process has phylogenetic meaning with regard to the CNS. The movements have destroyed the concept of a tube, although the cord has retained its overall oval shape.

Internally the grey matter has assumed in cross section the shape of a malformed 'H', tending indeed towards an 'X'. The horns of the 'H' represent the routes by which nerves of reflex type that are associated with each notional body segment connect with the grey matter of the cord. The two upper horns are the sensory routes and the two lower ones the motor routes, with 'upper' and 'lower' meaning 'dorsal' and 'ventral' in stricter anatomical terminology.

The spinal cord reflects the segmental nature of the vertebrate body, with this being indicated by the numerous grouped junctions of the various sensory and motor nerves. As with many invertebrates, the clarity of this segmental arrangement of man and other vertebrates is destroyed at the head end by development and fusion.

One such development at the head end was an early failure of the dorsal trough to close permanently and the resultant opening was merely sealed by the thin choroid membrane. The void thus created resulted in the two 'ventricles' being formed, one behind the other and joined by a thin canal. The fore-ventricle was associated with the forebrain and, together with the growth of the two cerebral hemispheres, expanded with the latter to produce two further cavities there, the left and right lateral ventricles.

The primitive brain can be thought of as the head end of the spinal cord, with various outgrowths from the top of it. In the human brain some of the structures have grown very complex and have tended to wrap themselves around the cord, but are still basically generated out of its top. One structure evident on the bottom of the cord is the *pons*, the 'bridge' of the cerebellum.

As stated earlier, the top or dorsal side of the cord is concerned with sensory nervous activity and the bottom or ventral side with motor activity. It is thus perhaps inferable that the major structures of the brain are developed from the sensory aspects of the nervous system, while the motor nerves have remained comparatively undeveloped, even in the brain. Motor activity in the brain is the concern of the "pyramidal system" (mainly for voluntary actions) and the "extra-pyramidal system" (mainly for involuntary actions), although the distinction between these two functions is not crystal clear.

With the failure to close the "dorsal trough" at the head or brain end of the cord (in order to form the ventricles), the motor nerve fibres of the white matter have generally been pulled higher up, but are still separated from the dorsal surfaces by the trough, while the sensory nerves are virtually in lateral situations, being pushed further apart by the interposed motor nerves and ventricle. The sensory nerve fibres are thus more specially separated here, with the raised motor nerves largely between them.

While the grey matter in the spinal cord can generally be termed internal and the white matter external, the cerebral hemispheres have developed in such a way that external surfaces are covered in grey matter, the great collection of neurones, while the white matter of the nerve fibres occupies the bulk of the enclosed volume. In stating this, it must be remarked upon that the internal volume thus created is reduced in size owing to the surfaces of the hemispheres being thoroughly divided up and penetrated to various depths by numerous chasms, the larger being called 'fissures' and the smaller ones 'sulci'.

The vertebrate brain has developed from various sensory outgrowths, originally of segmental nature, which have formed dorsally on the 'rostral' end of the primitive nerve cord and which became the "brain stem". These primitive outgrowths were responsible for the different senses and related activities.

The rearmost formed a pair that dealt with movement and muscular activity. This was also the seat of the primitive vestibular system, regulating balance and recording changes relating to this. From this rearmost pair has grown the cerebellum, which, together with the medulla oblongata, constitutes the hindbrain. The medulla itself is basically where the spinal cord itself changes its character as it reaches the point where it becomes the brain.

Anterior to the primitive rearmost outgrowths was a segmental pair concerned with hearing. These have now become the "inferior colliculi" of man, which have remained undeveloped. Anterior to these again is another pair of outgrowths that are also undeveloped. These are the "superior colliculi" that were the primitive centres of sight. Though not developed in the higher animals, they are retained as a reflex centre "related to visual happenings appreciated properly elsewhere in the brain". The superior and inferior colliculi constitute the midbrain.

Anterior to these was a pair of outgrowths called the "olfactory lobes", which primitively dealt with smell. In man these lobes still exist, but are overwhelmed by the enormous growth of the cerebral hemispheres that have evolved out of their rear ends.

A feature of the cerebral hemispheres has been a tendency to affect a dominance over segmental parts rearward (or 'caudal') along the brain stem and spinal cord. In this they have not always been entirely successful and for very good reasons. Thus have the cerebral hemispheres gained control over the motor nerves that operate muscles, but are unable to control them when an acute emergency arises. Likewise the cerebellum usually should refer to the cerebrum for guidance, but takes over like an automatic pilot once cerebral attention is diverted to other matters.

Another feature of the cerebral hemispheres has been a tendency to usurp the functions of the neural organs lying caudal to them. A major usurpation is the taking over of the sense of sight from the superior colliculi.

■ The Structures of the Central Nervous System

The Spinal Cord Apart from the simple reflex arcs of a single segment, nerve fibres run along the cord in the white matter. These can involve whole stretches of adjoining segments in the same reflex.

Such reflexes are nervous activity at an extremely primitive level. More advanced is the system that involves sensations transmitted to 'higher' levels of the nervous system and receives instruction from these. A feature of such nerve fibres is a crossing over to the other side of the cord, before forming a continuation of the relevant sensory or motor path. This is known as decussation and involves passing through synapses in the central grey matter.

One might well feel curious as to why decussation should be necessary. Well, it is clear that nerve fibres have to enter the grey matter somewhere, for that is where their own nuclei are to be found. Of course they could do this and still have all relays coming out on the same side as they entered, but if this were universally the case each of us would be basically two quite separate neural entities, even with minor lateral linkages traversing the segments. Creatures like annelid worms have separate twin nerve cords along their bodies and these are suitable for control of their segmental appendages, with their simple but co-ordinated movements. Even the segmental linkages between these appendages are simple and only needed to ensure co-ordination along each side and across the animal of a repetitive kind.

The trunks of the human Autonomic Nervous System are internally placed in a similar way and hence, like those of more primitive creatures, do not integrate the whole in the way that the Central Nervous System does. Complete integration of bilateral parts is a basic feature of the whole vertebrate CNS and is responsible for the difference between 'acting' instinctively in unison, and inherent co-operation while 'knowing' what one is doing and even why one is doing it.

The vertebrate nerve cord, with its bilateral arrangement, can be understood as former twin cords that have moved together as an evolutionary trend in the deep past, even though there is inconclusive evidence of this in the ontogeny of vertebrate species. Yet one might indeed claim that this merging of the twin cords to make one is the essence of being a chordate or a vertebrate, but the inherent advantages of the arrangement have not been exploited by all species and indeed these advantages have only been gradually dragged out of the great unexplored evolutionary pool in order to fulfil the requirements of natural selection.

The Brain As with all elongated segmental creatures the vertebrate nervous system is complex at the head end. Any former segmental repetition there has been lost in the complexities of the outgrowths, even though the segmental sequence of the various organs is still to be perceived in their order through the hind- mid- and forebrains. The basic repetitive arrangements of the spinal cord start to break up at the *medulla oblongata*, a transition section between cord and brain.

A feature with the evolution of the brain has been a tendency by forward or head end (rostral) nerve centres to take over the functions of rearward or tail (caudal) nerve centres in the brain. The cerebellum would thus seem to be the ancient vertebrate control centre, but the cerebrum, the bulk of the forebrain, has usurped its higher activities. The cerebellum has been left with handling and correlating automatic activities and must refer to the higher authority for changes in orders. However, if the higher authority embodied in the cerebrum is too preoccupied with other matters, it seems clear that the cerebellum still retains a modicum of ability to make modest decisions of its own.

Like the cerebral hemispheres, the cerebellum is a bilaterally divided organ, but the halves are likewise joined together by commisures (transverse connecting fibres, which cross in the organ itself and not through the brain stem). Under the cerebellum the brain stem is swollen by masses of grey matter to form the feature known as the 'pons'. The cerebellum is associated with the rear ventricle in a similar way to the hemispheres being associated with the forward ventricle and its lateral extensions.

An important function of the cerebellum is as the control centre for the vestibular system, the organs of which are in the ears, and which serves to relate the nervous system to acceleration forces such as gravity, spinning, etc. When all is well one is quite unaware of the activities of the vestibular system, but its inability to cope with circumstances can show up as imbalance, often accompanied by sensations such as dizziness or vertigo.

Although the cerebellum has come to be dominated by the cerebrum in higher mammals, this is not to state that it is degenerate. On the contrary, it has continued to evolve in its own right, but such modest growth has been eclipsed by the extraordinary development of the primate hemispheres.

The paired colliculi of the midbrain have remained aloof from the greater growth occurring in the hind- and forebrains. The rear pair have retained their association with hearing. Positionally these must owe their existence to the sharing of organs (the ears) with the vestibular system. The inferior colliculi are immediately in front of the cerebellum.

The two systems share the same nerve tract, which reaches and enters the brain stem near the pons of the cerebellum. It may be that the vibro-receptive auditory nerve was originally one of a group to do with the vestibular system of the ears, the others of which were absorbed into the motor activity that, as it developed, was the business of the cerebellum, while the spin-off non-motor function of hearing was left alone in its own segment. It can be noted that hearing does not need an ongoing automatic motor response, while vestibular balance sensations do. Auditory sensations demand specific responses, namely from the cerebrum, except for such reflex activity as a start(!) when hearing a sudden, loud noise.

The superior colliculi were ancient centres of vision, but now have been usurped by the cerebrum in this function. The main visual tracts now no longer pass through them, but are entirely within the cerebrum. The ancient visual centre of the colliculi now seems merely to serve some reflex function.

The cerebral hemispheres constitute the cerebrum or forebrain and are the seat of consciousness in man and higher mammals. This has wide powers to modify or override certain ongoing activities under the general control of the cerebellum. The motor nerves used by the conscious mind to exercise its authority are basically contained in the tract known as the Pyramidal System (on account of its shape).

The olfactory lobes are twin segmental organs that lie at the very front end of the brain stem and anciently dealt with the sense of smell. It is thus understandable that smell is one of the two senses whose nerve paths do not pass through the medulla oblongata. The other is sight, which is self-contained within the hemispheres and independent of the brain stem. Otherwise the cerebrum can be seen as a sort of ambitious, power-mad organ, seeking dominance over all others and demanding reports from all parts of the body, while issuing orders as it thinks fit.

The visceral activities of the body remain fairly aloof from the cerebrum. The Autonomic Nervous System thus reports in at the twin organs known as the 'hypothalamus' (below the thalamus) at the head end. The system works on a sort of "no news is good news" basis. The hypothalamus co-ordinates visceral activities, but provides no direct means by which the cerebrum can step in and exercise any authority, should there be signs that things are going wrong. This is all to the good, because the cerebrum, as a thinking organ, would be guessing, for it has no direct knowledge of what is going on within the body, only any effects. The body has its own separate agents for putting things right that may or may not be successful. Any reaction to internal problems by the forebrain is externalised - e.g. medical treatment - and done after consideration.

However, the cerebrum is a seat that has complete ultimate control over some activities and can interfere with others in varying degree. It has been formed by a tendency for nerves, with their reflex capabilities and relays, to send representative offshoots forward into the anterior extremity of the Central Nervous System. Here that great integrating body has been created, based originally on the lobes of the olfactory sense. Such 'encephalization' around the sense of smell is a particular development of higher mammals, for in other vertebrates the main co-ordinating centre remains the cerebellum.

So the origin of the cerebral hemispheres lies in the olfactory lobes, twin segmental organs at the very front end of the brain stem and anciently dealing with the sense of smell. In higher mammals they still remain under the cerebrum, but have been completely overwhelmed by its massive growth, especially in man. The fact that the most advanced part of the brain developed out of the nervous system dealing with the sense of smell of mammals requires explanation.

One might postulate that the brain developed from smell due to the nocturnal lifestyle of earlier mammals. Yet at the present time one might note that dogs and bears, essentially diurnal animals, have an excellent sense of smell that far exceeds that of primates. (So does one particular bird, the flightless kiwi.)

Primates have a poor sense of smell. They lost it as their eyesight got better as a result of changing lifestyle. A well-developed organ found something else to do with its surplus capacity: it improved the creature's capacity for intelligence. The evidence suggests that those primates that adopted an active arboreal lifestyle during the daylight hours developed the sense of sight in the forebrain. This change in the way of life induced a reduced need for a highly developed sense of smell. The redundant neural capacity for smell in the forebrain was free for conversion to a capacity for intelligence that was linked to a conception of the world that was achieved by improved daylight vision.

But the integrating brain in its most primitive form is represented by the thalamus – 'antechamber' – lying on each side of the forward ventricle and which now acts as a reception and correlation centre for the rest of the cerebrum. These masses of grey matter on each side of the ventricle would seem to represent the rootstock from which the great mass of each hemisphere grew.

The hemispheres of man are of considerably greater volume than nearly all other animals, but volume is not the ultimate objective of cerebral growth; that is surface area. The human brain achieves a massive area by being divided up by the various large fissures and small sulci that scar the surface. The 'fissures' can be identified individually, for they form very similar patterns in all human brains; but the sulci do not, for their patterns even differ between the two halves of the same brain.

■ The Mind

The cerebrum is the thinking organ that originally was 'non-thinking'. It still retains non-thinking functions. In order to understand what was involved in the change to a 'thinking' brain there is a need to return to neural basics.

Considering more primitive creatures, simple reflexes are contained within the affected segment on its one side. A sensation is received and provokes a motor response. Assuming the performance of a very simple action, such as the retraction of a protuberance on a single segment to protect it, the integration of the activity with other segments is not necessary, although they may have experienced the same independently. However, when it comes to locomotion the action of relevant protuberances needs to be integrated for greatest efficiency.

This need for co-ordination led to the growth and development of the longitudinal nerve cords, tracts of fibres of inter-segmental type that came to dominate the intra-segmental reflex nerves, except in instances of local emergency. The paired cords each had a junction point at such a segment – a ganglion. From these there also ran lateral nerves integrating the two halves of the segment, for the appendages on both sides needed to be co-ordinated for proper locomotion; each side had to be aware of what the other was at.

In earlier chapters it has been claimed that locomotor appendages are really descended from feeding appendages at the mouth end of earlier non-segmental creatures. Feeding is an activity that requires a greater degree of sensitivity than does locomotion among complex motile creatures and when these became food seekers, rather than food trappers, they were faced with a greater irregularity of circumstances.

The necessity arose to develop the ability deliberately to take advantage of such opportunities as arose. This demanded a more complex neural arrangement at the mouth. In addition, feeding encouraged the evolution of auxiliary functions at the head, such as vision, smell and taste. Thus could areas of sensitivity to light and chemicals be decreased or even become superfluous in the segments in general, but evolve to greater things at the head end. This again would provoke growth in the nervous system in this area, with the ganglia there not only growing in size and ability, but also tending to specialise.

The neural growth at the head end, being devoted to vision, taste, smell and detection of both food and danger, plus the seizure of food, necessitated that this leading part had ultimate control over the locomotor functions of all the segments behind it. It developed the ability to give orders that the other segments were generally unable to disobey. Yet the translation of visual sensations into locomotor activity in the segments involves the making of decisions – start, go faster, turn, go slower, stop, etc. This represents 'thinking' at a very rudimentary stage and can be classified as an enhanced stage of awareness, rather than consciousness.

The difference between these two concepts here depends on whether the activity is considered to be of a reflex nature or not. When neural activity is only reflex, awareness alone is present. However, if activity results after the consideration of options, then primitive consciousness exists. But consideration reflects some recording of experience, a degree of indecision and sensitivity to time; this implies a rudimentary memory.

The reflexes controlled by the brain are clearly more complex than the simple sensor-motor arcs on each side of a segment, but can reinforce the different activities relating to the sensations still available to the ordinary body segments, such as touch, pain and temperature. Touch normally does not require a local reflex, and is hence under cerebral control except when reactions to experiences are purely routine. But when extreme, as can occur with temperature, it will invoke local reflexes; yet if the sensation lessens it can revert to cerebral control once normality is felt to have resumed. This occurs in the worst cases when pain is present and this experience in general has this sort of effect. The difference in path between touch on the one hand and pain on the other would seem to be due to this. The latter decussates at the body section involved, while the former does so just prior to reaching the brain.

While one can comprehend with some ease that reflexes in the brain relating to touch or pain are essentially of the same nature as the simple ones of the segments down the cord, those due to sight and smell are more difficult to understand. However, one should endeavour to consider these sensations as being enhancements of lesser sensitivities to light and chemicals that all segments once had, before they degenerated and their relevant local reflexes with them. In these cases the local reflexes of the segments can have been lost, or only responsive to extreme stimuli, while reflexes seated more remotely, i.e. in the brain, have otherwise taken over their finer functions.

However, though in the brain, some reflexes will not be in conflict with local ones. It is 'thinking' that sometimes puts the brain at odds with reflexes. A crude example is, for instance, trial by pain, where the letting go of a red hot bar is adjudged the admission of guilt and ensures consequent punishment. It is virtually certain that a pain reflex will ensure one's fate and any insistence from the brain that the bar should continue to be held will be overruled.

At this level one can recognise three aspects of the brain. The reflex component gives instant and automatic response to stimuli received. The indecisive component can be confused by its own complexity and hesitate and wait for one of the conflicting sensations coming from different sources to become dominant. The reflective component has access to records of earlier experiences and refers to these before ordering activity. Only this last is independent in its activity and is the seat of consciousness, the others reacting quite passively to circumstances.

Here remains a necessity to consider how the passive aware brain was eventually able to develop the additional faculties of the independent proactive conscious brain. The reason would seem to be found in an explanation of the mechanism of the "conditioned reflex".

The brain can be considered to be dynamic, not only with regard to its activity, but also concerning its structure. The mass of nerve cells, with their axons, dendrites and synapses, can be regarded as a physical fully integrated structure, but functionally as selectively integrated as a result of the inhibitory nature of many cells. Simply stated, it is these inhibitory cells that control the pathways through which specific impulses pass and without them there would be chaos.

Two different stimuli can be given simultaneously to the subject of an experiment. Examples of single stimuli can be given to a dog. If shown food, it will have a visual reflex response to it in its salivary glands. If it hears a noise it will have an aural reflex response in its ear muscles, so that they will prick up. The two responses are not connected neurally because of the divisive effect of the inhibitory function of some cells.

However, if both stimuli regularly occur together, with the food being presented with a "come and get it" order, the need for inhibition diminishes. The brain somehow responds by selectively removing it. As a result, on hearing the noise only, the dog may give both responses, or even the wrong one alone. This is a conditioned reflex.

One can use a visual signal in the same way. Provision of food can be accompanied by the flashing on of a light. An animal in this situation will come to expect food whenever the light flashes. Alternatively a pigeon can be trained to access food in an experiment by pecking at the right place. A light can be arranged to flash on every time this occurs. Eventually the pigeon can be made to peck simply by flashing the light, whether food to be accessed is visible or not.

Such a conditioned reflex can be used as a basis for animal training whereby food is used as a reward stimulus. Furthermore, it is also fundamental to the way a human being is 'trained' during a lifetime. One can thus infer that, whereas instinctive behaviour is of a reflex nature, learned behaviour is a matter of building up large numbers of conditioned reflexes. However, the conditioned reflexes produced by experiments can constitute bizarre and immediate single-effect developments, while those produced by life are normally useful for survival in the real world. This is because of the effect of that agent of evolution called natural selection. The effect is modified and usually enhanced when the complexities build up and lead to reflection.

The Central Nervous System, and especially the brain, consists of a vast network of interconnected nerve cells. It is just as though neurones naturally reproduce so as to fill the available space in the growing skull of the young, without there initially being any specific use intended. New cells have always been kept in reserve by the inhibitory activities of others. They are chosen to be incorporated into linkages more or less arbitrarily by the mechanism of the conditioned reflex. Once a new linkage has been established and serves some useful purpose, it can be reinforced by repeated use, eventually to be perpetuated as a neural path. It has become a 'memory'. The final fixing would seem to be achieved by the brain section known as the hippocampus.

All the sense organs have the means to change physical and chemical experiences that reach them into a common neural form. Vision, smell, taste, hearing, touch, etc. are received in the brain as neural signals that relate to actual experiences. The physical brain itself cannot understand the signals; it is the ongoing activity itself that considers them and is capable of the necessary adjustments so that signals can be interpreted. They are recognized by the dynamic brain (or mind) because they are suitable to fit into neural pathways that are similar in varying degree to ones pioneered earlier.

Similarities invoke the memory; but deviation may be sufficiently outside of previous experience to remain unrecognized and hence be incomprehensible. However, repetition will again create memory by means of the conditioned reflex mechanism through being linked continually to something else that is recognized, even if logically unrelated. This is how the memory works and we consciously make use of this mechanism when we employ mnemonics.

The bundles of nerves representing one sense carry impulses in the individual nerve fibres that must be integrated at some centre for interpretation and comprehension. At the same time they must be reconciled with interpretation of outside experience coming from other sources, be they either other senses or memory references. All such interpretation results from a balance being achieved among the neurones, so that the correct neurones are transmitting and the correct ones inhibiting. When ultimate balance cannot be achieved confusion results, or blankness. Even the same sense can lead to confusion in normal brains, as with optical illusions. Continuous breakdown of the workings of the brain brings about the chaos of the mind that can be recognised as 'madness'.

The mechanism of the conditioned reflex explains how the brain builds up the 'databank' for use by the mind. Experiences that seem alike are built up in a similar way in the brain, although any differences are recorded by additional pathway creation. However, with the passage of time memory of small differences can fade and lead to loss of discernment. This can lead to inability to remember, but can also take the form of a reluctance to distinguish that can itself be a useful function, because it enables the mind to put things into categories, with experiences being grouped together on the grounds of similarities. This is the basis of the association of similar ideas, a necessary faculty.

However, in the wayward thinking of primitive man, either living long ago or in later savagery, it has led to the illogical concept known as imitative magic.

The conditioned reflex mechanism can also bring experiences together in the mind that are dissimilar. If dissimilar experiences reach the mind together on a regular basis, they become associated through the conditioned reflex mechanism and consigned to memory in this form, one might say as a 'package'. Thus things regularly found together belong together and this applies even after they have become parted. This is the basis of the branch of sympathetic magic known as contagious magic. However, this form of the association of ideas is vital for the functioning of the mind. Dissimilar experiences reaching it are associated by precedence, reflecting the appropriate neural formation in the brain. The association of bare twigs and leaves can be recognised even when these are viewed in disassociation.

However, one can logically claim that perceptions of the same experience by means of different senses are indeed themselves different experiences initially. Thus can a motor car be recognised by seeing it, hearing its engine, smelling its fumes and feeling it. All these perceptions are quite different in kind to the sensory equipment as and when recorded in the brain and are only made into a whole by the mind using the elaborate conditioned reflex mechanism. This is the method developed in the physical brain and used by the mind to integrate disparate signals that in fact emanate from an entity that can be recognised as a common source. However, in perceiving a 'car' one needs to recognize that sight is the key sense in that it works fastest.

The conditioned reflex mechanism is heritable only as a latent capability. The whole of learning depends upon it, while it can be regarded as a seat for instinct only in the way that it is ready for use at birth. One can notice that a baby is struggling to associate perceptions reaching it and get mastery over and co-ordination of its own senses and movements. Once it gets things right it reinforces the experience by repetition to establish the appropriate neural pathways. It is by this means that the laying down of memories is initiated and they are consolidated. Of course, a baby has already started to develop most of its senses for some time before birth. The one sense that is certainly completely unpractised at birth is vision and it is indeed a great moment for a mother when the offspring shows signs of distinguishing her by sight.

The brain of a newborn baby can be likened to a new-built library that is waiting to be filled with books. The building itself represents the 'brain' while the books form the 'mind'. A baby's brain is massive compared with what it can do and we can be certain that it 'knows' next to nothing at birth. All those idle (or rather 'inhibited') neurones are waiting to be put to use. During childhood brain growth continues to reach a maximum, after which a very slow decline in the number of neurones sets in. Yet this does not in itself prevent intellectual growth; such are the possibilities of the nervous system. Without some pathological condition setting in, it is normally only in extreme old age that the system breaks down. A major symptom of the onset of senility is difficulty in building up new memories, an indication that neurone loss has finally destroyed the "excess capacity" of the brain.

However, for the brain to have a greater capacity than it will ever need might seem an evolutionary impossibility. Yet the brain has indeed evolved to a size that embodies an expectancy that in adulthood all neurones provided will eventually be needed. Yet the fact that all neurones are intended to be brought into use does not exhaust the possibilities of the brain at all. Intellectual capacity can be expanded because the permutations of the network approach the infinite and growth occurs by the selective breaking down of the inhibitory nature of cells to create new neural pathways. A baby has immense overcapacity because it anticipates the learning needed to reach the eventual intellectual standing of an adult, but one must assume that a large proportion of the cells are as yet inhibited.

On the authority of these claims one might assert that a chimpanzee can never be taught to speak, since it has not been born with the neural expectancy that enables such to be acquired. This is in addition to its equipment to reproduce the sounds of human speech being severely limited. Yet it may be able to associate words uttered by humans with concepts; but never with meaning. Likewise, any bird taught to 'talk' has even greater inhibitory limitations with regard to understanding and meaning.

However, the human brain is not just concerned with external perceptions, but can stimulate itself by means of the mechanism devised to integrate these. Thus are the nervous impulses received from various sensors integrated, so that appropriate action may ensue. The result may be complex and the necessary reactions not clear-cut.

Such indecisiveness can be referred to as 'thinking' and the ultimate result will depend on the consultative body of the mind, i.e. the seat of precedence, the memory. Reference to the memory is of great importance to the human mind and this is due to the immense amount of information that is stored away as a result of the development of language. Because of this the human mind is able to stimulate itself internally and the great achievements of humankind result directly from this faculty, which is known as the imagination.

The human retrieval system would seem to be vastly different in quality and quantity, if not in kind, from those of other animals. With others there is mainly a matter of recognition involved. With man recognition is also present, but there is in addition a conscious search for information laid down in the memory earlier in the form of coded linguistic packages. Yet the system is far from being infallible and conscious attempts to remember something, especially if known to be 'there', may draw a blank, an irritating and frustrating experience.

The mind has been seen above to work at two levels, namely the recognition of similarities and the association of concepts that, although not necessarily alike, belong together. The pervading concept is one of comparison, in which incoming perceptions, together with self-generated ideas, are compared with precedents provided by the memory. The procedure of matching, though rapid, can of course be very complex, but once it has been followed to its limits decisions can be taken.

In the integrating activity of the mind resemblances lead to recognition, but the conditioned reflex mechanism will be needed to consolidate even small differences that begin to occur repeatedly. All repetitive differences, whether small or great, are eventually recognised by association, either as external fact or neural coincidence. Resemblances are immediately recognised and placate the nervous system. Unrecognised parts or entities irritate it and it is this that breaks down the inhibitory nature of neurones. In this way brain energy is channelled mainly into those areas of new or forgotten memories and differential associations are further built up in the memory.

The ultimate reflex nature of brain activity is disguised by the fact that immediate response does not necessarily occur. This is because mind-driven physical reaction is always diverted into memory building processes, which means that such conscious responses can be delayed indefinitely if required.

Motor responses due to brain activity have to be triggered off by changing circumstances and these are almost entirely external, even though primitively the prime movers are generalised inner drives to find sex, food, shelter, etc. The memory can absorb experiences, so that response can be suspended until called for. It is the human ability to make complex responses that gives one the feeling of "knowing what to do", a feeling that is impossible with simple reflexes.

■ Attention and Concentration

In beings considered to be 'aware' rather than 'conscious', one can assume that 'attention' is not possible; such are only capable of 'reaction' and the response of leaves to sunlight and coelenterate tentacles to food can be regarded in this way. Such are purely passive and automatic responses; 'attention' can only be present where there is a complete brain and this implies a memory. The simplest of nervous systems and more primitive brains lack a learning memory and hence will always be deprived of the benefits of precedent. Individual animals at this stage are unable to learn from experience and lack the faculty of making conscious decisions. Choice is not for them and their reactions are inherited ones that are not subject to adaptive changes during their lifetimes. Attention for them is neither possible nor necessary.

However, from among creatures at the purely 'aware and reactive' stage, some have indeed arisen with "conscious and attentive" capabilities, and such include all the vertebrates. Yet for them to reach such a level there must have been an earlier stage when consciousness and attention were just arising and the creatures among which such traits arose need identifying. One can speculate with some profit on the notion that purely sessile animals do not need to practise attention, and neither do those that simply drift with the current.

Hence can one associate early attention with motility and the deliberate search for food. Identified thus, attention had its beginnings among the invertebrates of higher evolutionary development than achieved by the coelenterates. Such a one as the jellyfish cannot give its attention to its surroundings; a sea-anemone is quite incapable of 'deciding' to trap and consume one of those remarkable clown fish that are adapted to dwell among its tentacles. Such creatures are entirely predictable; unpredictability is a trait that creeps in and grows with an increasing degree of capability to give attention, along with elements such as choice and distraction.

While attention is a clear-cut trait among the vertebrates, did it start up among some of the invertebrates? An insect such as the praying mantis seems to give a lot of attention to its prey. Yet it is very difficult to perceive attention among such creatures as sea-cucumbers, feather worms, sea-squirts, etc. and one is bound to exclude from possession of such all animals that feed by the fine-and-whisk method, but one can certainly see evidence for it among the molluscs and the arthropods. Attention is hence confined to animals that are coarse-and-grasp feeders, but to give it they need developed senses. The relevant ability reached its peak with the evolution of sight. Motility was also needed so that the attention could be actively rewarded after pursuit.

In the light of the above, one can clearly link the growth of attention with the evolution of the bilateral juvenile, especially those that came to dominate and finally eliminate their own radial adult forms. A sea-squirt cannot be claimed as able to give attention, but one cannot be quite so sure of its motile bilateral juvenile

The ability to give attention, along with some appreciation of time and a rudimentary memory, can be regarded as essential for a dawning consciousness. In addition, this primitive faculty then became a feature of the bulk of the life cycle of the creatures concerned. One might even say that in extreme old age human consciousness tends to be eroded away in the decay of the senses and the difficulty of making new memories and giving attention. Very old people may tend to sink back towards the aware state and such a condition can be described as 'vegetating', in other words becoming 'plant-like'.

Man shares the state dubbed here as 'attention' with other animals, but is capable of taking things a further step with what one might term 'concentration'. This differs from attention in that this is aroused, whereas concentration is deliberately given. Attention is generally still a passive state out of which concentration has grown and at its highest level is due to a state of consciousness that is engaged in by humans only. This achievement has been made possible by the reflexive qualities of language and the resultant Self Conscious Mind. Human concentration is so active as to be a fragile state and it is necessary for it to be continually reinforced by further orders from the brain, an essential quality not evident with attention, which is due solely to the attractiveness of the object of the attention.

Concentration is not necessarily the direct result of intending to fulfil some immediate aim. The latter may be distant and obscure and the mind orders itself to concentrate as some intermediate measure, even though the results may not give immediate pleasure and frustrate the giving of attention to other things that would do just that.

Concentration may alternatively be called "concentrated attention" and can easily be broken by situations arising that may only require attention, in other words 'distractions'. Only human beings have the reinforcing ability that allows full concentration. This is a function of the Self Conscious Mind, with its various unique facets that have grown out of the linguistic capability, such as reason and imagination, the ability to deal with abstractions, the choice of duty before pleasure, and indeed the triumph of mind over matter.

One can observe the stages whereby a human baby acquires consciousness and self-consciousness. The former, unlike self-consciousness, is a direct and instant relationship with the physical world. A baby develops consciousness out of prenatal awareness, which extends some way into the postnatal period. A newborn baby neither has nor needs full consciousness until it has built up enough memories that are useful for it to reflect on.

Attention to its surroundings and full consciousness are accelerated by improving vision, although blind babies can still achieve it by substituting hearing as the major sense facilitating attention. This can only be achieved because they are part of a species within which most babies are not blind and have evolved accordingly.

One might consider that a normal baby starts to become fully conscious by observing its hands and by coming to realise that they are part of itself. Then through the conditioned reflex mechanism they eventually gain control over them. This occurs in that the hands naturally come up to touch the face, especially the mouth. The observed movement of the hands, coupled with the felt contact with the mouth, will become neurally linked and the end product of the resultant conditioned reflex will lead to control over such hand movements and eventually to conscious recognition that they can be repeated with deliberation.

The transition from consciousness to self-consciousness can be associated with change from baby to infant. It develops because one comes to understand the concept "I am here", or "This is happening to me", not because one can form such in words, but because one is being bombarded with words by others that one eventually comes to understand.

Once self-consciousness has been established, the ability to concentrate can be initiated and consolidated. Thus is the transition from awareness to consciousness followed by the addition of a similar change allowing the possibility for attention to be overridden by self-conscious concentration. Once this has been established one might consider that the infant has become a child or juvenile and early educational work is possible. Even so, concentration in its earliest forms is not solely to do with the absorption of even oral instruction, but deep in man's prehistory it would be initiated by acquiring skill with the hands through observation of one's elders; namely it can be associated with tool making, the necessary intermediate stage towards the group sustaining itself.

When chimpanzees use stones as tools and modify twigs for use as such, they can be regarded as using concentration at its rudimentary level. However, even the earliest of human stone tools from the Palaeolithic indicate a degree of concentration far in excess of that of these apes. They indicate the birth of 'work', the performance of a manipulative task whose purpose only lies in the future. Work can seem to give pleasure, but this does not emanate from the task itself; it is derived from the future achievement that is unconsciously envisaged. Artistic activity works in the same way. More recently one can see the same mechanism embodied in the concept 'exercise'.

The Bilateral Brain and the Mind

The more primitive bilateral creatures have nervous systems whose halves perform identical functions for the relevant sides of the body. With such there is indeed no need for the sensory and motor apparatuses to differ on each side of the body, even if integration has been improved, firstly by simple transverse linkages and later by decussation. Where locomotion was mechanically repetitive, and perceptions and responses of an automatic nature, there has been no need to disturb the strict symmetry of the bilateral nervous system. However, for various reasons, a few invertebrates have modified their earlier bilateralism to produce lop-sided effects and the nervous system has of necessity changed to accommodate such asymmetrical developments

Snails and other univalve molluscs have lost symmetry in the shell-bearing area of the mantle and the nervous system perforce reflects this change. A nearer comparison to the human condition is that of the fiddler crab, whose one claw in the case of the male is grotesquely enlarged and used for signalling and fighting other males, but useless for feeding. Here the left and right claws are allocated quite different duties, the one having grasping powers used for feeding while the other 'fiddles' as required. An example of asymmetry among vertebrates is found with the flounder and similar flatfish. As the fish matures one eye moves round the head and the whole body distorts to suit a bottom-dwelling life-style.

But all these are unrelated developments, both to each other and to the neural asymmetry found among the higher vertebrates. The latter all maintain that strict external bilateral symmetry so typical of the more advanced creatures; yet inside their bodies lurks considerable asymmetry. This is primarily due to the convolutions of the alimentary canal and its laterally placed organs, such as the liver. This is served by the Autonomic Nervous System, which nevertheless contrives to retain a deal of symmetry itself and can hardly affect the Central Nervous System.

A form of asymmetry exhibited by man, and noted also among chimpanzees, is handedness, with the right hand normally being the more capable and more frequently used grasping organ. It is presumably due to chance that the right side is normally dominant, although not visibly distinct from the left. One can compare the dichotomy to the odd claws of the fiddler crab.

Oddness of appendage does not normally occur when grasping is accomplished by calliper-like paired organs, one on each side of the bilateral body, as are the jaw parts of arthropods. However, some arthropod limbs themselves evolved pincer-like ends and from these have arisen aggressive grasping claws such as those of scorpions and crabs. With these the need to work simultaneously with their opposite partners has become unnecessary, thus opening the door for bilateral specialisation. Physical evidence of this having occurred is rare, although clear enough with some, including the fiddler crabs.

When it comes to handedness among vertebrates one can observe that the same opportunity for specialisation has arisen as among the invertebrates, but it has been taken much further. Once vertebrates came onto land the alternating gait weakened the absolute symmetry when on the move, but gave no cause for any differential specialisation of the limbs across the body. However, one might note that tree-climbing mammals, with their clawed feet, were tending to develop grasping organs on the ends of their limbs (as noted above with arthropods, although of a completely different structure). The climbing itself induced no transverse specialisation; it was the further utilisation of the grasping function for feeding that led to this.

Most mammals are primitive feeders, simply applying their mouths to the food, although many do steady this with a forefoot. However, climbers often grasp the food between their prehensile forepaws and this is the characteristic attitude of a squirrel with a nut. Yet they always use both paws. The use of one forelimb alone for collecting food and passing it to the mouth is a speciality of the primates; it is common behaviour among monkeys and apes. Here there is a latent opportunity for this function to be favoured by one hand at the expense of the other.

283

However, the right-handedness exhibited by chimpanzees cannot be a purely physical preference exercised by each individual, but must be based on an evolved and heritable state in the brain. Right-handedness means that to a degree the brain is lop-sided. Yet one should understand that, while the right hand has specialised in food gathering, the left may be thought to have developed a sort of dominance of its own when it comes to supporting an arboreal primate by grasping.

There is no evidence to hand as to the general degree of handedness among primates, but it is clear that direct food gathering is not the only way in which chimpanzees exhibit it; they have been noted as 'tool-users' and even as 'tool-makers'. They may snap off long thorns to use for extracting social insects from their burrows. They may also snap off sticks to hurl at a threatening presence. All this tends to make them 'handed' and reflects a corresponding imbalance between the halves of the brain. From an evolutionary standpoint this could only occur when a natural imbalance, of however small a proportion, backed up the as yet untried advantage of right-handedness predominating.

The evolutionary advantage was inherited by succeeding generations and, by similar means, became more defined. The preference for the right hand, while originally being chance, must have later been reinforced by the apes observing each other and there may have been some evolutionary advantage for those who conformed. This still applies to us today (except for certain special circumstances, such as – say – playing tennis). What is clear is that left-handers are born as such and attempts to force them to conform to the majority situation causes distress. Handedness is located deep in the nervous system, neither being in the muscles nor the upper levels of the mind.

Should the observations made just above be considered just, then the conclusion is unavoidable that right-handedness arose quite early in the primate story and certainly with some common ancestor of man and the anthropoid apes being involved. A natural tendency among arboreal climbers was transferred to tool-users and then to toolmakers. This went together with a brain that was developing a basic yet ultimately advantageous imbalance.

Among simpler bilateral animals, whose nervous systems are restricted to motor responses after sensory stimulations, there is no advantage to be gained from a brain whose halves tend to differ. However, among more advanced animals, which have developed the capacity to remember and have acquired consciousness, the brain has internal workings that are quite distinct from those resulting in the motor responses of a simple reflex nature.

The Conditioned Reflex Mechanism was being built up and it was quite wasteful for both sides of the brain to be concerned with this advance and providing simultaneous and identical responses. This aspect of the brain deals with the whole animal, with no need to consider the needs of its bilateral halves.

Initially the primitive conscious mind would operate with the two sides of the brain happening to give slightly different answers to the same problem and activity might have gradually grown to depend on some kind of consensus. However, there is no need to assume that such a consensus was always superior to the opinions of either of its two constituents. The evolutionary prospect loomed that a real advantage could be gained from bilateral specialisation. As has been noted on earlier pages, when there is surplus capacity of a feature or faculty, it is open to favourable evolution by diversification and specialisation. The brain came to have one side too many when it came to thinking, rather than just responding. Initially the sides would specialise in different aspects of any one concept, but eventually certain functions would tend towards becoming the sole concern of one or other side.

The most obvious imbalance in the human brain is that the speech areas are concentrated on one side, almost exclusively the left hemisphere, while the other hemisphere can hence only be reckoned as being 'conscious' rather than 'self-conscious', at least when considered as a separate entity.

Human self-consciousness is unique, but can be claimed as being derived directly from the right-handedness evolved among arboreal ancestors and which was reinforced through the tool-using and tool-making phases. This pioneering right-handedness was a specialisation of the left hemisphere of the brain, from which the limb was controlled. Tools came to be of such cultural significance for man that one can see in this the confirmation and consolidation of the dominance of the left hemisphere over the right and it seems likely that speech would come to be located there when one side alone was selected for this.

Humans are mostly right-handed and have the speech centres in the left hemisphere, but the condition is not exact, so the link between them is not direct. It is the dominant role of the left hemisphere that throws them together. While it is usually harmful to try to train a natural left-hander to become right-handed, it has been observed that infants who suffer damage to the speech centres can compensate by developing them in the other hemisphere, but only at a cost to other faculties that are usually seated in this opposite area.

■ Specialization of the Hemispheres

The paired hemispheres of the cerebrum are joined together above the brain stem - and hence separate from it - by a massive independent neural tract known as the *corpus callosum*. There are similar commissures joining the smaller hemispheres of the cerebellum. Such commissures can be seen to serve a different function from the decussation (crossing over) of nerves in the brain stem and elsewhere in the nerve cord. One might well say that the nerve cord chiefly handles longitudinal neural messages, i.e. those that integrated the segments, and that lateral pairing of segments is only neurally equipped to handle reflexes. On the other hand, decussation has an integrating function, with the crossing over of neural messages - in both nerve cord and brain stem - serving to ensure that the bilateral animal with certain asymmetrical features can still react to stimuli as a whole.

The elaborate decussation at certain parts of the brain stem has led to the formation of great organs that go way beyond the ganglion stage, namely the cerebrum and the cerebellum and here it serves as a means to co-ordinate the neural messages of a reflex nature (both sensory and motor) with those emanating from the thinking parts of the brain, to some extent the cerebellum, but mainly the cerebrum. A commissure such as the *corpus callosum* can be thought to co-ordinate the activities of hemispheres in a way that is not possible by decussation in the brain stem, for this is only designed to handle 'physical' activities. It is commissures that are intended laterally to co-ordinate purely 'mental' activities, which without them would be quite isolated in their relevant hemispheres.

Without the commissures the left and right hands could still perform co-ordinated activities because of decussation, but the right side would only be aware of and not properly 'know' what the left side was doing and *vice versa*. Such bilateral ignorance is of no significance in creatures whose sides are completely balanced, but once imbalance of function set in, with induced imbalance in the brain, the need for the two sides of the brain to co-operate differentially became essential. This is the function of commissures; while decussation co-ordinates the function of the Central Nervous System, especially its longitudinal aspects, commissures ensure co-operation between the differential halves of the brain, an aspect that reaches its zenith in man.

Co-ordination by way of the brain stem and its decussations is thus involved purely with the sensory-motor apparatus of the Central Nervous System. On the other hand, co-operation is solely to do with the regulation of activity between the hemispheres and this involves purely mental matters, neural events that are confined to the hemispheres themselves. Those of the cerebellum can to a lesser extent be included in this use of a commissure. This is the stuff that higher levels of consciousness are made on. In the head commissures are separate from the brain stem because they are dealing with stuff that is of no immediate concern for that part of the CNS found in the body as a whole.

It has been claimed above that nervous activity in the thinking brain differed from that of simple reflexes in that constant reception of perceptions in close association with each other changed such reflexes into conditioned ones and from this grew the memory, plus the ability to refer to this facility and, in general, the ability to think. But thinking - no matter how complex or deep - is still just an intermediate stage between making perceptions and taking actions, no matter how sedentary man's lifestyle may become. The brain might receive a series of perceptions, they are mulled over and then the relevant activity is decided upon.

The thinking period can be long - the operation can even be deferred by transference to the memory – and this disguises the original reflex nature of thought. The need to take time is because the brain has to come to a conclusion or to make a decision and thus show itself as host to a mind. If this is immediately impossible the brain can contain the reflex activity involved within itself pending an anticipated satisfactory result and suspend any operations, which may only consist of instructions to the muscles of the mouth and larynx. This stifling of activity is certainly possible and may be a necessary intervention, but does not contribute to any feeling of well-being in the individual; on the contrary it can often irritate and be labelled 'frustration'.

CHAPTER 18

The Latest Achievement

■ Introduction

The trouble with thinking about thinking is that in doing so one seems to be trying to swim against the stream of one's own consciousness. While thinking thus one can be aware of the actual thoughts without having the slightest idea of the processes that put them there; and where is 'there'? Thoughts considered on their own have no body at all and only seem to be loosely "within one".

■ The Senses and Consciousness

The five traditional physical senses are touch, taste, smell, hearing and sight. Altogether they give the information one needs about the world outside of the body (together with some within it), and this is used by our 'selves' to take any necessary actions for self-preservation or betterment. The equipment we can use to record and react to circumstances, both external and internal, is the nervous system and it also serves to integrate all the various strands of information to give a sort of extra sense that we call consciousness.

Consciousness is a quality enjoyed by all the higher animals and is the condition from which has grown the highest known achievement of evolution, namely the human mind.

■ Sensory and Motor Nerves

These represent nerves at a primitive level and they can operate more or less on their own without any specific input from the Central Nervous System (CNS - this being the name given to the brain and the spinal cord in combination). At this simplest level the nerves are responsible for reflexes; a sensation is picked up that results in an automatic physical response. Expressed in a simplified way, receptors activate sensory nerves and these signals reach the spinal cord, from where they are passed back by motor nerves and bring about the characteristic muscular response. In bilateral bodies such as those of man such reactions are restricted to the side in which they occur. The other side, even of the spinal cord, is not involved. This of course has nothing to do directly with the sense of touch and its derivatives, which are registered in the brain; a reflex will work without being 'felt'. If you touch something very hot, you move your hand before you could withdraw it after realizing what has happened.

■ The Nature of Nerves

Nerve cells consist basically of pyramid cells (or neurones) from which there is a transmission branch (or axon) with branches to many other cells, plus numerous extensions receiving sensations from other nerves' axons and called dendrites. Impulses accepted by the dendrites may be passed on along the axon.

Nerve impulses are of electrical nature, the charge of which can be picked up by instruments, but must not be thought of as being the same as the current that passes along a wire from a point of high voltage to a low one. Nerve impulses are electro-chemical exchanges which progress rapidly along the fibre. The strength of a neural signal depends solely on the number of impulses that are transmitted in a given time, for the impulses themselves, by their very nature, are identical in both magnitude and velocity along any one nerve.

The initiation of nervous activity and the contact between individual nerves is basically chemical in nature. With regard to reflexes, an external receptor will stimulate electrically the nerve ending butted against it and initiate impulses. The mode of transmission between axon and dendrite is chemical and takes place at a junction called a synapse. The signal will not pass the synapse until the frequency of impulses along the axon has reached a certain level. A cell will only transmit impulses itself (or 'fire') when the weight of impulses received along its dendrites is sufficient.

However, cellular activity occurs in two different modes. In the transmission mode they pass on impulses that will induce other cells to transmit what they are receiving. In the inhibition mode their impulses prevent other cells from transmitting. Thus does a neurone's firing not only depend on the strength of the signals it is receiving from others, but also on the balance of the signals coming in along its dendrites, namely between orders to transmit and orders not to transmit. This has less significance where simple reflexes are concerned, but it is vital with regard to the pathways followed by signals through the vastly intricate volumetric network of the higher parts of the nervous system.

There exists a belief that nerves originated as a derivative of the skin and have developed their multitude of functions from this primitive condition. Thus can one indicate the primitive form of 'touch', as found in reflexes, as being the primary derivative of the earliest form of nerve. The nerve, with its electric qualities, could quicken up enormously the physical reactions of beings so equipped over those solely dependent on chemical responses. Yet nerves at the simple reflex level enhanced only awareness. Consciousness was to depend on further development.

■ The Central Nervous System (CNS)

This consists of the spinal cord, which has the brain positioned at its head end. The cord leads first to the hindbrain, which is concerned with involuntary activities, such as the beating of the heart and unconscious breathing. Organs found there include the *medulla oblongata* and the *cerebellum*. All sensations reach the brain through the medulla save those of smell and vision. The cerebellum controls involuntary activities in the body, but in certain cases, such as breathing, it may be overruled by further neural linkages to the higher brain mass. The balancing organs in the ears also work through the cerebellum.

The midbrain was primitively associated with sight, but in the higher vertebrates this has come to be the business of the forebrain.

The forebrain (or *cerebrum*) consists mainly of two masses known as the cerebral hemispheres. These again are rather arbitrarily subdivided into four further zones called the frontal, parietal, temporal and occipital lobes. Specific areas of the surface of the cerebrum are connected with specific functions of the body, such as voluntary movement, seeing, hearing, etc. The forebrain has ultimate control over all body activities to do with the external world, but has no direct control over most of the internal body functions, including the nervous system itself, even though it may be in possession of information relating to their condition. Many internal functions are governed by two separate nerve trunks that are only lightly linked in with the spinal cord and the brain and they constitute what is known as the Autonomic Nervous System (ANS).

The nerves are not the only agents by means of which our bodies operate. Primitive organisms can be seen to carry on with their simple activities without even possessing nerves and many functions of our bodies are instigated by those gland-produced chemical messengers known as hormones. These, like the ANS, are concerned with the internal operations of the body, about which one is quite unconscious until they go wrong and even then no personal means is available for determining the source of the problem. To pinpoint such ills external 'cultural' sources need to be consulted, the body of knowledge of the medical profession.

The primitive nature of nerves may be illustrated by recognising them as direct transmitters that do not employ ducts. In their case success has precluded any development of this nature.

■ Touch

This sense is generally felt externally all over the body as well as selectively internally: degrees of sensitivity might also be found within the body when abnormal circumstances arise. The nervous system, which operates touch, cannot itself feel anything, nor can the circulatory system that handles blood. Touch is a response to pressure and heat or cold, as well as certain other chemical and physical contacts. If the stimulus becomes too great, discomfort may register and eventually pain. Certain stimuli induce feelings of satisfaction or pleasure, but they too can turn to pain if overdone.

■ Taste

This sense is experienced in certain regions of the mouth, being a specialized development of touch associated with the ingestion of food and drink. The inside of the mouth retains all the other functions of touch, but, in addition, there is enhanced sensitivity to chemicals found in solid and liquid matter. The degrees of pleasure and distaste are stimuli towards the exercise of discretion over what should be consumed.

■ Smell

This is clearly allied to taste and the two were probably originally combined in a single sense. Now, however, smell is located separately in the nasal passages and deals primarily with gaseous contacts.

■ Hearing

The ear is a much more complicated organ than those surfaces that are concerned with touch, taste and smell. It deals with vibrations in the air (sound waves) that are far too minute to be picked up by other areas. The waves cause the membrane of the middle ear (the ear drum) to vibrate and the vibrations are transferred to the inner ear by means of the three auditory ossicles. Here there are liquid filled sacks and canals and it is the disturbances within these that stimulate nerve endings to give auditory experiences in the brain. Ears and hearing are a vertebrate speciality and the tiny bones known as ossicles are derived from redundant components of the jaw and the first gill bar of primitive fish, while the primary purpose of the liquid-filled vessels of the inner ear was the 'sense' of balance. The ultimate origin of the ears may lie in a specialisation at the head end of the two lateral rows of pits that constitute the balance lines of fish.

■ Sight

The eyes are organs that deal with waves of far greater frequency than those of sound and talk of the reception of vibrations is hardly apt. The front parts of the eyeball act as lenses to focus light rays on the back surface (retina). The 'seeing' of light that enters the eyes results entirely from refraction through, or reflection from other objects and gives a precise impression of them, which is projected by the focussing mechanism onto the retina. The contrasts and hues of this light are picked up by the rods and cones of the retina and the appropriate stimuli passed on to the adjoining ends of the optic nerve fibres.

All vertebrates have eyes of the same basic structure, but their more primitive chordate relatives are without vision. However, the principle of the vertebrate eye is phylogenetically transcendental, as evidenced by the molluscan cephalopods. Colour vision is lacking from many vertebrates and seems naturally restricted to some of diurnal habits. Lack of colour vision among so many mammals might seem to be linked to a nocturnal phase that earlier was perhaps universal among species with this characteristic.

■ The Higher Functions of Nerves

Muscles are activated automatically by nerves in reflexes. On top of these simple motor reactions, there is also the deliberate control of physical activity by the top end of the CNS, in which are also located those centres devoted to memory, thinking, instincts, emotions, intelligence, imagination, etc. All this is ostensibly of quite different nature from the activation of muscles, yet the whole business is contained within the most complex part of the CNS itself – the brain – and the type of cell involved. The neurones comprise about half the brain, the remainder consisting of various support matter. Like most cells, they are unimaginably tiny, and exist in the brain in numbers hard to comprehend. The higher functions of the nervous system are made possible by its complexity and by it consisting of cells differentiated into transmission and inhibition modes and their differing influences on synaptic activity.

These are the systems by which the brain works, even though they were originally evolved simply as a way of evoking rapid mechanical responses. The brain is an immense volumetric network of nerve pathways and without the cells incorporating inhibitory function it would be a place of shear chaos. It is they that guide a particular impulse along a certain path once it has been delivered with sufficient frequency to stimulate the various synapses. The routes in the brain are in a way like the footpaths that are to be used on a cross-country exercise; one is kept on the right track by "no entry" signs that have been put in place for the occasion. In addition the path, when trodden, gets to be worn and easier to follow, but if neglected becomes overgrown and obscure.

The pathways themselves are contained in the cerebrum and, together with the "no entry" signs, constitute the memory. The initial responsibility for the creation of the "no entry" signs appears to rest, at least partly, with the brain section known as the 'hippocampus', because surgical removal of such results in failure to build up new memories. Damage to the 'thalamus' has also been shown to have a similar effect. Yet sufferers from such defects are still able to recall memories from times before the occurrence of any such traumas. Older "no entry" signs elsewhere in the brain are unaffected by them.

■ Memory

The memory seems to work by a chain of external events being represented by impulses passing along certain neural pathways. The routes will to some extent be decided by the ways in which the information was gathered, e.g. involving the senses of sight, hearing etc., and be further complicated if there was also a linguistic input.

Otherwise the route will be governed by precedence. If the experience resembles a previous one, it will follow the same neural route through the brain, except for making diversions as found possible to account for divergences from that earlier experience. This "comparative method" would seem to be the only one possible for the building up of complex memories, a gradual extension of existing experience, without this being necessarily lost thereby, even though some confusion may set in as a result. The hippocampus, backed up by the thalamus, would seem to be the sorting house for new experiences. These particular parts of the brain may not themselves remember new experiences, but do remember just how to consign new experiences to the memory.

Memories appear to be recorded in three overlapping phases. Short-term memories seem to be reverberations that remain in the brain for a while after an experience, and do not involve the hippocampus. Medium term memories extend for several hours and seem to be an overlap phase during which the experience is being embedded in the memory, whereupon the short term memory of it goes into decline. Long-term memories are those that no longer occupy the mind significantly, but have been recorded for reference. The hippocampus mechanism would seem mainly to be involved in the middle phase. It would seem to instigate unconscious rehearsal of the experience to ensure that the neural pathways that it took have been well signposted for use when the same or similar is experienced again, or if recall is required.

The existence of memories is useless without means of retrieval. An animal is stimulated by current experiences, both inside and outside its body, e.g. hunger and the sight of food. The memory here seems to give answers to questions that have not been asked. Here the necessary action is basically instinctive, with instinct being a sort of inborn, genetically determined memory. There is also an element of learning. At a young stage the animal may have asked itself in its own non-verbal way, "Is this food?" and the answer would be consigned to the memory-bank.

With the very young, and most particularly with human babies, the retrieval mechanism is clearly very weak during the early months and years, while it is evident that the recording mechanism in the meantime is working away very effectively, yet secretly, so that adults are later surprised how quickly the child comes on. This is related to the rapid growth in the number of brain cells that occurs in early life, while later memories are set down against a decline in numbers and presumably depend more on some sort of synaptic development.

In a way all brain activity can be related to memory and its retrieval system. A situation arises, questions are asked, the memory is ransacked, answers are produced and action is taken: at the same time feedback occurs so that this experience too becomes a part of memory.

An aspect of a brain with a huge load of memories is the possibility of a number of them being aroused by an experience and then not agreeing with each other. Such clashes require the existence of a mechanism that can reject or override memories proffered by the retrieval mechanism. This "decision making" is the very basis of thinking and hence of intelligence. It might also be regarded as an important contributory factor in the difference between awareness and consciousness.

Awareness and Consciousness

Both of these can be thought of as existing at different levels. All plants can be considered as having awareness only, because their responses are automatic. The same can be said of the more primitive types of animal. Such can be thought of as not being conscious if they have nervous systems that are solely at the instinctive or reflex level. Consciousness begins when animals have nervous systems that can build up memories that are not biologically inherited and have centres that can disagree with, modify or override reflexes. In order to be able to do this at all they must indeed have access to precedent. For this, a working and expanding memory, however primitive, must be present.

For consciousness to be experienced there must also be an awareness of the concept known as 'time'. Memory itself is an acknowledgement that time has passed. Thus is the degree of consciousness tied up with the degree of evolution of the memory; without a non-instinctive memory-bank consciousness is not possible at all. In sensing the passage of time a non-verbal distinction between past and present must also exist.

The word 'present' can be used in two different ways. It can signify the short space of time of indefinite length that one is passing through and which we can term "the present time", or it can signify an instant at the time of its consideration, which can be termed "the present instant". However, the latter does not really exist for anyone, for the mind is always of necessity slightly ahead of such theoretical present. The present instant is like a staccato note; by the time the mind has appreciated it, it is already in the past; that is unless it has been anticipated. In particular situations, anticipation by way of memory can lead to prediction and this allows the mind to recognise a staccato note marginally more quickly when it comes.

So, in the case of the present instant, recognition can occur by way of the instant immediately preceding it. However, even a very short length of time like this may vary minutely from the expected and then adjustment made. Basic consciousness is seated in the anticipated next instant and unexpected variations found in it. It comprises a constant comparison of the sequence of such instants and assessment of any change. Yet the instants are not separate entities, but are overlapped and merged into a flow. As the flow hits the mind the memory comes to recognise the things floating in it, but may be confronted with variables that may raise problems of identification.

Appreciation of the present time at its shortest seems to be an essential component of consciousness. One might argue that a human at the 'baby' stage rises to this level of consciousness. It comes into being as a recognition of a short time span on either side of the present instant comprising both short time memory and short time anticipation. Basic consciousness is felt by the individual as the result of the fulfilment or not of ultra-short-time anticipations; its very existence is the appreciation of the continual barrage of such instantaneous sensations.

It is very difficult to appreciate which life forms have consciousness and which do not. One can at least feel sure that all mammals have it. It is also a fairly safe presumption that birds do too; but fish? As long as a life form is only aware of the present instant, it might be claimed that it does not possess consciousness. To be conscious a being must be able to appreciate the present time and project the present instant into this both ways. Until this happens young such as a human baby are only 'aware', but the transition may be quite quick. It starts when a baby begins to remember first what it has just done, followed soon after by the appreciation of wider experiences.

However, full consciousness is not just a recognition of fulfilled or unfulfilled anticipations, but results from a background comprising the integration of all perceptions, whether anticipated or not, both external and internal, and being derived from both senses and memories. However, recognition of the more complex forms of consciousness may be misleading. As just described above consciousness is a very simple process and supposed further forms of it are really resulting processes and hence not integral parts of it. One of these processes only significantly affects human beings, and that is self-consciousness.

■ Self-Consciousness

The type of consciousness experienced by human beings is vastly different from that of other animals and results from a single achievement of the human mind. As far as the body is concerned the human being is just another species of animal – another primate – but the brain is of greater size than with nearly all other animals, while being of far superior quality in particular respects to any others. The single cause of this is language. While ignoring artificial sign 'languages', like other animals man still makes audible and visual signals to others of his kind, but these are useless for passing on precise information, being simple indications of mood, from which imminent behaviour can be assessed and anticipated.

The first step towards language will have been an externalisation of the sound signals, namely their use to refer directly to things that are outside of one's self, a rudimentary practice that can be observed with birds as well as mammals. With man the process must have been greatly accelerated by the abandonment of the simple life in the trees and the taking up more fully the ways of the hunter. The functioning group would thereby get larger and the catching of food require ever improving organisation. The other aspect of life among these apes that were becoming men was the habitual use of objects as tools, followed by their deliberate modification into such. These socially complex tool-making societies evolved linguistically in line with their other strides forward, which were physical, mental and social.

Early language must eventually have included the concept 'verb' as well as 'noun'. It is the verb that provides and heightens the recognition of 'time'. Improvements in evolving language capacity also enabled man to regard the 'self' as though it were an external object, referring to it with the sense of 'I' (or the 'ego'). This recognition of 'self' had frightening consequences, because observation of one's fellows, plus identification of oneself as being one of them, led to the inevitable

299

conclusion that, like them, one was frail and mortal. At the same time the externalisation of the self by means of language enabled questions to be posed of the mind in linguistic form and receive answers in the same terms.

Originally oral communication was delivered by a mouth, picked up by the ears of another and interpreted by the brain of the same. The last facility enabled the brain to become adept at handling language internally as well as externally and this capability has now become the basis of the vast bulk of human thoughts. In the advanced thinking mode of humans their language is used silently. The words seem to be 'heard' in the head; yet the mouth and ears are completely bypassed.

While chimpanzees can hardly be thought of as self-conscious as defined above, when placed experimentally in front of mirrors they seem to recognize the moving image before them as themselves, but more certainly that they come to understand that they are the cause of the reflected activities. This can be compared with a bird like a blue tit sitting on a window ledge and attacking its own image in the glass. It is quite unable to understand that the moving bird it is trying to peck is a reflected result of its own activity.

The Brain and the Mind: Unconsciousness, Consciousness and Self Consciousness

The brain is the physical entity that in some way constitutes the seat of the mind. The mind is activity that takes place in the brain, but is in no way the brain itself. It is indeed not simply the activity there, but the summation of activities into a single whole. It integrates all those cerebral neural functions that are going on at one time, with the important exception of those restricted to reflexes, for the latter would extend the mind outside of the brain and into the spinal cord. The mind is more the tool of consciousness than of awareness, although it serves completely the latter among humans in the first years of life as memories are being built up with little outward sign. As the mind develops, much of the material stored consciously is lost to direct retrieval and this part of the memory serves the unconscious mind, which might only reveal itself selectively under special circumstances, e.g. in dreams.

The conscious mind provides a world of recognized precedents, these being selectively retrieved memories and the power to integrate such to give comprehensive solutions to current problems. Consciousness is a stream, because perceptions and the problems they raise constitute a continuous process. The conscious mind feels pleasure or displeasure, depending upon how well it is receiving the feelings and coping with any problems.

The self-conscious mind has one additional degree of consciousness and that is knowledge of its own existence, this being due to the reflexive powers of language. Whereas the conscious mind has intelligence, the self-conscious mind is provided with reason, because it can inform itself about and continually reinforce the validity of cause and effect.

Furthermore, while the conscious mind feels pleasure when it solves problems incidentally set by perceptions, the self-conscious mind can deliberately seek pleasure (and risk displeasure) by setting itself problems that would not arise from purely natural causes. The mechanism used to do this is a combination of imagination and reason. Such abstract problems must be initiated by imagination, before being formulated and solved by reason.

A feature of the self-conscious mind is that it 'knows' things, whereas the conscious mind can only be acquainted with them or, to use a dialect term, it 'kens' them. The conscious mind cannot deal with abstractions at all because this faculty depends on language. Indeed, abstract concepts do not even exist until we put a name to them. The verb 'know' has an abstract quality that 'ken' generally lacks. I can 'ken John Peel' without knowing his character.

The conscious and self-conscious minds can be in dispute, as in those optical illusions, where the conscious mind is tricked by the presentation of the pattern. The self-conscious mind knows that it is a trick, but in some cases can do nothing to alter the perception, or in others manages to succeed in this, but usually cannot hold on to the destruction of the illusion for long. This seems to show that visual perceptions, if not others, go first to the conscious mind and are then transferred to the self-conscious mind for final interpretation if necessary. Such illusions illustrate that the self-conscious mind has its limitations.

■ Language, Meaning and Writing

Language originated in sound signals formed by the mouth and able to be picked up by the ears of others. They anciently conveyed information about mood and hence one could judge intentions and anticipate actions; but they were useless for transmitting meaning. For this particular sounds had to be associated with specific concepts, rather than the wide generalisations and the elements of guesswork that go with mood. Objects had to acquire their own precise names, as well as to constitute units gathered under group names. More importantly for meaning, any senses of activity, movement and time had to be expressible and this was accomplished by distinguishing such insubstantial concepts by name. These aspects of activity in the mind eventually became verbs and abstract nouns.

In language the self-conscious mind co-ordinates and integrates meaningful sounds transmitted through the mouth for reception by the ear. This is language in the initial form of speech. The memory developed so that it could handle huge quantities of words and phrases and this was eventually helped by the formulation of grammatical rules and the evolution of rhythmic forms of retrieval, such as poetry. The latter in turn found more favour by the end product registering pleasure, both through its rhythm and contents.

In the meantime language had had immense social consequences. It had allowed folk to coach each other deliberately. The capabilities of the species were then not trapped in each individual member to the same extent. Language thus led to vast social advances and, with the growth of such intercommunication, much of the knowledge accumulated became the common property of the group. However, certain individuals or sub-groups were always going to seek prestigious advantages by knowing more and keeping such extra knowledge esoteric.

Large scale agriculture and pastoralism arose and allowed huge rises in population whereby massive surpluses could accrue that could be dissipated by the construction of prestigious structures, such as tombs, temples, palaces and cities; or in destructive dispute, especially in the form of warfare.

Names for various quantities would already be in existence before the need arose to count the vast surpluses. However, the necessity grew beyond the capabilities even of the human mind, largely because such accounts were by their very nature boring and not readily fixed as differentiated entities in the memory. In order to overcome the difficulty a system of tallies was invented. A simple drawing of it notated an object or contained quantity and the number in stock was given by strokes. This was at the beginning of both writing and mathematics. In this process the eyes also took part in linguistics, resulting in a great advantage over usage restricted to the voice and ears in that statements thus made were permanent records for those who could see and understand them, i.e. were able to read. A further step forward in writing occurred when symbols ceased to represent objects, but came to represent specific sounds that could be built up to form written words. Systems started to grow of which the alphabet is one of a persistent group of descendants. Very recent developments have been 'reading' by blind people through the use of Braille, whereby touch replaces sight, and 'speaking' and 'hearing' by deaf folk through the use of sight in their special sign languages and in lip reading.

CHAPTER 19

The Superego and Dreams

■ The Superego Concept

The term 'superego' was coined by Sigmund Freud to describe a hidden inhibitory process in the personality. He applied it to the repression of activity that showed signs of sexual sensations in very early years, a process that could later lead to unpleasant psychological effects in adult life. The repression could result from suppression of such activity during childhood by higher authority, namely the parents. The repressed activity was not destroyed, but was retained in the memory and waiting to be translated into action. The mind was generally frustrated in this prospect because of the inhibitory control of the superego.

This is hence a process by which values considered to be 'moral' are implanted in the mind of the child, which in later life cannot be disobeyed without feelings of guilt, even if these are subconscious. At the same time the unconscious mind is always aware of the fact that certain activities were once repressed, thus perpetuating a constant state of tension that cannot be eliminated wilfully, but, even so, needs to be alleviated, perhaps by indulging in diversionary activities that are not subject to inhibition. One can recognise this condition as a very elaborate kind of 'frustration' as mentioned in the previous paragraph. In extreme cases it can be a cause of irrational behaviour or antisocial activities.

Freud particularly restricted his superego to sexual matters, in particular the effects of living in a family, the observation by a child of the relationship between its father and its mother, jealousy experienced by the child of the parent of the same sex as itself, and the prevention of sexual interest in its siblings or preoccupation with its own genitals. It is clear that the circumstances most likely to cause such superego "personality damage" to a child arise when parents are ignorant or careless of the effect of their own activities on the child. Thus might the flaunting of sexual attraction between parents in the presence of young children stimulate jealousy that must be repressed if family life is to thrive. The brusque suppression of a child's own initial sexual awareness can produce inhibition scars that are much deeper than need be. Parents need to exercise discretion when showing 'affection' towards each other in the presence of infants and when correcting 'sexual' activity by them.

It might thus seem evident that in the repression of undesirable activities by an infant the process should be as gentle as possible if such activity is to appear natural at the appropriate age and undertaken in a proper manner, as is the case with sex. On the other hand one might argue that behaviour that is going to be regarded as antisocial in later life should be firmly suppressed in infancy so that the necessary inhibitory processes are set up.

However, the point at issue here is not a moralizing one, but rather an indication that sex is only one aspect of the mechanism that Freud called the superego. Sex is prominent because it starts so early in life as to be frowned upon severely when it shows up thus prematurely and hence the scars formed are deep seated and long lasting. It seems evident that all early unresolved frustrations fuel the same mechanism and contribute to our total behaviour in as much as they make it difficult to do many things that are otherwise appealing or even desperately desired, because of the generated sensations of 'guilt' that result, be these conscious or unconscious.

Even so, sex is a special case because its development is non-continuous. Three normal sexual phases can be identified that humans go through. There is an infantile sexual (but non-genital) phase in which there can be sexual experiences in both mind and body, but external sexual relationships cannot be established. There is an intermediate juvenile non-sexual phase that embodies a certain aloofness from the opposite sex. Then there is the post-puberty phase, which is both sexual and genital and during which full sexual and reproductive relationships with others can be established. It is evident that the deep-seated sexual inhibitions are set up in the infantile phase and lie dormant through the succeeding non-sexual stage. Although the latter is marked by a lack of interest in the opposite sex, yet other activities are pursued vigorously then that are characteristic of one's own gender at this stage. After puberty the division of activity tends to continue, but the erstwhile mild antipathy shown towards the opposite sex is normally replaced by selective attraction to members of it.

One is led to speculate as to the purpose of the intermediate non-sexual phase. It can be regarded as an insertion into the human growth and maturation period that started in distant prehistoric times and gradually grew longer as the millennia passed. A baby is born with a sexuality that should develop towards maturity in a straight and continuous manner. But it is in infancy that great strides are made in mental capacity. During this first period the child not only learns, but also learns how to learn. This is the organic phase of mental growth with cultural input at first of somewhat limited scope. Yet before the child can take his place in the adult world a huge amount of cultural matter must be assembled in the thinking brain, i.e. the mind. This would give a purpose to the intermediate phase and, to make way for it, any development towards sexual maturity is suspended, both functional and behavioural. After puberty learning does not cease, but then tends to become a cultural embellishment itself, dealing more with selected specialisations rather than encountered generalisations.

One can surmise that the imposition of the intermediate non-sexual stage is the reason why sex looms so large in the superego concept. Apart from the fact that earlier repressions are prolonged by this insertion and thus become more deep-seated, the occurrence of this phase is one great unconscious blanket act of repression by the mind, which differs only in being a heritable evolutionary development. Unfortunately the later cancellation of its effect, known as puberty, does not affect the earlier sexual repressions, which have been recorded on a different basis. Perhaps it is just as well, because one can maybe look to such unconscious frustrations as the very cause behind our drive to pursue the cultural aims that have led to our present and past civilisations and other achievements.

The "thou shalt not" inhibitions erected in infancy cannot be broken in adulthood without incurring an obscure feeling of 'unpleasure' in the mind. The unresolved situation may be alleviated by psychoanalysis; otherwise discrete intense mental activity is required, a diversion to blot it out. Thus are created your artist, your scientist, your 'workaholic', your civilisation builder, your man of destiny, your dictator, etc. Of course, all of this is additional to the basic and animal desire to be the alpha male or female, the "top dog". Man's inhibitions drive him into circumstances beyond any urge just to become leader of the pack or herd, which trait is simply an inherent selective breeding procedure with evolutionary ramifications that were generally advantageous for the species, at least under the more primitive of circumstances.

Sleep and Dreams

Normal sleep is a form of unconsciousness that is not deep-seated. The body does not become excessively torpid; stimulating the senses can readily arouse the sleeper. The external difference is that the eyes are shut; this serves to exclude the stimulus of light when applicable, but emphasises just how much vision contributes to consciousness. However, there are deeper forms of unconsciousness, such as those resulting from knockouts or anaesthesia, from which one cannot be immediately aroused. Hibernation is also of greater depth than normal sleep and during it the metabolism of the body is markedly slowed down. The question arises as to what purpose normal sleep serves.

Dreams accompany sleep, but as a rule such are only well recalled if one wakes up during one. Experiments have shown that if sleepers are continually woken up during dreams the results are deleterious to mental well-being. This suggests that one purpose of sleep, perhaps now the main one with us, is the allowing of dreams. Yet how did sleep itself originate?

Each day has one light and one dark period. During the daylight hours the relevant hemisphere of the globe is bathed in the Sun's rays. While many kinds of animal may have periods during which some of their external activities are switched off, among the land-dwelling vertebrates there is a clear tendency for creatures to specialise for either a diurnal or nocturnal life-style and to sleep during the other period. This is a simple matter of the maximum conservation of energy when there is nothing to do. This might even be regarded as the prime reason for sleep, but it has later come to be adapted and used for other purposes, these being beneficial to the mind rather than the body.

The puzzling aspect of dreams is their apparent senselessness, their obvious need for 'interpretation'. They are in fact not intended for understanding by the Self Conscious Mind and it is only by chance that they sometimes stray into the realm of consciousness. Or at least one might add that they sometimes are leading up to some either highly pleasurable or extremely horrendous event and the imminence creates such tension that the sleeper awakes and can recall the dream, either with disappointment or relief.

Characters or locations may be to a degree recognisable in dreams, but the action is not a replay of any occurrence in times past; it is an invention. So what purpose does it serve? On what basis is it created?

The dreams of young children are known to fulfil wishes. Something has been denied them during the day, but they can redress the balance in a dream, even if this is in a modified or generalised form. Wish fulfilment dreams can take place because inhibitory changes have not taken place in the brain. These are only established when the denial of the pleasure is continually reinforced. In adulthood, once the inhibitions have been set up, the denied activities not only cannot be consciously performed without unease, but the unconscious mind is also prevented from exactly reproducing them in dreams. However, the desired effect can be achieved by means of a subterfuge; during sleep a dream is created that secretly embodies the forbidden activity by disguising it with symbolism, which serves to hide the true meaning of the dream from those layers of the unconscious mind where the inhibitions lie, as well as from the conscious mind should it chance to become aware of it. The means used by the mind to create dreams is the conditioned reflex mechanism.

To recap, the conditioned reflex mechanism operates when events that are not necessarily conceptually close occur under circumstances of proximity, both in space and time. Yet repetition of the circumstances results in them becoming closely linked in the mind – the association of ideas. A dream may revolve around some activity or action that has been desired, but its realisation frustrated by external censure or inhibition. The mind, through its agent the brain, overcomes the difficulty in the dream by representing the circumstance in ways that are unrecognisable to intelligence or reason. In an associative way, as in symbolism, the mind actually arranges things so that to the baser levels of the mind, where the frustrated desires have been confined and lurk, the desired actions seem to have taken place and alleviation of the deep-seated tension results.

Apart from pleasant dreams, there are also unpleasant ones, the worst being known as nightmares. The last can hardly be substitutes for frustrated sex or other pleasure, but can perhaps be shown to be associated with emotional situations of a different kind. Sex substitution dreams are caused by denial of early sexual experience through external proscription. This tends to inhibit approaches to sex in dreams, even though most folk nevertheless do find themselves experiencing them, which in such cases are simple childlike wish-fulfilment experiences, as are most daydreams. Nightmares are connected with fear, an early entrant into the lives of humans and which, like sex, tends to get bottled up, because we are alarmed or even ashamed of the emotion it arouses. In a nightmare fearful circumstances are invented by the mind, perhaps with the intention of proving to itself that it is capable after all of standing up to that which is so terrifying. Perhaps one might understand from all this that all dreams are beneficial and conjured up to satisfy a pressing need.

This might explain why the watching of dramatic presentations can be so enjoyable; they are dream-like in that one seems unable to influence the action, which can contain elements like simulated sexual encounters, enacted terrifying situations, etc. They differ from dreams in that they purport to make sense. Writing drama may itself attract because it seems to allow one to control 'dreams', while acting appears to give the chance to allow one to take an active role in such. Frightening experiences can be enjoyed in drama, even when the content is as unpleasant as nightmares, but give relief to those parts of the brain, those hidden recesses of the mind, to which the real and unresolved fears have been consigned.

The components of meaningless dreams are based on the same covert ideas of comparison and association as lie behind sympathetic magic. The persons and things are symbolic, metaphors for concepts made unmentionable by inhibition. In addition the whole dream itself is a metaphor for the relevant repressed emotion, such as lust, fear, greed, etc. If one is aware of a derogatory word for some experienced feeling, then a dream can be had about it in disguised and meaningless form. The metaphors are constructed in the same way as one is able to think at all, using the comparative and associative powers of the so-called Conditioned Reflex Mechanism, as outlined above.

Here one can recall that civilized man does not believe in magic. He uses metaphors and symbolism to refer to matters and objects indirectly, with these constituting a linguistic device. The difference with magic in the anthropological sense is that here an object can be regarded as being truly identical in nature to the entity it represents.

■ The Wider Aspects of Sleep: Dogs

The preceding paragraphs refer to the occurrence of sleep in a human being. It is an explanation of the extra benefits of sleep in addition to the need to conserve energy.

Other animals also exploit it in this way, including those close companions of long standing – dogs. When dogs are asleep they can sometimes be seen to be dreaming, as indicated by eyes partially open, mouth in a snarl and legs twitching. The dogs may well be dreaming, but are its dreams of the same nature as those of humans?

Dogs may have a variety of dreams, but the state described above indicate that the dog is running in its dream; not only running but hunting. If an urban dog is taken to the countryside it will chase a rabbit or run at a sheep unless restrained. Because of their association with humans dogs are prevented from their natural behaviour, and that is to behave like wolves. Instead of being in a wolf pack with a leader, the actual leader of the pack they find themselves in is a human. In the early days of association with men wolf-dogs could still behave like wolves when they accompanied hunters.

Apart from certain breeds this pleasure is now denied dogs. They are trained to do other things, but deep down they still have it in their nature to hunt. The dog can overcome this denial of its natural instincts by indulging in dream hunts.

■ Primitive Traits in Dogs

The above indicates that dogs are still pack animals. In a wolf pack there is an order of rank. A dominant animal will expect one of inferior rank to be submissive. This is done by giving the right signals, such as exposing the throat or lowering the head and tail. The offer of a throat to a wolf of superior rank actually inhibits attack. In a human pack a dog accepts that it is inferior and does as it is ordered, especially by the one it acknowledges as the leader, its human master.

One way that dogs acknowledge dominance is by height. A submissive dog will lower its head before a dominant one. Humans tend automatically to dominate dogs by towering over them. However a big dog, especially one of a fiercer breed, may challenge the dominance of a small human child, who however not only does not recognise the signals, but is quite unable to give responses that the dog expects. This all suggests that it is always risky to allow a small child to play with a big dog, no matter how domesticated and docile it may appear. The dog may act badly in this situation and revert to dealing with it in a manner used in encounters with other dogs.

Due to their long association with human beings dogs have evolved behaviour to suit their generally subordinate situation. They have collectively adopted a life style that enables them to gain advantage from this relationship. Apart from inbred behaviour they learn to do things to please us. However, they are still sending out primitive signals that can be interpreted by other dogs, but are usually ignored by us, unless the import is obvious, such as whining and tail wagging, or may appear to be, as when growling, snarling or barking.

But dogs are forever testing each other out in ways beyond our ken. A big dominant dog will send out appropriate signals to a small subordinate dog, who will escape corrective punishment by appropriate behaviour. For example, licking of the face is less a sign of affection than of submission. But in the human pack a big dog may be unsure of his rank. There are members he towers over and he may under certain circumstances try to assert dominance over them. A baby has no idea how to react to such signals, while a small infant will also ignore them and probably continue to send off inappropriate ones of his own. This can sometimes lead to disaster. The implication is that families should never leave a small child alone with a big dog, even if apparently very docile, because the situation may trigger off some primitive behaviour in the dog over which it has no control. Such an impulse may completely override the cultural veneer acquired by its kind in the course of the enduring close association with humanity.

CHAPTER 20

The Significance of Human Characteristics

In the rest of this work obsolescent terms are occasionally used for the basic groups that humanity is supposed to fall into; this is a matter of convenience. This retention is done despite them being too simplistic, as well as allowing the opportunity for touchy people wrongly to infer derogatory undertones, despite these being unintended. They are used here simply as labels, terms of reference that are necessary to distinguish historical human categories. Thus, human beings are now all the same species, but not all of one race, both in the present as in the past; and whether we like it or not. However, for various reasons, racial blending has accelerated in recent centuries, which makes the study of racial characteristics more difficult among existing populations. Also, for historical and environmental reasons, races can vary in type of culture and, selectively, in ability. Tendentious use of these considerations has been disastrous in the past; caution is hence needed so that use of the evidence chosen is not excessive in nature, neither for political aims nor cultural purposes.

Physical Form and Behaviour: Movement, Stature and Stance

Man is the only species of mammal that habitually moves around on two legs and with an alternating gate, although this mode of locomotion occurred among the dinosaurs and was originally universal among the birds. The human gait is coupled with an upright stance that is also an improvement on the behaviour still observable as a tendency among the anthropoid apes and linked with a lifestyle that involves spending time down from the trees, even if the holding of such a stance is of short duration. The Asian orang-utans and gibbons prefer to stay up in the trees, but one might note that in Africa, the putative source of humanity, chimpanzees spend most of their time on the ground except when either foraging aloft or sleeping, while gorillas go even further; when food is available at ground level some adults may only use branches in the canopy as a place to sleep. They construct nests for this purpose. The largest male gorillas (silverbacks) may have even less need to climb trees since they sometimes build their nests on the ground.

Other types of animals have all tended to avoid the upright posture when using the bipedal walk because of the use of the tail for balance, these including the dinosaurs and birds, and to a lesser extent the monkeys.

As with other apes, in the trees the forelimbs of ancestral humans were used to grasp branches while climbing and feeding up there. At this earlier stage their food may have included smaller animals, such as invertebrates, as well as leaves and fruit. When they later descended to the ground on a prolonged basis the feeding function of the hands continued, but the climbing function of the forelimbs gradually became subsidiary to a capacity to grasp loose objects, such as sticks and stones. In parallel with the development of manipulation, use by the mouth as the primary instrument for the catching or collecting of food by primates in general has dwindled and finally ceased. It became the norm for one hand to pass food to the mouth. Except for dealing with large items of food, this tended to persist even after any need to cling on with the other hand ceased. The original development of use of the hands for feeding enabled food to be gathered that was awkward of access or even out of reach of the mouth. Use of the forelimbs to collect food and pass it to the mouth is not unique to primates. Squirrels also do this, but in their case, while actually eating, the use of both forepaws together is the norm.

Further evolution of manipulation associated with feeding led to the human precision grip gradually being added to the existing power grip. As it improved, the precision grip, together with parallel mental developments, allowed objects to be skilfully manipulated as well as just held. Man the tool user came to be in a different class to that tool-using ape, the chimpanzee. The human hand, developing in parallel with that unique human achievement known as 'language', led to the start of a decisive cultural evolution in addition to the biological evolution inherited from the animal world in general. The unique breakthrough was not just that man became a toolmaker; even some birds can be claimed to do this. Man's great distinction was that he started to use tools deliberately to make other tools. One might use this very important criterion to distinguish behaviourally the advanced hominins from more primitive hominids. Today it positively divides humanity from all other apes at the most basic level. While apes can be seen to handle things in a similar way to humans, there is a great gap in manipulative ability. An ape's hand performs rather like a human one when this is affected by arthritis. One needs only to consider the capability of a concert pianist to appreciate the gulf in dexterity that exists.

Some chimpanzees use stones to break objects. When such activity is also applied to other stones it is possible to perceive how the use of tools to make other tools might have begun. A further step was the deliberate production of 'edges' rather than surfaces that were merely useful for imprecise impact. Stone hand axes were an advance on stone hand hammers.

Although the human hand remains essentially the hand of an ape, it has indeed been subtly modified in ways that created the versatile organ that is so familiar and taken for granted. Arm and hand evolved together to improve movements that had originated to allow brachiation (arm swinging) up in the trees, together with more general grasping of branches and collecting food. Some twisting can be accomplished by the whole arm from the shoulder and this is supplemented by increased torsion within the forearm. The latter eliminates the need for twisting at the wrist. Here the hand can be bent all ways relative to the wrist, but without twist being involved, although further twisting is possible within the hand.

The human leg has a range of movement that is different from that of any ape: the hipbone and its joint have evolved to allow the particular fore and aft movement necessary for human walking, without the peculiar gait or pronounced body sway and the retention of bent knees seen with apes when they go into bipedal walking mode. However, the range is generally less than with the arm and the ability to twist, as observed within the arm, is reduced. The rotation that is available at the hip and the ankle provides the flexibility needed within the directional range normally required.

The foot of an ape has retained its primitive hand-like grasping qualities, with the big toe being thumb-like in appearance. Such a foot is not adapted for walking on the ground, although it does not prevent it. The human foot differs in that the 'thumb' is drawn up alongside the other toes, while its associated pad forms the major point of support for the foot's inner edge. In order for this to come into being it would seem that the human foot's 'thumb' must originally have been relatively large and perhaps less opposable than those of simians in general. In keeping with their changed functions relating to walking and balancing, all human toes have retained a capacity to bend back at their base joints, a facility largely absent and presumably a very early loss from the fingers. This might be thought to suggest that when the early human ancestors used all four limbs when walking, they may also have passed through a 'knuckle touching' phase. Yet it can be noted that man never does this now when reverting to 'all fours'; the flats of the hands are used. This echoes the mode used by baboons and indeed monkeys in general. This suggests that knuckle touching, the norm with other apes, was never a human trait.

The state of human feet indicates an evolutionary drive that was due to a thorough and enduring abandonment of tree climbing as a normal feature of society. Human feet may not be visually attractive body parts, but they are excellently adapted to supporting a person, whether in standing or walking mode. Yet, while apparently no longer assets when it comes to climbing trees, they restrict this ability rather than prevent it.

The difference between the human foot and those of all the apes can be regarded as of great significance. There is little sign that the feet of gorillas and chimpanzees are adapting in the same way to walking on the ground as ours have, although, if one thinks in terms of preparation, then their big toes are at least in a better position than is the equivalent digit on an orangutan. The great difference between the human foot and those of apes seems to indicate that its special and final development occurred after the ancestors of the relevant hominins left the trees on a permanent basis. The result of this can be seen in the remains of Australopithecines that have been found in East Africa and South Africa. This group that consisted of various small bipedal hominins provided the base from which arose the more advanced *homo habilis*. Footprints attributed to one kind of *australopithecus* suggest that the big toe was as yet not fully bound to the others. Even so, it might seem that feet possessing such a feature at all had been in some way destined to be suitable for walking on the ground

The (rear) feet of apes and monkeys are not ready for similar adaptation for walking at ground level, and apparently never have been. The kind of monkey that is least arboreal is the baboon. They are generally adapted for life on the savannah, an environment that other types of monkey and anthropoid apes mostly cannot tolerate. Some types of baboon use rocks or cliffs as places of refuge, rather than trees. Yet while they seem to reflect the same abandonment of woodland as the ancestors of *homo*, they have remained quadrupeds and their rear feet remain those typical of a simian.

A common factor with baboons and the ancestors of *homo* was that coming down from the trees led to a tendency to live together in larger groups, which in the case of humans and some kinds of baboon involved eventual leaving the trees altogether. It was hence the imperative to spend more and more time out in the open that eventually drove them towards the safety to be found in groups with increasing numbers.

Apart from apes and baboons, it can be noted that certain smaller species of monkey spend a lot of time down from the trees. Here one can consider examples like the vervet monkeys of southern Africa and some of the macaques of southern Asia. It was not necessarily increase in size that drove some primates out of the trees, but rather reduction in the tree cover and the bounty supplied therein, together with loss of the continuity of access provided by the canopy. It can be noted that primates that forage thus on the ground, such as some baboons and the vervets, are bush dwellers rather than inhabiting dense forests. With the passage of time, increasing descent from the trees allowed diets to be extended and eventually changed, but initially this would occur out of necessity rather than choice.

In the reduced areas where forest survived, the apes they supported did not develop human characteristics, such as those derived from standing and walking on the (rear) legs. Chimpanzees have moved towards this state, but ongoing restriction to a wooded habitat slowed down and even inhibited progress in this direction. It was the ancestral forms of *homo* that were involved and as they found themselves in an environment more and more denuded of trees they had to evolve both culturally and physically, or die out.

The earliest known apparent ancestors of *homo* are the diminutive Australopithecines of southern and eastern Africa, who were distinguished from simians by their habitual and persistent standing and walking on two legs. They were of smaller stature than all later hominins. But who were their ancestors? These are elusive and this is perhaps due to them not only being originally rather rare, but also occupying areas elsewhere, from which they were driven as the woodland was diminished. Their smallness was probably another factor. When forced to survive without forests they presumably had to evolve a larger, more robust structure, or they would have been eliminated. Even so, their frames seem to have remained rather slight compared with other apes. All this would be in parallel with improvement of their social organization and a more developed stick-and-stone based technology. The driving force behind these developments would be the existence in the more open landscapes of some larger animals, some of them hunters and others that eventually would be hunted. To perceive the conditions of the small ancestors of the Australopithecines one can imagine them as becoming socially organized for existence in the bush as are the troops of meerkats found there today.

The structure of the (rear) feet of anthropoid apes, such as chimpanzees and gorillas, makes it difficult to envisage that feet like these could ever evolve in the same way as have those of humans. It seems unlikely that their opposable big toes could ever fall into a row with the other toes, as is the case with man. Perhaps these apes have long been too big for this to happen. The gist of the matter might well be that, while the earlier ancestors of men (who were smaller) developed arms and hands similar to those of gorillas, i.e. designed for swinging from branch to branch, their (rear) legs and feet were more like those of some monkeys. They could make progress up in the trees in a way different from other apes. Their rear feet evolved away from suitability primarily to grasp branches, but rather for running along stout boughs (as indeed do the smallest apes, the gibbons) and hence better adaptable to further evolution as organs more suited to moving about at ground level.

This all points to the earlier simian ancestors of *homo* being among the smaller members of groups who were to evolve into the various hominins. These smaller apes were pre-adapted for life on the ground because of their rear feet. The ancestors of the gorillas, chimpanzees and bonobos retained feet that were more suitable for grasping boughs than for walking on the ground. Yet they did come to spend more and more time below, but this, together with the availability of food there, was combined with a tendency to increase in size. The gorilla, the largest ape, is a more lethargic mover in everyday activity, especially when compared with the chimpanzee. One might also note how the smaller 'pigmy' chimpanzee (bonobo) spends more time aloft than his larger relatives. A gorilla on the other hand likes nothing better than to lie around among his food, as found among the luxuriant growth forming the underbrush of a reduced tree cover, which last feature, high above the ground, is now of less interest to him now than it used to be to his ancestors.

A major difference between human beings and other primates is that none of the latter kneel. When not just untidily lounging about they often rest in a squat. Some weight may still be taken on the soles of the feet, but the bulk of the body rests on the rump, since this touches the ground or other surface. The animal's knees can be drawn up and may be held apart to flank the body. Human beings may also adopt this mode of sitting on the ground, especially among more primitive cultures, but modern chair-sitting man is generally not comfortable in this position. Various sitting positions may be assumed, either because they are more comfortable or more convenient. Under certain circumstances men may simply extend their legs forward or squat on their heels, a position that in 'pitman's' parlance is known as 'sitting on your hunkers', this being reminiscent of Danish *sidde på hug*. In this the weight is taken by the pads behind the toes, the heels being lifted. For those not used to it this squat soon becomes uncomfortable and use it only as a temporary measure, if at all.

The human frame has been further modified to suit bipedal walking and a completely upright stance, in the latter case with vertical straight legs under the body and with the pelvis exhibiting major changes. Associated with this are the large muscles that constitute the buttocks and these again have been given further duties as support pads when sitting. Habitual buttock sitting (as opposed to squatting) is a human specialty and the rounded features themselves are often found attractive both aesthetically and erotically, presumably because they comprise appreciated smooth curves and thus anciently differed radically from homologous areas on apes and monkeys. Rather than just being sources of speed and strength, the huge muscles of the buttocks may initially have developed because of a need for greater staying power among hunters once they started to track larger game with the intention of wearing it down during a long chase, after which it could be overwhelmed by numbers. One can readily observe that Bushmen have relatively large rumps. This reflects a phase before more sophisticated hunting methods were developed, such as organized ambushes and trapping. The stamina needed for hunting by tracking will have contributed to the ability of earlier hominins to travel far when food gathering and eventually undertake long migrations when the need arose.

Otherwise modern folk – when away from furniture or other raised ledges suitable for sitting – may resort to kneeling on the floor. But why are humans the only primates to do this? It is not a natural position, yet it clearly has long been a feature of the human condition: the knee has been modified to support the human body weight, even being provided with a pad. Yet it is not all that well designed for eventualities; it is rather vulnerable to damage by impact.

Squatting is suitable for doing tasks that involve the holding of things fairly close to the body, such as nursing babies or handling or working on light objects. But when it comes to working with heavier objects or the application of downward pressure, e.g. using a saddle quern, kneeling provides a better position. Kneeling has become habitual for humans because of advances in cultural evolution that have not been approached by other primates, even chimpanzees, the most advanced ones.

Kneeling is just one of the activities that humans can undertake that other primates cannot. Such exceptional abilities have resulted in the need for performances that have led to both behaviour and structures that have been made permanent as the result of evolution. Human babies embody the evolutionary destinations of their species. Apart from being born into an environment that accommodates such developments now, they also are pre-programmed to follow the human evolutionary path, which is the specific ontogenesis.

All young primates are born helpless and each mother needs to hold her baby close to her body, not just for suckling, but for safety. But newborn simians soon learn to grip with their four feet, so that they can cling to their mothers' hair, this providing a backup to being held and is advantageous for creatures that spend much of their time foraging in the treetops. It seems safe enough to infer from this that ancestral humans behaved similarly before they left the treetops and that at the time they retained a good covering of hair. Another form of behaviour that developed from clinging to the motherly front was back riding, clinging by grasping the hair on the back of the mother and, especially when grown larger, other adults.

There is something in the upbringing of human babies that physically prepares them for the ability to kneel. All such babies initially go through a phase when lying supine is the norm for them, before a state is achieved where crawling is possible. In the initial condition the baby is laid down on his (or her) back and cannot move the body as a whole, neither from this condition nor away from the spot. However, a degree of rolling can eventually develop, which can result in an ability to turn over. Once lying on the front is achievable on a regular basis a degree of uncontrolled change of position occurs through random movements of the arms and legs. This seems to awaken in the baby the desire to explore his immediate environment, which is encouraged by the prone position making this more easily visible and which in turn reinforces the ongoing desire to get off his back, while acquiring the ability readily to do so. His parents may also recognize this change and then habitually put the baby in the prone position when laying him on the floor so that the tendency towards crawling can be exploited.

A baby wants to get its head in a higher position. As opposed to lying on the back, when lying prone the head can be lifted and this enables better all-round viewing than is possible right at floor level.

Eventually the activity known as 'planking' may then take place. In this the head and body are lifted from the floor with a bridge being formed, this being achieved by pushing upwards from support points at the hands and feet. This has been initiated by the arms forcing the head upwards to enable a higher viewpoint to be achieved for scanning the immediate environment. In the meantime the baby may become aware of the body angle being better and the span reduced if support at the rear is provided by the knees, rather than the feet. In this way an early form of kneeling appears.

But this is not enough: more height is required. The random movement realized by the friction produced when moving the arms and legs against the floor has been replaced by this becoming more organized so that purposeful forward movement can eventually be achieved (crawling). This purpose is then better fulfilled as a result of the planking exercises: forward motion is then performed much better by progressing on all fours, with the body being supported by the hands and knees (or feet) and movements of these points - as already practised in crawling - being naturally translated into 'creeping', with a tendency to use the co-ordinated alternating pattern that is so instinctive and efficient for most quadrupeds.

The next stage in the raising of head height is 'standing'. This is at first difficult to accomplish without support. But creeping allows upstanding objects, such as furniture, to be approached and, after initially being able to haul himself into a sitting or kneeling position, perseverance by the baby results in the ability to stand. What follows then depends on the perfection of 'balance'. With this stage reached, standing without holding on is possible, followed by ever improving bipedal walking. By this time the baby prefers to sit or kneel on the floor, rather than lie, and comes to be able to stand up from such positions without using supports and with great ease. The mentioned use of 'furniture' in this way has replaced the primitive use of the mother's body or nearby natural objects for this purpose.

The behaviour of simian young is quite different. They are never laid aside by their mothers while they are still helpless, and they only put them down after they show themselves as capable of secure physical activity while able to support themselves by their four feet, a necessary stage to be passed through when living aloft in the trees. Even when still small they may also independently clamber off the motherly breast onto a supporting branch or the ground. In doing this their first contact is with the rear feet, which provides them with early acquaintance with bipedal standing. But once they relinquish hold with the hands they then move away as quadrupeds. The simian ancestors of *homo* and the predecessors of the Australopithecines must have gone through this stage before change set in and a sequence gradually adopted that involved the laying aside of young by the mother while still at this helpless stage. Such behaviour indicates a serious change in circumstances that exerted a great deal of evolutionary pressure. What caused the human baby eventually to be so different?

The human being is the only primate that habitually stands and moves around on two feet. While apes such as chimpanzees and gorillas also do so occasionally, monkeys hardly ever do: this reluctance is tied to their more arboreal lifestyle and the retention of a tail. The behaviour of all simians on the ground is related to their mode of progression in the treetops. Indeed, the 'more primitive' ring-tailed lemurs of Madagascar have developed a bipedal way of progression on the ground. It is a skipping motion that relates to the way they leap about when aloft. Monkeys leap between trees in a different manner. Their way was never going to lead to habitual walking on two feet. This even applies to those monkeys that are least dependent on a habitat where trees are found – the baboons.

The tendency towards bipedal walking on the ground arose among those simians who had lost their tails and had arms that could reach more easily above their heads, and this had evolved with the acquired ability to 'swing' through the trees. This activity favoured an upright posture, even when just clambering about among the branches, and this could then be transferred to other situations, e.g. when on the ground.

With regard to locomotion at ground level, the difference between apes and human beings is that the former can and sometimes do stand and walk on two feet (albeit with bent knees), whereas the latter not only can, but under normal circumstances they always do. After the baby stage, creeping on all fours became difficult for them long ago and is now only undertaken when special circumstances call for it. The difference has arisen because other 'apes' have kept their dependence on trees and with all of them the ability to climb trees well has been retained. This is reflected in their physique, including the grasping function of the rear feet and the inability to straighten the knees. However, one kind of 'ape' broke away from dependency on woodland and this opened up new opportunities when dense forest was first replaced by scrub and eventually by grassland. These creatures led to the Australopithecines, who were like little men, but with the heads of apes. Yet though the style of the head was initially conservative, once evolutionary change was well on the way to perfecting a man-like body, the head was destined to evolve in its wake to resemble the feature that we all bear atop of our bodies today.

The Australopithecines were descended from some kind of African ape, but no trace of any such early creature in a satisfactory intermediate form has been identified. While still up in the trees they would be like some short-armed gibbon in size, with monkey-like hind feet, but with a genome that is now most nearly represented by the chimpanzee. That the early and arboreal form is elusive in the record is to be expected, but why have no examples turned up of the subsequent and diverging forms leading to the Australopithecines? Since they were forced out of the trees due to environmental change they were probably under considerable evolutionary pressure and such vulnerability would initially keep their numbers small. Over an immense time span it was only after suitable physical and behavioural changes became established among select populations that they were eventually able to exploit the environmental drifts that were taking place in Africa, rather than be constrained or overwhelmed by them.

It is obvious that those monkeys and apes that were deprived of trees had to change or they would die out. It is clear that one type of simian was poised to become a super-ape in a way that the equivalent monkey, the baboon, could not match. The great future changes were initiated by the way they handled their babies.

The remote ancestors of *homo* that were destined completely to abandon the arboreal lifestyle had physical characteristics that distinguished them from other apes. Apart from bearing a covering of long hair on the head (a mane), they had hind feet from which it was possible for feet as found on *homo* to evolve. For convenience they can indeed be regarded as the equivalent in Africa of the gibbons of south-east Asia.

Apart from physical considerations, leaving the trees allowed certain behavioural changes to occur, especially with regard to the treatment of babies. One of these was the custom of laying babies down while still helpless. This behaviour was not possible up in the trees, unless secure nests were involved, which seems unlikely. The possibility would arise when the troops of such ground-dwelling hominids carried out their foraging operations from 'secure' (albeit temporary) bases, which might be regarded as homes. At such bases babies did not need to be held securely to their mothers' bodies except when suckling, although body nursing could still be practised because it was a source of pleasure, rather than necessity. Yet the advantage of occasionally avoiding nursing was that it freed the mother to do other things, such as adult social activity, especially involving sex, or work requiring manipulation. This still applies today with us the latest representatives of *homo*.

The advantage to the mother of not having continually to nurse the baby probably came later, at least where everyday activities are concerned. A specific result was that, with a child reaching the laying-on-the-floor stage, the next baby could be produced earlier – or, if occasional twins be considered – could be better cared for. The possibility also opened up for the elder or other child to be attended to by other members of the family. However, the particular advantages felt by the baby who had been laid aside were perhaps of greater significance, since this practice was to give the whole species an evolutionary boost based on the freedom thus provided to move experimentally.

How a baby primate grows into an adult depends on the development of the senses. As with other apes the human baby is born with a relatively poor sense of smell. Because it will experience nursing from the point of birth any baby ape does not need its sense of smell to find a teat, as is the case with some other mammals; its mother leads its mouth to the nipple. This has contributed to a decline in this sense among apes and the vacant capacity thus produced in the forebrain has become available to be taken over by something else: it has enabled this zone to be the kernel from which has grown the cerebrum, which in *homo* has led to a massive brain.

In humans the need to test food by smelling has been catered for by the nostrils being positioned immediately above the mouth so that smells are detected involuntarily or voluntarily while eating, whether pleasant or foul, and whereby the substance can be confirmed as being edible or otherwise.

As with other apes, being a creature of diurnal lifestyle and arboreal habitat, the sense of sight became more importance than smell early in the line leading to *homo*. A newborn baby takes some time before his vision becomes clear. The first persistent visual experiences normally involve his mother's face and he begins to associate this with pleasurable experiences such as feeding and care. At first a baby does not mind being laid down on his back so long as he feels well fed and has no bodily discomfort. From this position he can freely see the faces of other family members, and especially that of his favourite, his mother, once he is capable of making such a distinction.

■ Perception and Language

But as his eyesight further improves and the capacity for random head and body movement acquired, a baby also becomes aware of its wider environment. Such observation arouses an interest and desire to explore and this will culminate in the acquisition of some motility (as already described above).

However, once acquired, vision does not change in essence, and man's initial perceptions are of the same nature as those of other animals, as well as those of the primitive ancestors of his own species. Indeed, his grasp of his environment only changes and improves when he comes to realize that perceived entities have been given names and that this enables him to utilize things to his advantage in ways additional to those that come naturally to him.

A human baby is born with the ability to hear and is probably familiar with certain sounds heard while still in the womb. At the time of birth a baby often needs to be encouraged to start breathing. This can be brought about by a suitable slap on an appropriate part of the body, whereupon the newborn not only shows he can breathe, but also howl.

So a baby goes from simply being able to hear sounds to eventually reach that great achievement of mankind, the ability first to receive and understand, then transmit in kind. It has then taken the first steps in the use of language. How is this initiated in detail?

The human baby's brain, like those of other species, is not just a piece of matter that starts from complete unawareness. Evolution has ensured that it is equipped at birth with the neural infrastructure that will enable it to follow the growth in perceptiveness that is normal for the species and the ability to build on this. It is thus ready to cope with such primitive concepts as cause and effect and the association of ideas. It first realizes that when it cries it gets attention. From this the corollary is eventually understood that if attention is required, start to howl. Even when the requirement is not urgent, the notion of 'control' is born and the activity deliberately pursued for the feeling of pleasure that is engendered. This association of ideas – i.e. howling with attention – can later spread to different areas of contact with other human beings. This is indeed the inception in humans of the conditioned reflex mechanism as discussed earlier and leads to automatic reactions that are slower than other reflexes, but likewise without conscious intellectual interference.

The foregoing is very important with regard to the importance played by the sense of hearing, in that it enabled the human species eventually to reach its pre-eminent position among the other living beings who came to occupy the planet Earth.

A baby becomes familiar with certain sounds before it becomes obvious to attendants that such is the case. Eventually it will not only come to recognize nearby objects and entities, but will also indicate this by using its vocal apparatus to utter sounds that represent a version of the names of these. This usually starts with that entity it feels closest to – its mam or mamma. Having once learned that this sound is associated with its mother, by a process of trial and error it succeeds in uttering an ever improving version of the sound itself. This is helped in the case of 'mam' by it being associated with the use of the lips when suckling. Simply by raising a sound while repeatedly opening and shutting the lips produces a 'mam-am-am...' effect. By means of the incipient conditioned reflex mechanism it can be noted that this process is primitively the cause for the association of the sound 'mam' with mothers. It is also prominent among the first steps on the way to being able to understand and use language. A baby comes to associate specific sounds with objects it is aware of and then realizes that - using the brain power being acquired - it can enhance this idea by using its own voice. This is very satisfying, especially when encouraged, and the practice develops, even eventually to encompass the wonderful world of abstract ideas that the human intellect has created.

The conclusion of all this is that the main sense that allowed the growth of the human intellect to a position of such superiority in the animal world is that of hearing, and that this occurred because particular sounds acquired specific meaning and eventually became organized into the oral/aural systems that we call 'languages'.

◼ Skull and Jaw

Unlike gorillas, chimpanzees and baboons, early humans lacked the biological means to defend themselves against dangerous animals, such as having great strength and fangs. Even so, vestiges of the fearsome canine teeth of other apes and many monkeys can still be seen in present human populations. In all cases these flank four incisors in the upper and lower jaws. The reduction in size of human teeth and jaw accompanies the fact that they have long abandoned biting as a major ploy in both defence and offence. Apes and other simians often use the hands as offensive weapons, with a form of slap being common.

The human skull contains basically the same bones as the skulls of other animals, but continues the trend towards retracted jaws and enlarged cranial capacity that is observable among some other primates. In comparison with modern man, earlier hominin skulls are characterized by having a more massive structure, receding foreheads, brow ridges and large yet chinless jaws. Evolution towards *homo sapiens* has shown a tendency for these aspects to be eliminated, although some modern races still exhibit vestiges of such primitive traits. Yet the retention of such in the skull hardly presents an obvious practical drawback, especially when the cranial growth has allowed an increased brain capacity. However the chin, with its association with advancing linguistic powers, has become more prominent with the passage of time, and this has been in combination with a general loss of robustness in the skull; and indeed elsewhere in the skeleton.

■ The Effects of Paedomorphosis

Among features of the child, one that can be observed is that the brain case tends to be larger in proportion to the face. With this in mind, paedomorphosis was among those trends that allowed the adult brain to increase proportionally within its containing head.

The occurrence of aspects of paedomorphosis, albeit selective in nature, can be perceived among extreme types such as the Eskimos and the Bushmen as mentioned above, among such features being the reduction of beard and body hair. More generally, selective paedomorphic differences in form may be taken by the human female as an aspect of sexual dimorphism.

Paedomorphosis works in parallel with the development whereby the reproductive system becomes active at an earlier date against general body development. The breeding adult is then physically at a more child-like stage. It was not so much a harsh environment that brought this about, but the changing physical requirements of the improving hunting-by-tracking way of life. Such primitive hunters were also able to achieve the brain they needed by relatively increasing its volume against the size of the body (and perhaps absolutely), while at the same time resisting the need to increase the length of adolescence. Paedomorphosis would progress in a population when the members could reproduce while still remaining fit to survive even though having reached the reproductive phase at an ever more adolescent stage of their development.

However, people whose culture was becoming more complex found themselves with an ever increasing trend towards a longer adolescence.

Even so, the Caucasoids of the west today might be considered to have sustained among them a greater tendency towards some aspects of paedomorphosis than is evident among the Australoids. Even so, as a generalisation, it might be assumed that all kinds of *homo sapiens* to some degree owe their structural appearance to the effects of paedomorphosis.

However, there would seem to be an anomaly. Although the Neanderthals of Eurasia had a large brain capacity, they maintained the primitive characters of massive build of skeleton and skull, including pronounced eyebrow ridges. They bore fewer signs of paedomorphosis, than the *homo sapiens* types that impinged upon them in Europe from the south at the east end of their range. This was to do with their generally more conservative culture. More specifically, their hunting method can be examined. The Neanderthals apparently were equipped to hunt larger creatures by rushing at them while wielding thrusting spears. These men needed to be very strong and fast rather than cunning, and their skeletons reflected this. This general robustness was also preserved in their skulls. But the newcomers to Europe had throwing weapons, with some eventually being equipped with spear throwers as examples of their more diverse toolkit. Such advances in their culture were reflected in the paedomorphic features they had acquired long before, such as lighter skeletons and lack of brow ridges. Along with this went more pronounced chins, which development indicated advanced language skills. It might well come to pass that in the future the Neanderthals will come to be regarded as brighter than we now give them credit for and they did possess rudimentary chins; but, even so, there seems little doubt that culturally they lay behind the level reached by their contemporaries, i.e. *homo sapiens sapiens,* the folk who came to share their territory for thousands of years before eventually becoming completely dominant.

The advent of paedomorphosis, casually considered, only seems to affect the physical appearance. However it would be improvements in the workings of the brain and concomitant improvements in culture that would be the driving forces behind the changes in appearance that occurred. Changes in lifestyle, including diet, meant that later *homo* would not need the heavy bone structures of his predecessors, and a leading means by which these were discarded was paedomorphosis.

Eventually a great breakthrough could be achieved when such cultures overlapped and racial mingling occurred. In such a situation, within the limits set by the way that genes work, those hybrid individuals that showed the best of both worlds would tend to thrive and multiply. Paedomorphic developments can be seen as being derived from adults who had become practitioners of higher forms of culture. These paedomorphic hunting cultures relied more on throwing missiles or the use of projecting equipment to kill their quarries, this aspect eventually being dominated by bows and arrows. This suited paedomorphs but necessitated a greater use of technique, rather than brute force. This does not imply that the intruding types of *homo* were culturally more advanced than the Neanderthals in every way, although their way of hunting might require greater co-operation, since a greater diversity of methods became available. Yet one can speculate that this led to social advances that did not occur with more primitive hunters, who tended to be set in their ways and hence less resilient in the face of threats to their existence. The end of the Neanderthals should not be regarded as a quick process. The most successful populations would arise where hybridization had bequeathed greater efficiency to their successors. The folk groups who were gradually whittled away by competition would be those at the bottom level, particularly when conditions for survival deteriorated. It would be unchanged or little changed Neanderthal populations who would lead the way on the path to extinction. However, it was the later arrivals on the scene that had the combination of characters that suited the way that European conditions were heading: the resultant populations may well have consisted of the better hybrids, yet in which the proportion of Neanderthal features acquired by their genome in the end remained quite small.

The physical progression can be understood thus. At one stage the core type of early humanity comprised *homo erectus* derivatives and exhibited heavy cranial features. However, there was a tendency to increase the cranial capacity while at the same time reducing the bone structure through paedomorphosis and selection. This development originated in Africa and from there spread along the southern shores of Asia. Their descendants can be recognised as the pockets of small hunting folk who have survived in these areas, usually in forested areas.

Apart from their material culture being different in many respects, the Australoid way of life can hardly be regarded as being on a significantly lower plane than that of the Bushmen of the Kalahari. The observable difference between Australian Aborigines' and African Bushmen's heads is that the latter are more child-like in appearance. In other words human cranial and facial evolution has to a large extent been affected by paedomorphosis, with Bushmen being more affected in this respect than Aborigines. Thus, while human evolution must certainly have been imperceptibly affected by the beneficial appearance of small chance mutations, more rapid evolution has been made possible by exploiting the separate evolutionary pathways followed by the various immature stages of the species. Over the generations members tended to reach puberty at a younger stage of growth and development. Thus can one argue that the children of earlier hominins naturally lacked the massive cranial developments of their adults and it was these features that were shed by the trend towards sexual and mental maturation at an earlier physical stage of that development - i.e. ontogeny - of the individual.

The last vestiges of the heavy browed and jutting jawed structure that was so typical of all earlier forms of *homo* were exhibited by the Neanderthalers. These inhabited Europe and Western Asia, but were excluded from the northern regions. Over the years their northern extent will have been governed by the state of the ice cap. In Asia to the east a similar but genetically distinct population has been discovered in recent times. They are known as the Denisovans. This realization has given rise to the suspicion that there are others waiting to be found. Indeed, the groups of *homo sapiens* who migrated out of Africa and impinged on the Neanderthalers would contain just such extra population elements. They are not recognized as such because their superior survival culture has since enabled them either to eliminate or absorb all others.

◼ The Sequence Leading to Homo Sapiens

Although there is still much ongoing debate on these matters, here for convenience the various stages are presented in simplified order using established nomenclature. The first recognized representative of *homo* was *h. habilis* and he appeared in Africa as a result of evolution among the hominins known as Australopithecines. After this particular advance, cultural improvements went hand-in-hand with physical evolution to produce related types who can be dubbed collectively as *h. erectus*. Further evolution then led to similar beings who can be grouped as versions of *h. heidelbergensis*, who were large of stature and from out of whose ranks eventually emerged folk at the Neanderthal stage of evolution. The typical body and skull of *h. (sapiens) neanderthalensis* remained significantly more robust than the usual anatomy found with modern man. The latter is designated *h. (sapiens) sapiens* and all 'men' found on Earth now belong to this single species.

Beginnings in Africa. While the whole of recent humankind can be seen as having origins in Africa, it is not clear as to how the present inhabitants of that continent arose, except for those resulting from recent colonization. These last arrivals cloud the picture in Africa, as indeed elsewhere in the world. Since the last Ice Age, to what extent has there been reflux from coastlands lying to the north (Europe) and to the east (Asia)? Apart from recent sea-borne Europeans and Arabs, with their commercial and colonial activities, the picture that presents itself is still obscure.

However, both physical and DNA evidence indicates that there has not been any ancient reflux of *homo* since the various occasions that have arisen when he spread into south-east Asia. This is the model to be used until research happens to prove otherwise. It is proposed that *homo erectus* was the first to migrate thus and his descendants found their way along the southern coastal areas of Asia where they evolved in their own way to form a population of rather diminutive hunters. The driving force behind this need to get out of Africa was presumably habitat deterioration there initially resulting from climate change and periodically making the Middle East a more attractive place to live.

■ Aspects of Evolution Leading to Homo Sapiens

The above argument can be used to tackle various problems, including that of Neanderthal Man. This Palaeolithic Eurasian type was heavily built, and showed some retention of the heavy cranial features characterized by *homo erectus*; yet his brain could be large. Here we apparently have an Ice Age hunting type that was restricted from evolving paedomorphically because of the lack of need to abandon his own cultural level. Yet eventually, and where circumstances allowed it, varieties evolved that had arisen from Neanderthal-Paedomorph crosses and tending to take advantage of the best features from both groups. Such advantages including the tendency drastically to reduce the heavy Neanderthal cranial features, but retain the advantages of some of his other physical developments and even cultural achievements. Changes in appearance were probably accelerated by the fact that some features were regarded as being ugly and their frequency diminished, if not eliminated, by sexual selection. (Yet others, such as hair colour, may have been preferred.)

Such hybrids benefited from combining the mental achievements and physical make-up of both parties. The Neanderthals contributed an increase in robustness that could be useful in individual and intertribal conflict. But the overall prettier appearance of the paedomorphs was preferred and dominated the selection process. This could explain the patchy survival of the paedomorphs across much of northern Africa and southern Asia. The African hybrids developed separately from the Asiatic ones and the balance of physical features was different. However, one might well assume that among the tropical paedomorphs both in Africa and in southern Asia, some shared a dark skin and curly hair. Hence any hybridization between evolving descendants of *h. erectus* and *h. heidelbergensis* would produce folk with combined features and they can be thought of as being represented today by the Negroes of Africa and Asian islanders like the Fijians. While their dark skin and curly hair would indicate discrete origins and give an appearance that was different from Eurasian crosses, they too will have tended to eliminate any external features considered ugly.

In the Far East the dominant type of paedomorph is the Mongoloid and one can see that it was probably originated from a degree of hybridization with Neanderthal-like folk (Denisovans?) and that these at some time had taken part in a persistent migration southwards. Here again the paedomorphic element has produced populations that have eliminated undesirable heavy bone structures. Along with the paedomorphic development, features have been preserved through association with populations that were built to withstand the fierce cold found towards the Arctic. This applies particularly to the face. There are of course intermediate zones between the western and eastern hybrid forms. The dominant strain among Afghans is clearly western in appearance (Caucasoid), but the adjacent Tibetans exhibit features that indicate evolutionary links with populations further east (Mongoloid).

It would seem that further to the west paedomorphosis occurred among folk who lacked typical Mongoloid features. Such eastern traits seem to be quite absent from those incomers of advanced *homo sapiens* type who, according to the records, eventually impinged on the extreme Neanderthal folk of western Europe. These intruders bore all the advantages of previous hybridization and it was perhaps inevitable that descendants bearing their genes were to form an ever-greater proportion of the population, eventually eliminating the Neanderthalers of the west altogether, at least in their original recognizable 'purer' form. The interaction took place over several thousand years and study of DNA has shown that some interbreeding occurred, but it may have been limited in that the new folk mostly found the Neanderthalers unattractive. Such sexual contact as did take place may not have been under congenial circumstances.

There is cause to ponder over why the *homo sapiens* who penetrated western Europe were apparently quite lacking Mongoloid features, considering how persistent such traits have been further east. One must presume that these were not derived from crosses with Neanderthalers, but rather from discrete hybrid strains as occurred in the orient.

The 'Negroid' hybrid type is basically a west African sub-Saharan development, but in argument below is claimed to result from migration and to have origins further east within Africa. While the smooth Negro skull can be regarded as progressive, the tendency of the jaws to jut seems paradoxically to be a physically conservative feature. One way in which this might be explained is that such a jutting jaw and the associated thick lips became regarded as desirable features because they differed from the equivalent ones found among the earlier Bushmen/Pygmy populations. Protruding lower jaw and thick lips are also noticeably less in evidence among some of the tribes of eastern Africa, reflecting the proximity of Eurasia, in which adjacent area the jutting jaw has been thoroughly eliminated, as exhibited by the fate of the Neanderthalers.

From the above it would seem that, apart from the slow change due to mutation, there were also two quicker driving forces behind the evolution of *homo sapiens*, as he is now found across the world – paedomorphosis and hybridization. However, one might well bear in mind that the latter requires social contact between forms that had not drifted too far apart from the evolutionary point of view. Any interbreeding would lead to intermediate forms, some of whom would be of superior nature one way or another. Such superiority could then lead to the emergence of dominant strains that in the course of time would tend to eliminate all traces of their immediate predecessors if they remained in contact. Hybridization requires that contact between relevant races occurs. This suggests the existence of relatively dense populations of *homo*. The effects of such contacts could appear and become dominant quite quickly. This is in contrast with the effects of paedomorphosis, which always involves a slow evolutionary process. All this indicates that it has had millions of years to take effect and such a process has been taking place since the very beginnings of *homo* and even beyond, i.e. in the preceding *australopithecus* stage. Throughout most of this time populations were very small in number and the likelihood of suitably related groups coming together before they had drifted genetically too far apart for interbreeding to take place was slight. With the increase in populations the occurrence of evolution through paedomorphosis due to groups coming together who were still able to interbreed accelerated. Indeed, its effects are still being felt today: in recent decades it has been noticed in modern societies that there has been a tendency for puberty to be reached at an ever earlier age.

■ Physical Adaptation to Environment and Way of Life

Many human groups still embody a stature that suits their present geographical position and traditional way of life. The Eskimos (Inuit) have short squat bodies that exhibit the best shape for conserving heat, since resistance to cold is far more important to them than athleticism. Some African tribes are very tall and lean. This is the best shape for heat loss, while also providing a good stature for vigorous yet sustainable outbursts of energy, as in chases, fighting and vigorous dancing. The length of leg and arm is also proportionally great. This is the shape of people who evolved in a less demanding natural environment. It was better suited than were the smaller hunters who inhabited deserts and forests for dealing with the larger animals that congregated on the grasslands, many of them dangerous both as predators and prey. Such taller folk eventually came to adopt pastoralism on those same grasslands. Their stature was selected for and perpetuated.

Folk who are heavily dependent on hunting lead a life in which the food supply is restricted to limits beyond their control, while the environment may also be harsh. Yet even the most primitive of human hunters catch food by cultural means. As time passed, numbers of hunters may have become of more importance than the stature of the individual. Indeed, a small man can set a trap competently or shoot an arrow accurately just as well as a large one. The Pygmies and Bushmen of Africa have hence resisted growth in stature because it held out little or no advantage to them. Their small size only became a drawback when they were confronted by bigger human intruders from elsewhere in Africa.

The 'Indians' of South America are hunters and farmers in varying degrees, but the hunters in particular have tended to retain the fairly squat body shape that relates to their remote Asiatic ancestors. There has apparently been neither time nor need for them to evolve to anything else. In North America the plains Indians often display traits that diverge from the typical eastern Asian. They are hunters who may even include some separate influx from Asia that never penetrated as far south as other incomers. As a group the Eskimos represent the latest of such arrivals. While varying degrees of interbreeding may be assumed to account for differences, it would seem to have been migrants lacking the more extreme Mongoloid elements who colonised South America right to its tip.

Farming arose and was developed among Americans in a central zone, and they benefitted from the advances associated with it. By the middle of the second millennium AD – when their civilisations were disastrously impinged on by the Spaniards - progressive folk of this type were occupying a strip from the east coast of North America through Central America to the west coast of South America.

The Polynesians have responded to the plenty provided by their insular paradises, but the heavy build in their case can facilitate the development of plumpness. They do not display Mongoloid traits.

The Mongoloids of central Asia inhabit harsher regions and this exhibits itself by a tendency to leanness. A non-Mongoloid sub-strain in all these populations may have contributed to the prevalent stature. The folk who inhabited the bleak regions of Asia north of the Himalayas (such as Siberia) developed various pastoral lifestyles that could be supplemented by hunting. They herded animals such as yaks and goats in the south and reindeer in the north.

As a means of fighting the cold, any tendency to grow fat among these Asians has largely been obviated by a need for occasional vigorous physical outbursts and a specialized culture developed to cope with environmental challenge, and much the same can be said of the most northern Europeans, e.g. the Lapps (Sami).

Way south of the last are the Mediterranean types among whom one can detect a natural propensity to put on weight, which is now to some extent unconsciously controlled by diet. They may be due to vestiges of an Ice Age population that was largely reluctant to move north once the ice retreated, even if the possibility beckoned.

The land freed from ice further north was occupied by folk who included taller and naturally leaner folk who generally must have come from further east. However, one might note that Palaeolithic cave art and more widespread contemporary figurines generally depict persons of a certain rotundity. Yet these are usually women and they are perhaps symbols of fecundity rather than accurately portraying actual stature.

'Hamites' and 'Semites' would appear to represent populations of lean stature, reflecting the adoption of a nomadic pastoral life style. The desert Arabs of today preserve the build, this adaptation being suited to much moving about by them in a harsh environment. There are of course plenty of plump Jews and Arabs now, but the Near East especially has been penetrated in historic times by folk from lands further west beside the Mediterranean. One can mention "sea peoples" such as the Philistines and, especially, the Greeks.

Extreme human groups can be typified by the Siberians and Eskimos in the north and the Pygmies and the Bushmen in southern Africa. Of these extremes, one has a short physique suitable for a harsh cold environment and the other has a short physique suitable for a harsh hot environment. The former is squat with straight hair, while the latter is lean with curly hair. Between them is found much diversity of physique, with plenty of hybridisation having taken place throughout the past. Stature has tended to become greater among progressive groups and this has been developed by migration driven by changes of climate. Yet one can see disparity everywhere in the form of marginal types due to different adaptive hybrids in different areas. Where hunting has remained the mainstay a shorter stature has generally been retained, and especially where the habitat is dense forest.

It was when it came to strife between groups that larger stature emerged as an advantage. To some extent this would involve the throwing of missiles like spears, but especially so with hand held weapons such as clubs and lances. Such squabbles were the forerunners of warfare, which developed when folk had more worth stealing, as did pastoralists and agriculturalists. However, any disadvantages that accompanied lesser stature could be mitigated or even overcome by cultural advantages: the Romans of the early empire depicted their victorious legionary soldiers as being comparatively small when engaged in battle with larger northern opponents.

Over much of the mainland of eastern Asia the Mongoloid element has apparently moved south and formed hybrids in which vestiges of this type are still evident. This incursion has caused disturbance among folk who were originally there. Mongoloid influence did not reach the far south. It is difficult to see among the Polynesians and in the disturbed populations that reached Australia it is surely absent. Others were constrained to press westwards and caused waves that reached northern Europe. This should not be regarded as a single event, but as a trend that took place over a long period of time and often accompanied by climate change. Thus a drying out in the middle of Asia would tend to cause folk to seek refuge in the damper climes of western Europe. Such movements can be observed in historical times with folk like the Huns and Scythians. The spread of Indo-European language family can also be attributed to such activity at an earlier time, including its arrival and eventual dominance in Europe.

■ Colouring

The various human races exhibit differences in the colour of skin, hair and eyes. To state that dark hair, skin and eyes go together is adequate only as a rough generalization. One might also harbour the opinion that dark people inhabit hot climates in the south, while fair people dwell in the colder north. As a generalization this may represent some truth, but the distribution has been much disturbed by geography, past climate change, and, especially in recent centuries, exploration and settlement,

Skin colour is less stable than that of hair and eyes, being susceptible to seasonal and geographical variations in the amount of sunlight, as well as geographical situation and life style. All skins are subject to sunburn, but 'white' ones are readily damaged by exposure to the rays of the Sun. For northern Europeans time spent in or near the tropics (or even just in more southerly parts of Europe), accelerates the loss of light colour and rosy cheeks, which changes are otherwise delayed to form part of the normal aging process. Truly 'black' skin is rare anywhere, while 'white is really either 'off-white' or 'pink'. Three basic colours can be identified for convenience and they are brown, sallow and whitish, although such distinctions are not absolute. While brown skins have the most pigment, it is only in extreme cases they can be considered to be 'black'. They historically comprise majority populations found in Africa south of the Sahara, as well as being widespread in southern Asia. 'White' skins are properly at home in northern Europe, where a historical rarity of sallow elements has led to skins that are 'pinkish', especially with local features like the cheeks, ears, etc. Such 'fair' skins can be susceptible to the pigment running together into small clumps or 'freckles'. Brown and sallow skins are generally unable to exhibit the involuntary reddening of the face known as blushing, usually an indication of embarrassment. Sallow is the most usual skin colour, and many Europeans can vary between light or dark sallow, especially in the south. It is to be noted that brown skinned folk, such as sub-Saharan Africans, retain hairless areas of pink skin where historically unaffected by the sun's rays, most noticeably on the palms of the hands and the soles of the feet, zones which are also in frequent abrasive contact with hard or rough surfaces.

In the Americas the sallow skin colour of Asiatic derivation was carried to the extreme south of South America by such 'Indians'. Even such 'Native Americans' inhabiting the tropical zone have not generated darker 'brown' skins. This resistance to being brown reflects the lack of tropical Asian elements among the migrants to America. But, as with other sallow skinned folk, older members of the population can be dark in the face, this reflecting a lifetime of this feature being exposed to sunlight and weather.

Fair and light brown hair also occurs among a small proportion of Australoids and Melanesians, suggesting that their ancestry may have contained strains of lighter skin colour.

Mammals and birds all have light coloured skin under their hair and feathers, where these are dense enough to be light exclusive. Elsewhere on their bodies different colours are in evidence, including black, as with gorillas. It follows that pigment in skin results historically from exposure and that the ability to generate it reflects a sometime lifestyle among humans with hairless bodies in which the wearing of clothing was absent or minimal. The whitish skin of Europeans implies that they of all peoples over the years have most avoided any tendency to strip off in the open air. By the same argument the sallow skin of Mongoloids suggests that they comprise a deal of ancestry that at an early stage dwelt in a clime that allowed more nakedness away from shade and shelter.

The import of the preceding paragraph is that the 'whitest' elements among the Caucasoids are due to a cultural response to the cold that was discrete from that of the Mongoloids, who had more opportunity to respond in a biological and physical manner. With the former folk, extreme cold was withstood by using shelter, fire and clothing to the maximum. Clothing, however crude, was worn on a regular all-year basis when away from home in the open air. Their bodies still retained a deal of hair, this applying particularly to the adult males. The latter were the ones who were resisting most the onset of paedomorphosis and their features tended to retain primitive elements. As for the ancestral Mongoloids, they eventually came to rely even more on clothing, especially furs, to withstand greater lower extremes of temperature. In response to the cold their facial features needed to be more covered when abroad and this tended to favour the setting in of paedomorphosis to a greater degree with their males, as well as the females. This showed in the unprotected features, namely the eyes, nose and mouth, which all tended to change in order to resist the effects of bitter cold. The first acquired a fold of skin on the eyelid, while the second and third features both became smaller. It was these cultural and physical trends that contributed to an ability to live on the bleak wastes on the fringes of the Arctic, or even on the icy lands further north still.

Those hybrid Caucasoids, who came to inhabit Europe and western areas of Asia were habitual wearers of skins and developed a very light colouring – fair or red hair, pale skin and light eyes. Was this solely an evolutionary response to them seldom being exposed to solar rays? Perhaps, but the light colouring appears to have become prized, especially where women and children were concerned and this may have led to sexual selection and hence preference being acquired by the relevant genes. One can note that in European folk tales and mythology heroines and goddesses usually have descriptions like 'fair' and 'golden' applied to them. Such manifestation in its extreme blond form seems to have been inherited by the Mesolithic hunters who historically inhabited the lands around the southern Baltic Sea after the last Ice Age. Further west, in the lands lining the Atlantic seaboards, blondness of hair tended to be replaced by redness.

As already asserted above, this move away from darker colouring embodied visible traits derived from the Neanderthal side of hybridization. The observation can be added that the lighter hair colouring found among such as the Melanesians of New Guinea and the Aborigines of Australia may have a similar cause. It could be due to an admixture of Denisovan genes and due to a population that was analogous to the Neanderthalers of the west.

While developments in Europe and along the northern seaboards of Africa may seem complex, the situation in Asia is even more so. It can be postulated that the various types of *homo sapiens* who spread from Africa eastwards into Asia attained certain cultural levels such as wearing clothes when necessary to keep warm. However, according to this argument it would seem that the Mongoloid folk should also have developed 'white' skins, but they are universally sallow. This suggests that the ancestors of the Mongoloids originally lived in open areas for a long time, but not ones of unrelenting fierce cold. Penetration of the Arctic came later, along with habitation in higher, hostile places. They were originally dwellers in scrub, something like the savannah of Africa. The climate was regularly sufficiently warm for their bodies to be exposed to a certain degree of solar radiation. One can compare the occurrence of such a skin colour to that of the African Bushmen in modern times.

Light eyes, such as grey, blue or hazel, are generally linked with fair colouring. They can involve disadvantages. Such irises do not exclude unwanted light as well as brown ones. Like all light colouring, blue eyes have been helped to persist by being selected for as an aesthetically desirable feature. Light colouring may historically also have attained a class significance in northern Europe. A darker substratum may have slipped into a position of social inferiority at an early date, where it would be more susceptible to the vagaries of life, with a resultant gradual reduction as a proportion of any population.

■ Evaluation of 'Race'

Genetically speaking, the physical differences between current human races are relatively trivial and have not resulted in any noticeable interference in the ability to interbreed, except for exclusion by choice. In the recent past obvious differences of culture linked to racial diversity has become blurred due to the increase in mingling due to the ease of travel. The different kinds and levels of culture hitherto achieved by different races did not necessarily imply different levels of general intellectual ability. This is because the intellect itself is the result of an evolutionary trend towards extremely wide mental adaptability and this is never fully exploited in communities as long as they are relatively isolated. The differences indeed largely depend on whether historical and geographical factors have given rise to opportunities for broadening the capabilities of the human mind and if these opportunities have been exploited. For example, cultures with writing have huge intellectual advantages over those that do not. This is irrespective of the calibre of the brain that individuals may have inherited.

The human brain is an apparatus that is capable of handling circumstances quite different from those for which it was originally evolved. Even so, observations indicate that races can specialise in certain intellectual strengths, in a similar way that some are better equipped to excel at specific physical activities than are others. For instance athletes of west African extraction dominate sprint racing, while those of European extraction dominate swimming.

The fundamental point is to avoid indulging in any fruitless exercise intended to 'prove' that any one race is superior to another. One needs always to bear in mind that in matters to do with evolution there is always a degree of overlapping: although not necessarily much. It is possible that the bulk of any racial population may not generally reach excellence, but a relatively few may be so good at some subject or capability so as to reach levels not attained by the best members of other racial groups. Indeed, not facing up to the contention that races do have their physical and mental strengths and weaknesses can create its own kind of problems in society. The desire to be 'fair minded' can drive one into a state of 'denial'. This in itself can drive problems 'underground', the realm of psychological uncertainty.

That generalized racial differences are observable in common spheres of activity should not surprise us, for it would be extraordinary that races could evolve with diverse physical features yet have brains that show absolutely no racial characteristics whatsoever. Hence, as a generalization, there are bound to be some differences in how the mind works between different races. By the very nature of the matter, such variation is most likely to show up and be contentious in the case of achievements made among the complicated state of affairs found at the higher intellectual levels. This can be held to be true even though such observations are open to exploitation by bigots and denial by the over-sensitive. Ultimately individuals should be judged by how suitable they are to perform in the circumstances they find themselves and not from any racial characteristics they may display, which may wrongly be perceived as being 'alien'. Cultural traits are far more likely to cause minorities actually not to fit into the standards set by a majority, than do perceived racial differences.

Yet, unfortunately, racial groups may carry with them cultural baggage which can alienate them from their neighbours. One cannot choose one's ancestry or appearance, but there is an imperative to accommodate the background of other folk groups if friction or more serious strife is to be avoided. The fear that others are threatening your lifestyle must be based on actual hostility caused by real zones of contention, rather than blind prejudice; otherwise humanity will continue to experience serious disputes and armed conflicts in unnecessary excess and leading to the disastrous disruptions that are still afflicting many communities in the 21st century. This is asking a lot, since the pervading aspirations of humanity are still mainly based on the choice that it is preferable to try to get ahead of the others and stay there, rather than try to get on with them. Of course, racism is not the only cause for people to be subjected to blind hatred. It is one contributory factor to the struggle to get on with other members of one's own species because of perceived differences that are often relatively petty or even quite false.

Newcomers seeking to adapt to a strange culture may or may not seem to do well. They may succeed because they try hard. Yet hidden difficulties may eventually manifest themselves and lead to frustration that good will and aspirations alone cannot overcome. Yet as time passes some hybridization of life style is bound to take place between disparate cultures coming to occupy the same space. The distribution of genes will then tend to become more eclectic, so that different characteristics become more blurred, with social relationships changing as a result. Differences in standing then revert to being more a matter of class than ethnicity.

CHAPTER 21

Primate Hair

At the time of writing the study of human evolution is in a state of flux due to ongoing developments in knowledge about the genome. As discoveries about the DNA of various human types turn up the picture we receive of our ancestral types changes and even the prospects for the discovery of yet unknown hominins is ever present. This makes any discussion that is based on such matters liable to become outdated as more evidence emerges; it is hence safer to make observations based on the more readily perceivable human physical traits. One such is hair

■ Primate Hair Types

Certain anthropoid apes, namely the chimpanzees and orang-utans, now have not very much more general body hair than humans, at least when age asserts itself. Older individuals show a significant reduction of hair quantity. In the long term their hair amount would seem to be on the wane and they lack the dense insulating layer so characteristic of many mammals. The thinning cover of long hair may be quite attractive to them as adults, but its causes are surely evolutionary. Among chimpanzees the dark body hair makes a contrast with the bare pink area at the rump whose fluctuations in size and colour have sexual import.

Monkeys in general retain an overall covering of body hair, as do the majority of the smaller active types of mammal, even in warm climates, as, for example, the lemurs. Gibbons are apes, but their body is comparable in size to a big monkey. It would seem that this is why they retain a covering of quite thick hair, i.e. because, being so agile there is a necessity for heat/energy conservation. Large, thickset apes like chimpanzees and orang-utans do not need this facility, at least when adult. So why are gorillas covered in denser hair? Even though large silverbacks can be seen to be generally less hairy than juveniles, they are still considerably hairier than orangs and even chimps at every stage of growth. This seems to indicate that the ancestors of today's gorillas were all adapted to live in forests that clothed high mountain slopes and hence needed better insulation against a degree of cold and damp.

The above confirms the generality that heat is better preserved when the body is large and the build heavy. It also illustrates the need to remove heat from the body when activity generates too much or the temperature rises too high, and the need to conserve heat when faced with cold and damp weather. Different methods to expel heat have arisen: dogs are hairy creatures that breathe over their tongues, whereas *homo* has lost his hair and dissipates excess heat automatically by way of widely distributed sweat glands.

■ Occurrence of Human Hair

Since monkeys and some apes are still quite hairy, yet live in the tropics, it can hardly be a hot climate in itself that caused the loss of human body hair. From the observations above one might postulate that, even while still in Africa, the various types of hominins grew less dense body hair as body size increased. Even so, it seems likely that at the time when the primitive *homo erectus* spread north out of Africa, in general the body was still more hirsute than that of any modern human race. Could it actually be the case that moving into a cold climate created conditions that favoured further loss of body hair? Having been lured into the north, some derivatives of *homo erectus* were eventually overtaken there by the onset of Ice Age conditions and must either adapt for the increasing cold or retreat to the south.

At this juncture a problem presents itself, namely: what was the state of the body hair of *homo habilis*? Had the ancestors of *homo* indeed lost some of their body hair at the *australopithecus* stage? From the examples of modern monkeys and apes, especially chimpanzees, it would seem that body hair thinning could take place when large apes spent more time on the ground, although they still remained very hairy at least until well into adulthood. Against this one can note that ground-dwelling monkeys like baboons have retained hairiness throughout their lifetimes and irrespective of habitat.

However, this comparison with apes may be invalid. A noticeable trait with all kinds of *homo* now is that, no matter how primitive, they adorn their bodies with gear, usually including garments, even though of the most simple kind such as animal skins. This implies that at sometime in the past *homo* began habitually to cover himself up and that this would eventually supersede any residual body hair. This would work in parallel with the increasing use of sweating to attain loss of heat. Evolutionary pressure would ensure that these trends increased chances of survival.

It could also have been to some advantage to achieve an increased body size even in a hot climate. Some other kinds of animal have lost virtually all their body hair, including very large creatures like elephants and rhinoceroses. (However, in the past, when such could be found living in a cold climate, hair was retained, as with the mammoth and the woolly rhinoceros.) With these types of animal the skin has compensated for the almost complete loss of hair by becoming tough and thick, both to control the passage of heat and as protection against the Sun's rays, as well as other harm. With man this has not occurred, but where protection from Sun radiation has remained a basic requirement, as in Africa and southern Asia, the skin has responded by the addition of pigment to its natural light colouring. Otherwise adjustment to prevailing heat and radiation conditions could be made by the donning and doffing of animal skins. Even so, use of animal skins in the earliest phases may not originally have been entirely for protection, but rather to increase success at hunting by affecting a disguise to resemble the quarry, either visually or to hide the hunter's own smell.

Darker coloured skin is a response to exposure to sunshine, the cause being loss of body hair and a degree of nakedness in a clear and hot climate. It has been shown recently that migration northwards into a temperate climate can now be harmful to dark-skinned folk due to vitamin D deficiency caused by reduced ability to absorb light. This can result in an inability to build bones properly, a condition known as rickets.

It is hard to decide at which stage of human development the advantages of covering up began, but it seems safe to say that it was already practised by *homo erectus* when members of this species started to expand in Africa. One might likewise assert that such cladding was rudimentary, with little (if any) by way of fastenings, although still-attached leg skin could have been retained to secure covering by twisting such extensions into primitive knots. Among the early forms of *homo* were some who started to make such use of skins to control the loss or retention of heat when at rest, as well as when out hunting or foraging. But at first the head would not be included, because, at least for this purpose, this was unnecessary. All the early hominins would have good heads of hair and the men would display some facial growth (beards), while they did not go bald in the way of modern men.

As the relevant interbreeding populations of *homo erectus* type evolved towards *homo sapiens* they spread out and took with them certain differences in physical appearance that had already become evident among some *homo erectus* populations. Among these would be hair type. One might then postulate that the further away from Africa *homo erectus* travelled, the greater the likelihood for straighter head hair to be retained. Yet as progress was made towards *homo sapiens*, some genetic contact remained with other advanced but divergent populations of *homo erectus*.

While it can thus be suggested that there have been variable climatic and cultural reasons for a reduction in hair on the human body, there remains to be considered why hair was retained in certain places on the body, together with the causes for differences in texture, form and colour between the hair of various modern races.

There is some hint above of natural selection having been at work, but it might seem that sexual selection played a major role in the loss and distribution of human hair. Mankind tended to get the kind of hair desired, rather than that needed. Hair evolution was used unconsciously by progressive man while still habitually naked to distinguish his kind from other animals, such as monkeys, apes and those hominins more primitive than himself. It is generally accepted that man originated in Africa, which he shared with a variety of simians, all of whom had straight hair. With his dawning self-consciousness man started collectively to resent everything about himself that was ape-like, even though he was not necessarily individually conscious of this. The effect on his evolution was that such attitudes were to accelerate any tendency for changes, such as to make the hair less straight and to reduce the body hair amount to a less than ape-like level. At the same time the amount of head hair grew, so that it outdid in quantity and length those crests that might be born on any simian.

The advanced hominins in Africa apparently fell into two kinds, the smaller *homo habilis* and the larger *homo erectus* and with a common ancestor among hominins of more simian form. Mammals generally have straight hair. Monkeys and apes certainly do and one can assume that such was the case with the earliest still hairy types of *homo*. On the grounds that he wanted to be different, one might surmise that at some stage progressive man in Africa preferred to have wavy or even curly hair as further evidence of his distinctiveness, especially from the more closely related primates. Thus was its evolution away from straightness subliminally encouraged.

From this it might be claimed that curliness is later than straightness, and that curly people emerged in the wake of straight haired predecessors. It can with some confidence be claimed that curliness originated in Africa south of the Sahara where its presence until modern times has been virtually ubiquitous. At the same time it might be noticed that straight or nearly strait hair is the norm across the more northerly tracts of Eurasia and across the American continents in their entirety, even as far as Tierra Del Fuego, the southernmost tip. Against this can be noted that head hair between northern Eurasia and sub-Saharan Africa is much more variable; it can vary between straight and curly, with various degrees of frizziness and waviness between. The inference one can make from this is that the earlier emigrants from Africa were all straight haired, while the later ones, irrespective of other characteristics, could be recognized as curly, even if this exhibited itself in ways that differed for other reasons, one being density. It seems evident that curliness developed in Africa among populations of *homo sapiens* who were straight haired and who were already constituents of migrants penetrating south-west Asia and spreading out from there. The curly heads eventually followed them on such enterprises.

It has been claimed that the vestigial human hair tracts are actually denser than those of chimpanzees. Given this, it follows that man's more immediate ancestors were much more hairy that he is now. This suggests that the drive to be less hairy among our ancestors was also sparked off by comparison with another kind of hominin, rather than solely with apes. Hairiness would come to be (wrongly) associated with inferiority. Initially the inferior hominin in question would be *homo habilis*, who we might hence presume to be hairier than the more advanced *homo erectus*.

When someone is cold he might display gooseflesh. This is apparently a residual ability to raise the body hair to increase its insulating qualities. Why this characteristic has not fallen into disuse after all these millennia is odd; perhaps it has been retained as a useful communal visual indication that it is time to seek warmth or cover up before the actual need was felt.

None of the apes exhibit clumps of hair in the pubic and armpit areas, as does man. The pubic hair is not designed to hide the genitals, but serves rather to draw attention to their location, for it only appears once puberty is reached. Since pubic hair is concentrated at the front, it would seem to be associated with the upright stance. Armpit hair appears to have assumed the role of pseudo-pubic hair that reinforces the signal of the real thing. As with apes, the amount of any kind of body hair is eventually reduced as an effect of age.

One can hence attribute the development of pubic hair and the decline of general body hair with progressive man's desire to diverge from the appearance of monkeys and apes, as well as from that of other hominins. Yet he remained unconscious of this ability to fulfil such a desire by way of sexual selection and evolution. Those others were hairy, but had no specific genital or pseudo-genital clumps. Man's general body hair was hence reduced, while the hair of genital areas was retained as a contrast, but with its occurrence linked with puberty. Although the hair tended to obscure the genitals to some extent, it also drew attention to the human pubic area, which seems to have been its prime function.

Cultural adaptations included cave dwelling, shelter building, fire making and the wearing of animal skins. Man's relatively meagre hair covering on his body was largely irrelevant as a means to combat the severity of a northern climate. Yet the thermal benefits of wearing animal skins could not have been realised until later. Initially the practice may have arisen for other reasons, either during primitive ritual or, more probably, to enhance the chances of success while hunting by stealth.

Human hair has peculiarities that seem to be associated with man's rise to pre-eminence in the animal kingdom, but also exhibits traits that are due to simpler animal evolutionary factors. Except for residual wisps in the case of the male, hair has virtually been lost from the human body, except for the head, the armpits and the crotch. On the head the female face is totally exposed, while only on males does hair extend down the sides of the face (sideburns), to overhang the mouth (moustache) and cover the chin (beard). In contrast to body hair, head hair can be quite dense and grow to great length, especially in the immediate post-puberty period.

The distribution of hair on humans is evolutionary in origin and can be regarded as the combination of two factors, i.e. natural and sexual selection. Even in a cold climate, because they were clothed when outdoors males did not need their covering of body hair, which in any case was inefficient. As a facet of the fight against the cold, the necessity to wear a cowl also usurped the function of the hair on top of the head, so it too tended to be superfluous. Yet the chin and mouth were exposed so that the growth of beard and moustache was encouraged. However, there were also air spaces at the back and sides of the head and these too encouraged the retention of hair in these areas as some protection against the cold. Originally the head hair of humans had taken on mane-like qualities. The preference for this was hardly for practical reasons and the cause would lie in other evolutionary pressures. From beginnings serving the needs of thermal insulation, the selective retention of hair around the cheeks and chin in the males for similar reasons would be further enhanced by sexual selection.

◼ The Beard

The beard is a male character. It was perhaps originally considered desirable due to its absence from other primates (including more primitive hominins) and the way it emphasises the human chin, this itself being a feature unique to *homo sapiens*. Beard growth is restricted to males because body hair in general represents maleness, while a larger jaw is also a masculine feature. The development of facial hair may also have been encouraged selectively because of the chin's exposure to the cold during prolonged hunting expeditions, while most other areas of skin were well protected by clothing.

The growth of the beard on men was a specific human trait divorced from precedent among the closely related simians. While one can see analogous male head adornments on the heads of African animals such as the manes of lions and certain baboons, excessive chin hair is specifically human. One might note that the mountain gorilla does develop a crest in the male, but no beard.

The appearance of facial hair is a sign of the onset of puberty in males. Although not universally popular with modern women, the beard can be regarded as being at least partially due to sexual selection, because it indicates that the boy is no more. Beards grow at their thickest on the men of Europe and nearby parts of Africa and Asia. Women do not need a beard as a visual signal for such a purpose, for they have other developments, such as the breasts. Additionally there is onset of menstruation, with the changed pattern of activity this entails. With men a slight amount of general body hair also tends to appear at puberty. The main truly hairless areas of all humans are the forehead above the eyebrows, the ears and their setting, the palms of the hands and the soles of the feet.

◼ Baldness

Baldness is a loss of hair that affects men only, but to a very variable degree. It seems to have its main centre of occurrence among the Caucasoid folk of central and northern Europe. Baldness has to be distinguished from the thinning of hair that affects both sexes with the approach of old age. Hair thinning occurs over the whole of the head, whereas baldness is restricted to a well-defined area of the scalp, this following a curved line a few inches above the top of the ear and with a pronounced dip at the back of the head. The occurrence of baldness in European populations can vary greatly. Its beginning can be noted as early as the twenties and develop into a shiny pate over a few years. Others may start to go bald at a greater age, while there are also those who look as though they will never go bald and, if anything, only show the thinning of old age. With those who start later the onset of baldness is not as rapid or complete as with some of the young sufferers, but takes the form of a more prolonged decline before a bare pate is achieved, possibly over several decades.

Baldness might be staved off ill-advisedly by absorbing female hormones, thus implying that it results from a change of gene activity at a certain stage of life. Baldness can occur while quite heavy beard growth is maintained. (From the appearance of the great man one might well call this the Darwin Syndrome.) This anomaly is very noticeable with the early acquired complete kind of baldness and emphasises its selectivity.

Seeking the cause of baldness, one might note that it starts either at the crown, or at the upper temples, or both. From these positions it spreads into the rest of the vulnerable area of the scalp. Early occurrence may be at both locations together, but occurrence can be staggered. Should one imagine that hair could be worn off, or inhibited from growing through being stifled, then this is just the kind of effect one would expect from the wearing of a loose hood or cowl of fairly stiff material and the weight of which is mainly taken on the crown of the head, but with further touching just above the temples. The predominance of crown based baldness suggests the prevalence of skulls that tended to slope downwards from back to front,

But baldness is heritable and appears now irrespective of headgear. One might then assume that there was a tendency for men in a certain climatic zone to lose the hair in this way because it was superfluous to requirements. But there again this is hard to justify as a proposition on evolutionary grounds unless the trend at the same time proved advantageous or was considered desirable in some way. Male baldness has never diminished survival prospects and the presence of selective pressure for it not to occur varied, being dependent upon the circumstances that populations found themselves in.

Thus might one consider that in the Ice Age of northern Europe and adjacent areas of Asia groups of men habitually left their womenfolk and children in warm shelters, while they went on hunting forays into a harsh wilderness. They were skin or fur clad, including a crude hood that pressed down on the scalp and temples, but left the face and chin exposed. It made no direct contact with the sides or lower back of the head. When the cowl was doffed certain areas of the hair on the top of the head would be shown to be to some extent flattened and less than vigorous, especially with men of more advanced age and many seasons of hunting behind them. The process would indicate men with plenty of experience in the field and hence of survival capability due to prowess. Such men might be venerated and become leaders. They would presumably have special rights among the group's women. It could thus have occurred that men with hair prone to "cowl damage" would marginally have a better chance of passing on their genes than those more resistant, even if their actual prowess was not greater. One might also suggest that a natural evolutionary trend did occur to accelerate the condition, this being derived from the fact that the limited insulating properties of hair was less required where the efficient furry skin of the cowl was in contact with the head.

The fairly stiff material of the cowl ensured that above the forehead contact was made at the high temples, while centrally the edge formed an arch above the brow which enabled the forelock to show itself. This formation is evident today with some Europeans as a crest stretching back over the crown of the head, this resulting from the forelock being left after the adjacent areas on each side of it started to lose hair.

Men of later times came to resent the onset of baldness, for they no longer went on prolonged hunting trips in the cold or wore heavy and primitive skin cowls. Whether or not today's women are generally repelled by baldness is doubtful, but this in any case is irrelevant to the progress of the species, for many men acquire a sole mate before the onset of balding becomes serious or even apparent. There is little selection for or against it, while the sexual advantages that used to be the prerogative of an experienced male no longer do so to quite the same extent.

Baldness has spread into many races, perhaps by diffusion of genes from the northern Eurasian zone, for the further away one gets from this the less it is apparent. Modern hunters of the Arctic, such as the Eskimos, are resistant to balding. This should not be taken as proof that dress has no effect. Perhaps their rather coarse hair has never been so prone to damage by fur hoods, which in any case are more advanced in design and in the more extreme climate fit snugly enough to be self-supporting and, being more flexible, do not press selectively on the scalp. This would then obviate the onset of baldness being made more likely by sexual preference in the manner described above.

■ Hair Traits and Tracts

General hair colour is related to that of skin and can best be considered along with that. Greying however, like baldness, is a feature of ageing. (It affects other animals, e.g. dogs.) As with balding it can start as early as in the twenties, but in some individuals is delayed until old age. Unlike balding, the loss of pigment affects both sexes. Greying may set in about the same time as balding, even though the two can hardly be closely genetically linked. Both the effects result from aging, but whereas balding can be attributed to differences in the genes found on the female 'X' chromosome from those on the male 'Y' one, the genes leading to greying are shared by chromosomes of both sexes. One might postulate that greying, like balding, started off as a male-acquired character, but that it was transferred to the opposite sex. However it is preferable to regard greying as unrelated and consider it as a universal timed signal to others of the species that this individual is going "over the hill" and is near to being "past it". As with balding, greying became less important as man's lifestyle became based more on culture than on physical activity. The great time differential for the onset of either of such features may have cause in such changes of custom and in the degree to which one's various genes have developed along different time scales during one's ancestry. The signals given off may originally have had different import. Balding, when or if it occurred, carried the additional indication of physical experience, whereas greying was more a variable sign of the effects of age, but with the acquirement of wisdom that could go with that.

The Eyebrows It is normal for each human adult to have a distinct bow of hair above each eye. These eyebrows are generally of coarser hair, with their presence normally being a bit more substantial in the male. Above them is a hairless area that is clearly defined at its upper and side edges by a usually symmetrical but possibly erratic frontal limit to the head hair. This is the forehead.

Modern apes generally do not have these features, but it is safe to say that earlier hominins like *homo erectus* and Neanderthal Man did. The differences from them was that their skulls had bony eye ridges, behind which foreheads sloped upwards and backwards at a low angle, whereas with modern man, i.e. *homo sapiens sapiens,* it is much steeper and can be more or less vertical.

The purpose of the eyebrows may be obscure, but it can be claimed that they were expressive, rather than functional. It is noticeable that they can be raised - either voluntarily or involuntarily - when conversing with or simply confronting others. In either case a wide-eyed expression results from raising them that delivers the import "trust me" or "this is true or important". The voluntary eyebrow lift can be used for deception. The involuntary one is often little more than a twitch.

Among earlier hominins the raised eyebrow gesture would originate as a signal of friendly intentions. It worked by contraction of the muscles behind the eyebrows, thus causing these to rotate backwards over the bony ridges upon which they featured to give a flourish of expression. The muscles were housed in an area of bare skin behind the ridges that could be wrinkled. The effect is still similar today, with the result contributing to the production of the "furrowed brow".

The opposite effect will also be ancient, whereby the ability is used to produce the frown of displeasure or hostility. Now the eyebrows are pushed downwards by forehead muscles, usually with some vertical and even slanting furrowing above the nose. Anciently the eyebrows would be rotated forwards on the ridges and thus contribute to the narrowing of the eyes. This gesture was to convey feelings of dislike to a companion or hostility to a rival.

The Hairline Above the eyebrows the forehead is defined by this feature. With the male, from the level of the top of the ear it continues the line of the sideburns on each side in a generally upward direction before eventually crossing over the head.

Although the pattern varies considerably, a feature evident with every individual of *homo sapiens* is the "temple point", although in some cases it is vestigial. This is formed by a sudden change in direction in the extension of the sideburn line. This edge of the hair (here usually well defined) would appear to be starting to turn inwards across the brow, but this is arrested and the line changes direction so that it runs upwards again, and generally backwards from the vertical at a variable angle. It is from the top of this alignment that the line direction changes again to form an angle: at this higher level it turns to cross the brow.

So, what is the reason for the temple points? They are of significance in that they reflect the developments of both brain and skull above and behind the eyes. The postulate is that the early man *homo erectus* had eyebrows on his significant brow ridges, behind which was a narrow bare area where there were muscles to operate them. Beyond this reduced forehead area was the hairline which at this early stage crossed the skull to turn down at each temple to pick up the line of the sideburns, but with no sign of points. The points started to form when the skull widened and assumed a more vertical and higher front. This was a result of the need to accommodate the growing size of the frontal lobe of the brain. The skin of this bigger forehead was thus evolving solely out of the muscle area behind the eyebrow-bearing bony ridges and would hence also be hairless. The expressivity of the eyes could actually be enhanced when the brows became less rugged and the requirement was for the upper lids of the eyes to complement eyebrow activity by being pulled further open or pushed further shut. The area containing the brow muscles and wrinkling skin was enlarged by occupying the space created by the increased size of that part of the skull that housed the frontal lobe. The increased forehead was hence hairless and the temple points mark the particular time when this brain development and activity began. The lateral change in direction of the line of the sideburns at these points suggests the situation of the earlier hairline at the time of *homo erectus*.

So what was the event that caused these developments? It must have had something to do with the use to which the frontal lobe was being put. This excludes the use of language, since this is controlled from elsewhere in the brain. The frontal lobe concerns itself with the process of thinking. "The main functions of the frontal lobe are to control attention, abstract thinking, behaviour, problem solving tasks, and physical reactions and personality." It can safely be assumed that the main driving force behind all these traits came to be the acquisition and development of language.

The sudden change in direction of the temple points suggests a quite sudden evolution of brain activity. This is not necessarily so. The forehead grew by the gradual and simultaneous increase in its height and width: the points are indications of how far this process has occurred in the case of each individual (as a representative of his kin, tribe or race). There appears to be no difference in the development of the height of the forehead between the sexes, but the female head appears more often to taper inwards somewhat at the front. However, this is mainly due to the temple points being less evident owing to modern female hair styles and a tendency for the hairline to continue above the points with a different (i.e. inward) slope. This means that the indents where frontal male baldness begins are usually absent, while the bare forehead can be considerably narrower. The female brow is also less likely to show vestiges of the erstwhile bony ridges above the eyes. This is all to do with the female head's tendency towards paedomorphosis and generally to be of lighter construction.

The occurrence of paedomorphosis in females is perhaps most evident in the setting of the eyes. The brow overhang is reduced from that of males, who are often regarded as "beetle-browed". This causes the male eye to appear narrower even when at rest, as though crushed from the top. Because of this difference, the female eyebrow is set higher above the eye and contributes to a more bland expression. Because of the lack of overhang with the brows there is a much smoother transition upwards from the bridge of the nose to the forehead, sometimes lacking overhang completely. It can be noted that these paedomorphic characters are more in evidence with some races than with others. Beetling brows are most obvious among Caucusoids and least so among Mongoloids.

CHAPTER 22

Preparations in Africa to Reach Nearly Everywhere

■ The Three Aspects of Evolution

The generally accepted view that evolution was initiated internally by mutation and driven externally by natural selection can be modified to include two further aspects, both of which have already featured above. In one case a species may come to reproduce at an earlier stage in its development, so that such younger form becomes the new adult and the old adult form no longer exists. This is paedomorphosis (or neoteny). In the other case speciation within membership of a genus has not been quite completed, so that the production of fertile offspring (hybridization) is still possible between creatures that could otherwise be suspected as being discrete species and externally appear to be physically distinct to a significant degree.

Evolution based on paedomorphosis is a slow and deliberate process that involves internal and external developments of an animal to take advantage of changes exterior to it, if that is at all possible. A major effect was that *homo* was able to prolong the lives of individuals after they had lost their fertility. Signs of this can also be seen with other gregarious primates, who live in troops that can protect aging members and prolong their lives. Those human beings with extended lifetimes might still be of value to the troop because they could become sages, i.e. depositories of the collective culture of their kind.

Evolution due to hybridization is haphazard in its effect, but when the result is appropriate it can occur quite rapidly. In the case of *homo* successful results from such unions may have been rather rare, but when they did occur they could be very significant with regard to directing him onto the right course towards being *sapiens*.

Yet as a generalization, for either of these changes to be successful they still had to pass the tests set by natural selection. However, an exception arose where it concerned *homo*. Members of this genus were to develop a special trait with which to resist the destructive demands of nature, and that was culture, which itself provided greatly increased adaptability. Eventually this grew to such an extent that it provided the means to modify his circumstances in order to overcome the normal consequences of evolution due to natural selection, even should the results sometimes be initially restrictive in some new way.

With regard to matters of sex and hybridization, the bonobos are apes known for their promiscuity. Yet this occurs with virtually no violence. They are a community in the Congo that is isolated from related primates, such as chimpanzees, and this effectively prevents any resulting hybridization. But with *homo* there appears to be a violent aspect that can raise its head in sexual situations, whether a relationship exists or not (a trait shared with chimpanzees). Brutal circumstances can arise that induce serious frustration and lead to resolution by attacks on females. Apart from the general brutality involved, such violence sadly occurs commonly when armies lacking discipline overrun civilian populations. This could have applied from early in the Pleistocene when hostile contact was resolved in a similar way by any group with overwhelming physical advantages, and perhaps also cultural ones. In such circumstances normal selective sexual attraction played little if any part (as is still the case), but the result could be the same as if it had – offspring. Such encounters could have occurred among intermediate (yet still differing) versions between *h. habilis* and *h. erectus* versions of the genus, and inescapable servility may have been the fate of females, who became slave wives. This sort of situation developed in more recent Africa when tribes of larger stature, such as the Bantu speakers, moved south to impinge upon the pygmies and Bushmen they found there. Then European colonists arrived, who could use their technical superiority to dominate the native tribes and form imposed unions with their women. The result is a blended community which was eventually separated out racially as 'coloured' and became socially distinguished from both 'whites' and 'blacks' in the oppressive system known as 'apartheid'.

Despite the drift towards the status of subspecies, sexual unions occurred between some of the various 'species' found within the genus *homo*, which often continued lines that were fertile. As already discussed, such offspring could contain members who benefitted from both parental lines. Eventually such superior successors, because of their advanced survival capabilities, would tend to eliminate the original separate strains, as well as less satisfactory hybrid forms if social contact was maintained. Because of these circumstances the constituent members of *homo* came to be reshuffled biologically, as well as culturally. There was one factor that was incidentally giving preferment to certain lineages and preventing extinction, and that was being carriers of the genes that were on convergent critical paths that were all leading towards *homo sapiens*. They would be recognizable in the future in the emergence and eventual complete dominance of this single species, albeit still being visibly divided into many physically diverse races.

Hybridization was able to occur more frequently among types of *homo* than with other vertebrates because of his increased adaptability. This led to him becoming more able to escape over hostile territory when his habitat went into serious deterioration. This was due to his physical and cultural developments making it possible to migrate to an extent that exceeded all other earthbound creatures and thereby increased the likelihood of contact with other members of his genus, who due to earlier circumstances had become separate and to a degree different.

 ## Limits on the Presentation of Ancient Human Evolution, Culture and Expansion

Owing to the state of knowledge about these matters that still obtains in the early decades of the 21st century it is impossible to make firm statements on many aspects of them. Archaeological evidence is still sparse and new discoveries are ever being made, any of which could upset any dogmatic positions that one might want to take up. It is hence better to keep any claims down to trends, rather than to express oneself with a degree of certainty about details that discoveries actually suggest rather than prove. Yet even the most basic of trends can still be in dispute, such as establishing the zone where the genus *homo* first appeared.

■ Remotest Human Origins

To digress to earlier times, the difficulties that are evident when trying to pin down the evolution of *homo* are nowhere more frustrating than in the study of the emergence of mankind from the realm of the apes. This has long been understood as having occurred in Africa; yet some schools of thought try to throw doubt on this, and suggest China instead. The argument is very forced, since nothing has been found in Asia to serve as a candidate to compete with the evidence of the Australopithecenes who inhabited Africa over immense periods of time. It would seem to be irrefutable that it was an advanced version of these that gave rise to the eventual *h. habilis* and his putative larger successor *h. erectus*. There is even an intermediate form of hominin whose remains have been found at the right time and in the right places. Paranthropus has been found in eastern and southern Africa. This has been identified in three forms, contained somewhere within a time span of 3 and 0.5 mya. Otherwise it has been claimed that *p. walkeri* (or *aethiopicus*) ranged between 2.7 and 2.3 mya, *p. boisei* between 2.3 and 1.0 mya and *p. robustus* between 2.0 and 1.2 mya. The last has only been found in South Africa.

It is to be noted that *homo habilis* was present within a very similar range of both territory and time (c2.4-1.5 mya). While *paranthropus* may have made a contribution, it would seem that this hominin was really a dead end form and that it was *h. habilis* who was the major contributor to the emergence of *h. erectus* in the areas concerned, i.e. East Africa, in particular in his more specifically African form *h. ergaster*, within the time span 1.9-1.4 mya.

As things stand, the best hypothesis arises from accepting that East Africa was the original centre from which radiated various early forms of *homo*, forms that continued occasionally to make fruitful contact with each other. Rather than proposing that *h. habilis* was succeeded by and generally replaced by *h. erectus*, it would seem likely that he existed in Africa alongside a distinct contemporary *h. rudolfensis*, in accordance with interpretation of skulls found in Kenya. While versions of *homo* remained in east Africa the distinction between them has become blurred. Otherwise it is a question of who stayed and who left that was the driving force behind diversity and a matter of when they left and which direction they took.

So, while it is proposed that *h. rudolfensis* evolved in east Africa as a contemporary of *h. habilis*. It was from this background that derived forms were able to roam and instigate emigration, one direction being towards the north-west. *H. ergaster* seems likely to have provided the antecedents of those who crossed over into Europe at the western end of the Mediterranean, to become *h. antecessor* and eventually *h. neanderthalis* (by way of *h. heidelbergensis*) as discussed below. The start of these processes is illustrated by the finds of Oldowan tools from Ain Boucherit, Algeria, which have been dated from as early as 2.2-2.4 mya and suggest the presence of an ancestral form of *h. ergaster*. Later this 'migration' route can be seen in reverse with eventual return of *h. heidelbergensis* from western Europe via the Maghreb to the areas of their origins and to emerge in the form of the Heidelbergers that have been found in east Africa.

So in times long ago the ancestors of *homo* commenced to peel off from the existing simian world and apparently existed in a variety of forms at any one stage in the ensuing process. Yet since Palaeolithic times all these have been reduced to one species, i.e. *homo sapiens*. So why did this one line succeed while every one of the others failed?

The available evidence points convincingly towards humanity having originated in Africa. Fossilized bones of creatures showing the most primitive human traits have been discovered at various sites on that continent, but nearly always in east Africa (in areas immediately west or south-west of the Horn) and in South Africa. Such persistent and slowly evolving beings are generally known as Australopithecines. Yet this view of origins depends heavily on the lack of convincing evidence indicating something else, which may be unrecognized or still lie hidden while awaiting discovery.

While the claim can be made that man is descended from some kind of ape, anything like such early primate is not readily recognizable among known species of simian; man and ape share an unknown ancestor that was more primitive than both the Australopithecines and any present-day anthropoid ape. Both chimpanzees and modern men, on their own time scales, have evolved away from such an unknown extinct animal along divergent lines, to produce these two present day species that are relatively diverse in structure, appearance and abilities. Certain beings that are found to lie more directly between that obscure ancestor and the beginnings of *homo* are the variously discovered but rare Australopithecines, some of which extinct examples would be closer to the actual human ancestral line than are others. It follows that the divergent elements in their DNA profile would be even fewer than those shown between man and chimpanzee.

DNA analysis is advancing so fast at the time of writing that one needs to be especially wary about being too dogmatic on the subject of the emergence of the various strands of the genus *homo* from remote origins in a primitive primate pool, followed by their subsequent distribution throughout the world. Here one can provisionally note that researchers claim that only something like 1% and 2% of their DNA differs in the case of chimpanzees when compared to humans. This does not so much indicate a close relationship as animals, as that the "1-2%" represents the differences that have emerged during their evolution as discrete primates; the common rest of the DNA represents all of their previous evolutionary past as living entities. In consideration of the divergent evolution of humans and chimpanzees from one ancestral ape it is the "1-2%" that is important; the rest would seem to be irrelevant when it comes to specific differences between these two particular forms of life of relatively recent divergence, whether these are obviously recognizable or not. One should be wary against using DNA to exaggerate the closeness of the relationship between human beings and chimpanzees. The degree of similarity and difference that we can readily appreciate that this still basically depends on what it is possible to observe in their physical structure, appearance and behaviour

The suggestion made by scientists is that the human and chimpanzee lines diverged between 7 and 5 mya (i.e. at the end of the Miocene, the duration of which epoch is considered to be from 23.03 to 5.33 mya), but such a timespan shows how tentative this is. In 2020 the claim of Africa to be the motherland of humanity has been challenged by finds in Greece and Bulgaria. These comprise a jawbone and teeth that show human characteristics, but dating to 7.2 mya. This appears to predate the earliest finds from Africa, which have been limited to 7.0 mya; so at the very least this approach depends on the accuracy of the dating. The European specimens have been dubbed 'Graecopithicus'. Discussion on the implication of these finds is deferred to a later page.

The Australopithecines were clearly modified apes that over an immense period were in the process of acquiring certain characters that can be regarded as human. Although they were smaller than later human types (1m+, i.e. somewhat over 3ft), their skeletal remains indicate that they walked upright. Yet despite that their feet were gradually evolving towards suiting this, relatively their legs remained short and their arms long. These are ape-like features that still facilitated climbing, but in the way of apes, rather than monkeys. With the exception of the dentition, from the neck up they were still veritably apes, but from the neck down they were starting to show features that were to become characteristic of more advanced creatures.

All this suggests that early evolution that was specifically human with regard to structures and nervous system (i.e. in the way that the body formed and functioned), started with changes affecting the feet and legs. With some overlap, this then passed upwards to the hands and arms, then to the jaws and mouth finally to affect seriously the form of the skull that was needed to house the enlarged and hugely complex nerve control centre, the brain. All this commenced with a lengthy transition stage. At first limbs and other features would lag behind and not be fully changed to suit an evolving life style. Yet some of them would already be heading along the critical path that was leading the body towards suitability for the life style that had arisen. Indeed, those that did not follow the right trend were on the way to becoming extinct. Thus was clearly the fate in the case of the various types of *paranthropus,* as well as most versions of the Australopithecines.

The reason for the ancestral creatures giving up a life style that had been mainly arboreal for one that was mainly ground-dwelling needs to be considered. Collateral evidence indicates a long presence of Australopithecines already in Africa more than 2,500,000 years ago (2.5 mya) and which continued there from that time for up to a further 1,000,000 years. During this hard-to-grasp huge period a variety of forms existed that did not necessarily ever contact each other due to the immense length of time available for each line to exist, evolve and generally reach the dead end known as extinction. This could occur through inability to cope with environmental change. Even so, such early evolutionary selection might sometimes also have occurred due to competition arising between differing yet related hominins.

These various Australopithecines were the most relevant hominins that have been found in Africa at this earliest stage. With the passage of time and accompanying changing environments, these have been created as a result of divergent evolution. Small yet vital changes to their genetic structures were occurring that ensured eventual inability to interbreed in the most affected cases, and thereby multiple species of the genus would be created. Nevertheless, before such full speciation happened it is possible that divergent forms chanced to come together again, with successful hybrids sometimes being produced. "Success" implies ability to survive long enough to reproduce and the offspring to be fully fertile. Yet as is the case with dogs, mongrels can be very variable; some are aberrations, while at the other extreme there can be produced delightful creatures that may be physically attractive or with above average intelligence, or both. This illustrates that hybridization can be a good thing. The results of genetic contact between different yet related forms could lead to some 'mongrels' that were superior to either or both of their parental lines. Natural selection would see to it that inferior forms would die out, but the superior ones would provide the opportunity for the survival of themselves and produce races that would out-compete their own parentage and eventually eliminate and replace all conservative ancestral lines because of their acquired 'superiority', the ultimate sense of this word in this context being 'better ability to survive within a species'. In short, chance hybridization could speed up evolution. (In modern times the deliberate application of this process can be seen in accelerated form as one aspect of the selective breeding of domestic animals. Human evolution has also been speeded up, albeit in ignorance of the true cause, by the selective processes used in the choice of partners by mankind.)

While searching for causes for these developments, it can be asserted with some confidence that around 2.5 mya environmental changes had long been occurring that had forced certain types of ape in some areas either gradually to leave the trees or become extinct. The vital variation would involve a general retreat or breaking up of woodland due to drifts in the climate, especially when these included reduced rainfall. The manner of getting about in the trees used by these apes was distinct from that of monkeys, and this suggests why they developed the bipedal gait as the habitual means of locomotion when moving about on the ground, while baboons did not.

The great apes of Africa are gregarious, unlike those of Asia, and there is every reason to believe that this applied to early hominins like Australopithecines. Moving away from the densely growing trees must have been possible as an elaboration of the rudimentary communal foraging life style that had already been developed in the forest that went with an omnivorous diet, while the tightening of existing group-living became a growing necessity. While still forest dwellers they could frequently have been in contact and competition with other simians, such as contemporary chimpanzees. Their slighter stature would put them at a physical disadvantage in such a struggle; they could have been persecuted by the latter apes and perhaps even hunted.

But unlike chimpanzees and gorillas there were aspects of their physique and traits in their behaviour that enabled them to adapt to the expanding savannah landscape, even though fewer and fewer trees made this an increasingly dangerous place for ape-men. Open grassland was even worse. These landscapes could however also be exploited as hunting, scavenging and foraging areas, but it would seem that these early Australopithecines would need cover to retire to and this would best be provided by river valleys or lake shores, where lusher vegetation could still flourish and cliff faces provide shelters. The residual tree-climbing features of their build would give them an ability to clamber about on such rough terrain, which was likely to include rocky outcrops.

Another way of looking at this is that they did not so much voluntarily leave the trees; the trees left them. Some communities may have found themselves to be isolated in islands of deteriorating habitat when fragmenting areas of forest first shrank and disappeared. It was only because their peculiar evolutionary trends that they happened to be preadapted to the new circumstances and thereby had a chance to survive.

There can be little doubt that these early hominins were omnivorous feeders. This probably was the characteristic that enabled them to survive in environments they were forced into but otherwise not particularly well suited to at the time. Their survival strength lay in adaptability. They would seek vegetable matter in the form of fruits and, especially, roots. Animals they could catch would at first be small and include invertebrates, amphibians and reptiles. The regular inclusion of larger species in their diet would eventually develop from the exploitation of relevant types of carcass, such as those resulting from the activities of other predators in the more open environment. Such scavenging would lead to the need to confront dangerous animals, and a strategy must have developed to achieve this, as well as an improved weapons kit. As red meat eating became more prevalent the Australopithecines would go about seeking larger live prey for themselves, rather than leaving it to chance. It was also less dangerous than trying to steal the kills of fearsome hunters such as big cats and hyaenas. Even so, the hunting side of their lifestyle would lead to an advantageous increase in robustness, especially in the case of the males.

Among the techniques developed would be stalking before laying an ambush, i.e. group attacking out of cover where the quarry could be overcome by surprise and weight of numbers, a method that usually results in a high failure rate when used against fleet and alert prey. A more dependable outcome might result from the digging out of their refuges such burrow-dwelling animals as wart hogs and aardvarks. Another method may also have been developed from the experience of trying to catch fleet ungulates, such as antelopes, especially the more solitary ones. The adoption of the bipedal gait had resulted in a type of animal that was slower than quadrupeds, but this was compensated for by an increase in stamina. This has allowed human hunters to develop at some stage the method called persistence hunting. In this the chosen quarry after being spooked is tracked determinedly until it is finally halted by overheating and exhaustion, whereupon it can be dispatched relatively easily. While being pursued the animal would not be allowed time to recover. The inability to continue flight largely resulted because furred animals have difficulty in disposing of excess heat in a hot climate. At some stage *homo* developed a method to do this by means of sweat glands in the skin. Evaporation of perspiration helped him to cool down. This points to a diminishing of body hair as a parallel development.

In the course of several million years there were always some Australopithecines who were in the process of slowly evolving towards being true human beings at their most savage stage. Being gregarious would help towards being able to survive when roaming on more exposed areas away from dense forests, but they would still need to develop weapons to protect themselves and overcome prey, as well as using "tools" to deal with carcasses and dig up roots. Originally their build inherited from an arboreal past would serve them well in the rocky terrain that would have suited their physical abilities. Even so, the increased abandonment of tree climbing would free both hands for other tasks and, together with the acquirement of the bipedal gait, this would allow them to bear sticks that they could pick up or break off, as well as shorten to suit the required purpose. Otherwise they would pick up stones likewise to use as tools or weapons. This can be assumed since even monkeys such as types of capuchin in the New World use stones to break open nuts or split flakes from other stones, while in Africa chimpanzees behave similarly as one of their advanced activities.

The battle between the inherent ability to climb and the ability to walk well on two legs would eventually be won by the latter achievement. The benefits of being able to roam far and wide on the ground in search of food would be the decider. As a consequence evolution ensured that legs became longer and arms shorter, while the buttocks were developed to house large muscles. Together with the ability to dissipate heat, all this would lead to a being with a lot of stamina.

Chimpanzees cast stones (sometimes mysteriously at tree trunks), while others brandish sticks. In disputes both between and within groups sticks may be thrown, although with little skill and without great accuracy. With hominins improved methods of production and application of equipment made of such materials would have been useful for hunting, but also as a means of defence against predators and rival groups. However, no defences are perfect and losses would certainly occur, as happens even with baboons, the adult males of which are especially formidable. Yet, apart from technical and social developments there would be an evolutionary drive towards physical change, such as increase in stature and the ability to perform better and better on two legs.

■ The Rise of Homo

All the Australopithecines except those destined to become *homo* went extinct, but the next phase of human development also occurred in Africa. This suggests strongly that intermediate examples between them and a later form like *homo habilis* have either not been found or not recognized. This emergent early form of man could result from the evolutionary improvement of just one strain. If so numbers of this would at first be small and survival uncertain. However, the possibility also exists for the appearance in Africa of other forms of early *homo* from slightly different strains of *Australopithecus*, with the timing of their emergence as "ape-men" being staggered. What they would all have in common was that the main stream of evolution was transferred from the body to the head. The question to be answered concerns the identification of which among them contributed to later mankind; did this involve more than one type? If so, this would indicate a retention of some reproductive compatibility between types that were otherwise slowly diverging.

One way in which the head developed, alongside a general reduction in robustness of the whole feature, was a lessening of the ingestion mechanisms. The jaws were reduced, so that the associated muscle anchorages on the skull became less massive. This was accompanied by changes in dentition, with the crushing power of molars not being required on the same scale. The fang-like quality of the canine teeth was eliminated; they ceased to protrude significantly beyond the adjacent incisors and premolars, if at all. They were no longer needed as weapons and to overcome food species. This all suggests a replacement of their functions by tools and to some moving away from the consumption of vegetation, or at least its tougher aspects such as many types of leaf and stalk, with increasing preference being for suitable fruits and roots. It also hints at the use of fire and that meat came to be softened by being roasted before being eaten. This realization probably developed accidentally from recurring incidents. The eating of the meat of animals that had been killed by natural fire would lead to the deliberate copying of this occurrence by the creation of hearths as the ability to use 'tame' fire grew.

Adoption of life on tracts of land that were low in tree cover meant a reduction in the amount of edible vegetable matter available. Primates generally do not and indeed are not equipped to eat grass. The emerging members of the genus *homo* filled the gap in their nutritional needs by hunting and devouring those that did graze – such as ungulates. The development of such pursuits required much more organization and co-ordination within the members of a group than did the collecting of edible plant matter. This created an evolutionary drive to improve the performance of the brain and the capabilities of the body, especially the hands. The changes in brain-hand co-ordination led to an ever improving ability to fashion tools and the discrete use of both hands in this practice. Eventually the use of a tool in one hand to make a tool held in the other tended to displace the more haphazard results obtained from the need to use stone anvils. Precision striking enabled even a hard material like flint to be knapped into accurate shapes. Improved technology meant that larger game could be hunted more successfully and along with this an increase in stature would also be helpful before the use of hafted projectiles was developed and became widespread and transformed the ability of smaller men to kill game. (The capabilities of some communities of smaller men were eventually further improved when they discovered poisons that could be applied to the tips of their hunting points.)

Early African finds that are recognized as human (and later ones elsewhere) are linked together under the genus *homo*. While it would seem unlikely that the extremes placed under this label could have interbred, should contact even have been possible, fertile hybridization would surely sometimes result when groups who were in the process of diverging happened to come into contact again.

The ancestry of modern man passed through forms found among the ranks of earlier hominins most often classified as *homo habilis, homo rudolfensis* and *homo erectus.* All of these emerged in Africa and some of them, escaped from this continent, assuredly in forms derived selectively from one or more of these three designated types.

■ The Development of Homo in Africa

The Epochal Background of Homo Towards 6 mya the Miocene epoch gave way to the Pliocene, which then continued for nearly 3my. The latter epoch became a time of warmth with temperatures averaging some 2-3°C higher than the present and it has been suggested that sea level could have even have reached at least 25m higher than now, this being accompanied, among other factors, by a reduction in the amount of ice near the pole. The Pliocene was a time when much of Africa was covered with forest. The rise in sea level eventually led to the Atlantic breaking down the land block at Gibraltar about 4 mya. The creation of the Mediterranean Sea as we now know it and an accompanying enormous rise in the level of the Red Sea would severely decrease the possibilities of migration beyond them.

But before this, during the preceding Miocene, significant tectonic activity had still been going on in the Earth's crust. From c25 mya the African and Arabian plates were drifting apart along the line of the Red Sea. To the north was the so-called Tethys Sea, which comprised a zone that included the present Black, Caspian and Aral Seas and the Mediterranean Sea and the Persian Gulf to the south. At the end of the Miocene, say 5-6 mya, the original Mediterranean dried out and the Persian Gulf retreated, this being a heralding in of the Pliocene. The

Black Sea still stretched as far as the Aral Sea to the east. Then, as the Atlantic broke through at Gibraltar to flood the Mediterranean basin about 4 mya, the west end of the Black Sea became land and thereby created much of the Balkans in south-eastern Europe.

It is would seem that the Australopithecines never got out of Africa and that the conditions that first brought them down from the trees and then away from them arose there, as has been presented above. Any use of objects that they had developed while they were becoming bipedal apes and still dependent on the tree cover must have been increased by the need to adapt to the new conditions. In the end physical adaptations were not enough; any divergent strains that did not also evolve mentally could neither survive adverse conditions nor serious competition and would sooner or later die out.

Although hard to contemplate now, during the Miocene 'apes' lived among the fauna north of the Mediterranean basin. Dryopithicus inhabited the forests of western and central Europe between say 13 and 8 mya, while in Greece and Bulgaria, as noted earlier, jaw parts have been found of an ape exhibiting human traits and dating from some 7.2 mya. The name given to this type is Graecopithicus.

In the earlier Miocene much of the present land area north of Africa was inundated and formed the Tethys Sea. This was progressively reduced in size and resulted in the Mediterranean drying out nearly completely some 5-6 mya. Its bed formed saline lakes, while, as already mentioned, the Atlantic was held at bay by the land block at Gibraltar. So, at this time, Africa and Europe were united by low lying land and the Sahara had not formed in the state we recognize now, while at the same time movement by animals was clearly possible between the two continents. But this was brought to an end when about 4 mya the Atlantic broke through at Gibraltar. This means that any traces of any early *homo* over a great area thereabouts now lie on the sea bed and are lost to us. Later similar evidence from much earlier times was also lost in north Africa because of the geophysical effect from the start of the Holocene c11,700 ya (11.7 kya), when the Sahara Desert was beginning in the form we know it and obscuring ancient evidence under its sand dunes. One site that seems to have avoided such obscurity has been discovered in Chad. Remains of *sahelanthropus tchadensis* have been discovered on the southern fringes of the Sahara and are dated between 7 and 6 mya. Until the inundation of the Mediterranean c4 mya it was possible for descendants of Graecopithicus to exist in northern Africa. After the flood this creature became extinct in Europe to the north but could persist in Africa to the south as antecedents of creatures akin to those found in Chad.

It would seem rather unlikely that the man-like dentition of Graecopithicus could have evolved quite independently of that of Australopithicus. The corollary is that the ancestry of the latter arose while in contact with the former in the course of genetic descent, which process provided elusive later versions derived from this that were still occurring in north Africa after the formation of the Mediterranean about 4 mya. The inference is that one line on the way to *homo* by way of Australopithicus - and perhaps the main one – was in the form of Graecopithicus and originated in a zone that included southern Europe. The finds at Chad suggest an intermediate presence dating from 7-6 mya, but before the Mediterranean basin filled more advanced creatures appeared in Ethiopia that have been given the designation Ardipithecus. *A. kadabba* has been dated from 5.6 mya (late Miocene) and *a. ramidus* from 4.4 mya (early Pliocene). In east Africa around the time of the flood the discovered forms have been designated Australopithicus, hence *a. anamensis* from the date range 4.2-3.8 mya and found in Kenya and Ethiopia. In the later Pliocene in east Africa the type takes the form of *a. afarensis* 3.9-2.9 mya. From this same time span *a. bahrelghazali* (3.5 mya) has turned up in Chad, thus suggesting the possibility of a reflux of Australopithicus towards the north, or rather north-west, when conditions allowed it.

Stone tools have been found in Kenya in East Africa and dating from 3.5 mya. These and somewhat later ones are evidence of tool making and use by some Australopithecines from a very early date and for a very long time indeed. But eventually a certain section of them made cultural advances that included the use of specific tools to make tools of a different kind and with this came increases in brain capacity and improvements in mind function. These developments are in evidence as a range of fossilized skulls that once enclosed brains of ever greater volume. These trends eventually resulted in the emergence of *homo habilis.*

Through Australopithecus to the First Homo At the beginning of the Pliocene in Africa, apes with primitive human tendencies were already in evidence (5-6mya). The remains of these have been found in Kenya, such as *orrorin tugenensis* (c6mya) and *ardipithicus ramidus* somewhat later, with indication of bipedal gait and reduction in size of the canine teeth. An even earlier find was that of *sahelanthropus tchadensis* in Chad, quite a distance west of the usual sites, these being centred on the great African Rift Valley. Such finds of early date initially suggest that the absolute beginnings of humanity were during the later Miocene in that part of Africa whose latitude was near that of the Horn (Somalia, Ethiopia, etc.). Unless equally early finds turn up in southern Africa, the implication is that the *australopithicus* finds made there result from later southwards migration down the east side of Africa in the Pliocene. From this one might conclude that the long drive towards abandoning the trees commenced somewhere further north in Africa, where apparently the most progressive results of this eventually became concentrated in the area of the East African Rift Valley. As discussed above, recent impressions suggest that these hominins were from parts of Africa lying even further to the north and west and with an ultimate origin represented by the terminally extinct Graecopithicus of south-east Europe, with this having been the northern extremity of their range.

Late in the Pliocene (duration 5.33-2.58 mya), types of Australopithecine eventually evolved cerebrally and developed culturally to give rise to variations, one being known as *homo rudolfensis*. The increase in cranial capacity of such hominins indicated cultural advances that possibly included a capacity to forage further away from their home territories based on rock shelters, scrub and the availability of water. But the type of landscape that had emerged to the north was unsuitable to allow them to escape out of Africa.

The various kinds of environment that existed in Africa were so vast that they acted as barriers to migration for creatures who were adapted to survive in a particular habitat. The role of deserts in this respect is obvious, but the vast expanses of forest, savannah and grassland that might be entered to a limited degree for foraging and hunting may not have been traversable either. Other impediments to travel must have occurred when stretches of water needed to be crossed, such as wide and dangerous rivers and also seas, no matter how narrow. Like chimpanzees and gorillas, these early versions of *homo* still had not yet evolved a comprehensive means to escape from any habitat they had become so dependent on.

In general the Pliocene climate probably remained fairly stable, although somewhere about its midpoint a reversal occurred that was initially relatively rapid and led to a climate that was cooler and drier. The landscape barriers that existed in the north part of Africa would remain, even if there was some fluctuation in the distribution of the various environments. The route out to the north still remained impossible for the earliest *homo*. Even the more advanced *h. rudolfensis* was still just as trapped by his enforced habitat and was still as dependent on the presence of adjacent dense forests as chimpanzees were trapped within them. However, evolutionary improvements did enable advantage to be taken of a way out when eventually it became possible.

Homo Erectus Another facet of early man's containment in Africa was his dependence on his life style and equipment. If different areas were to be occupied they had to supply the resources that were necessary for survival, at the very least nutrition, wood for digging sticks and suitable stones for tools and weapons. The way out towards the north-east was through deserts, barriers that were obviously impenetrable. This made the crossing of the Red Sea at its southern end the most likely initial route. (During these earlier times it may have been less of a barrier.) Otherwise one might think in terms of the northward course of the River Nile as sometimes providing a passable route towards Asia. Even so unsuitable areas would exist on the way. The problem of survival when travelling into new areas depended on having the means to transfer your essential equipment during an interim period of want, until you reached some distant place where you were again able to find the raw materials to operate your life style and would sufficiently provide other necessities to serve as "home".

Homo habilis probably operated a system whereby material suitable for use as equipment was picked up as required and discarded after use. Yet reuse could occur when daily life involved places of retreat, even if new materials could also be worked on as required. This was probably the case with *h. erectus*. But penetration of unfamiliar territory could only occur when a means of carrying enough equipment was devised so that normal operations could continue on land that was materially unproductive. This seems to imply rudimentary thongs and bags made from animal skins or strands of fibrous vegetation. As techniques improved great skill was put into beautiful and often symmetrical tools, i.e. Acheulian hand axes that were produced by early mankind over an immensely long period; it seems unlikely that well-made objects would lightly be abandoned unless they either got mislaid or that damage seriously lessened their suitability for continued use or repair.

■ Migration

On foregoing pages reference has frequently been made to 'migration'. This word carries different meanings within a general sense of 'movement of people or other animals'. More basically two kinds of migration can be recognized. Firstly there is a regular exchange that occurred between fixed locations where the journeys are usually governed by the yearly cycle. In human society it often refers to the reciprocating movements of herding animals that are performed on a seasonal basis and in order to reach traditional areas where good - but temporary - sustenance could be expected to be found. Such nomadic events are usually fixed by geography and governed by climate stability. Some wild animals perform similarly and with birds such activity can involve enormous distances. Before pastoralism was adopted, man could be constrained to follow such movements by wild mammals, and some populations still are.

Then there are more sporadic movements usually caused by gradual climate change. These could lead to permanent displacement of folk groups and were caused either by it being impossible to stay in the current location, or were due to serendipitous discovery that conditions for survival were better somewhere else. Such 'migrations' in prehistoric times were really more erratic. The actual details of such drawn out drifts of populations into new areas during immense periods of time are now usually indeterminate except in the vaguest of terms.

Human hunter-gatherer communities in the past would always be affected by such stable or fluctuating conditions, and over the huge length of time involved they grew to be much more adaptable when it came to facing adversity. Indeed, this flexibility allowed an ape-man, such as *h. erectus*, to escape from Africa and in the course of many generations penetrate Asia all the way to China.

■ The Sequence of Migration

Time and Space in Africa In recent times stone tools have been discovered at Lomekwi in Kenya that have been dated to about 3,300,000 years ago (c3.3 mya). Since the descent of man from ape began 5-6 mya - from a common ancestor he had with the chimpanzee, and followed by the appearance of *Ardipithicus* 5.6 mya, *Australopithecus* c4 mya and *homo habilis* well before 2 mya, these tools would appear to have been made by Australopithecines who had made a start – at least in a cultural way - to becoming *homo*. Before the above Lomekwi finds, the earliest stylistically fashioned stone tools had been discovered in Africa from c2.5mya and called Oldowan from the find site at the Olduvai Gorge in north Tanzania.

Homo habilis habitually used the bipedal gait, a condition that had been pioneered long before by his ancestors among the Australopithecines. The appearance of *h. habilis* was stimulated by the climate change that introduced the Pleistocene Epoch – c2.5 mya - and usually referred to as the Ice Ages and with the ice growth causing increased aridity and loss of vegetation in the distant zone inhabited by early *homo*, i.e. in Africa. At this commencement, as a type the Australopithecines were dying out in Africa and in human terms it was the start of the Palaeolithic period, or Old Stone Age. Although the times had generally remained challenging, windows of opportunity did keep opening that members of the genus *homo* occasionally found they were in a position to take advantage of.

As the Pleistocene advanced, some *homo habilis* were evolving and the earlier type was being gradually accompanied in Africa by advanced forms designated *homo erectus*, earliest traces of whom appeared c1.9mya, and whose improved stone tools (Acheulean) have been dated from c1.7 mya. These hand axes remained as part of man's tool kit until c250,000 ya. In the wider world that was otherwise devoid of recognizable human ancestry (in the form of hominins), it was the appearance in Africa of *homo* that provided forebears who were capable of penetrating new habitats, either when under severe environmental pressure, or even simply when opportunity in the form of emergent possibilities beckoned. An early form of *homo,* resulting from such wandering northwards out of Africa, appears to feature in finds in the Caucasus, namely at Dmanisi, Georgia, and dating from c1.8mya.

The Pleistocene has been conventionally divided into 4 cold periods in which the ice caps at the poles spread and - at least where associated land masses were covered over - grew to enormous thicknesses. These were interspersed by 3 warmer interglacials, periods of fluctuating but less extreme climate and wherein the ice was in recession. During the latter conditions, the rainfall and the area of forest in Africa both increased, while the Sahara desert shrank, with such changing territory being added to the area of the intervening landscape of savannah type. But when huge areas of the hemispheres were again covered with ice, aridity returned and the deserts reclaimed much territory. At the extreme occurrence of glaciation, forest in Africa was restricted to a relatively small area in the west.

While noting that the above account represents a simplistic version of the effects of the Ice Ages on Africa, the question arises as to which climatic condition either allowed or forced *homo* to leave Africa? The indication that some of his type had already moved northwards out of the continent by c1.8mya suggests that they had been unsettled by conditions arising there by the first phase of the Ice Age (traditionally called Günz) and that he took advantage of the relaxation of conditions that occurred in its aftermath with the onset of the following interglacial. However the move meant that any later advances northwards into the Asian hinterland would always be curtailed or repulsed by the three other major icy periods that were to come during the remainder of the Pleistocene. Apart from residual populations of *homo* in Africa, this would result in the other major survival areas being limited to the southern fringes of Asia. However, the Pleistocene lasted for a period in excess of 2my, and during this, apart from times when ice dominated, there were minor fluctuations in the climate. Set against this huge stretch of time, the finds identified as versions of *homo* are very sparse indeed. It seems certain that during this time span a certain amount of resulting - though very slow - migratory response must have been generated, generally of a reciprocal north-south nature where this was physically possible. Under certain circumstances this may even occasionally have included some return towards Africa.

Physical and Mental change: the Drive of Evolution In the period when populations of human ancestry consisted of divergent but related forms of *homo,* the finds suggest that differing conditions suited differing physiques whereby divergent statures were favoured, i.e. tall and short races, or robust and gracile ones. This is what the sparse finds indicate, but future discoveries may drastically refine this simplistic impression. However, using the evidence as it stands, the picture emerging is of taller more robust types who could dominate their smaller relatives if they made contact. To survive the competition, the latter were constrained into living in remote areas where conditions could be extreme and the prospects hostile to comfortable human existence. Hence spearhead evolution was apparently more likely to occur among larger races, who had greater freedom of movement and less need to become specialized for a particular habitat. In actuality, the proposition is that it was under encountered difficult circumstances that innovation was likely to occur. But in the end, as human society became less primitive, evolution depended less on physical necessity and more on a need for increased intellectual achievement, while any progress could be accelerated selectively by hybridization. Yes, despite discrepancies in size and culture, the successors to *h. habilis* and *h. erectus* were all still contained in a single genus *homo,* some of whose otherwise divergent species remained sexually compatible and capable of producing fertile offspring.

The adaptability of *homo* and his superior ability to migrate in the face of environmental difficulty meant that total geographical isolation was less likely than with other animals, while the opportunities for hybridization were thus increased. These circumstances acted to quicken the pace of human evolution to a rate - and to comprise a kind - that exceeded the possibilities of any other kind of animal.

Africa Reconsidered Before coming to grips with the problems presented by Asia it is necessary to explore the situation in Europe after considering Africa again, where diminutive races of *h. sapiens* persist in the 21st century. Why do the pygmies of the Congo basin and the Bushmen of the Kalahari exist as wee folk? They are good examples of small people being confined to difficult areas of an Africa where tribes elsewhere are usually of greater stature. There is evidence that the Bushmen used to be more widespread, as for instance at the headwaters of rivers in the Drakensberg Mountains, to the east of their present range. This took the form of rock art dating from 4000 years ago. They have since shrunk back to the Kalahari, which until recently served as a harsh enough habitat to protect them from encroachment by others. The dense forest of the Congo Basin similarly used to protect the pygmies who dwelt there. Thus can one observe that small folk could best survive when dwelling in habitats that were unsuitable for larger men, even though these were physically more robust and, in many respects, of more advanced culture. The question arises as to whether this situation is a relatively recent development or does it somehow reflect a dichotomy that was already evident early in the Pleistocene? It is provisionally suggested here that the latter case applies.

The Background of Early Homo

The emergence of modern man appears to have been driven by the difficulties imposed by the Ice Ages and the effects these climatic fluctuations had on habitats everywhere. It is reasonable to suppose that it was only the intermittent retreats of the ice that provided the necessary opportunities for human advancement. Yet the times of apparent increased hardship when the ice was in ascendency may have contributed in greater degree to the driving of humanity into more advanced forms. The implication is that it was not the easy life that drove the evolution and distribution of *homo*. Primitive man was usually driven to change and make changes when he found himself outside of his comfort zone, this being the one in which he was at the time best equipped to survive, until needing to amend himself culturally or even physically when more challenging circumstances arose.

The following is a postulate scenario. In Asia Pekin Man was found in the north-east of China. He was a form of *h. erectus* who perhaps could only have survived at that latitude during an interglacial. But when the temperatures dropped and the frozen wilderness encroached he would be faced with the alternative of moving south or risk being eliminated. This sort of situation must have been faced by any variety of *h. erectus* that had got out of Africa and some would be lured northwards when the climate was favourable and where the habitat was becoming more attractive and accessible than the more southerly situation wherein he found himself. Later advances of the ice would tend to force such northernmost communities of *homo* back into territories since occupied by other hominins and thus creating competitive situations with increased evolutionary possibilities. Yet *homo* was well along the road leading to a state of ever increasing adaptability, and the genus was now being led along this way by two inseparable companions, physical and cultural evolution. An improvement in one would lead to a compatible response in the other. This led to enormous adaptability, which enabled *homo* to escape from damaging changes in climate and habitat that the World could throw at him. No other form of life came anywhere near such capabilities and in the 21st century even the greatest depths of the oceans and the vastness of space are not safe from his penetrations.

The Ice Ages of the Pleistocene had cosmic causes resulting from changes in Earth's attitude to the Sun. Apart from these, there were ongoing geological activities over these immense periods that were affecting the Earth's surface and hence the habitats in which early men were striving to survive. Sudden tectonic movements were ongoing and causing violent events like earthquakes, volcanic eruptions and sea surges such as tsunamis. Old and new mountain chains were being eroded on a longer time scale, but the more recent formations had peaks that were tending to be pushed to even greater height, both for tectonic and volcanic reasons. Stretches of water were widening or narrowing and some were even disappearing.

In the Pleistocene, however, of greatest significance were the persistent effects of the ice on the types of landscape available and suitable for habitation by primitive humans. Of specific importance were the changes induced by the state of the ice caps on the level of the oceans. When the amount of ice increased, areas of land would be exposed that had been inundated during interglacials. Apart from the ice reducing the amount of water available, a further effect was that the sheer weight of the ice caused some of the adjacent land towards the equator to rise considerably due to counterpoise. Aridity was also increased because of the area of ocean available for evaporation being reduced. The root cause for such effects on the size and distribution of the various kinds of habitat lay in the changes that occurred in the Earth's attitude as a member of the solar system: and indeed still occur.

Those areas that slowly alternated as seabed or littoral as the Pleistocene progressed may have been the zones that supported or allowed the most extensive of man's activities for the duration of the high and dry phase of such fluctuations, both for habitation and migration. As sources of information about man's past they are now mostly beyond our reach due to submergence caused by the present interglacial state of the world's climate.

The origins of *h. sapiens* are generally believed to be found in Africa and it can be postulated with some confidence that some of the natives of that continent have achieved this status while remaining within it. Any contemplation of modern man having originated elsewhere and then migrating to Africa where any earlier versions of *homo* were all completely eliminated presents an unlikely scenario. For example, any notion suggesting that the bushmen of southern Africa represent a form of *homo sapiens* that originated as a finite entity somewhere in Eurasia and somehow made their way to their present situation must be resisted. A more reasonable hypothesis is that ever changing forms of *h. habilis*, *h. erectus* and later forms of the genus were being created somewhere or other in Africa and spread elsewhere when circumstances both demanded and allowed this. Later emigrants contained modified versions of any earlier types of such beings, and on their travels they encountered populations descended from earlier migrants, whether elsewhere in Africa or in Eurasia to the north or east. Such encounters would result in some degree of interbreeding and sometimes the hybrid offspring would carry improved characteristics in their genetic structures and such would then be in the best position for evolutionary, physical and cultural dominance and thereby survival.

■ Heidelberg Man

The first signs of further significant evolution in Africa are provided by the remains of *h. heidelbergensis* that have been found there. Yet this type - evolved from *h. erectus* as already presented above - was named after a site where similar human remains were found in Europe, namely at Heidelberg near the River Rhine in Germany. The presence of such a being in this northern location demands consideration. In the south remains of *h. heidelbergensis* have been found in eastern and central Africa, namely in Ethiopia (c600 kya) and further south in Zambia (c300kya). The vast intervening distance from the European places, apparently without known examples, is curious and requires explanation. One might just infer from this that movement into these two African areas has been out of some elusive intervening core zone generally to the north of them, and was hence southwards.

As it happens human remains of earlier origin have been found at Gran Dolina in the Ebro Valley, northern Spain. They have been identified as intermediate between *h. erectus* and *h. heidelbergensis*, while being dubbed *h. antecessor* and dated to 900,000 years ago (900 kya). Even earlier human presence has turned up in the form of finds of teeth at Sime del Elefante (1.2 to 1.1 mya) and Orce (1.4 mya). These dates are significantly earlier than the ones for the forms of *h. heidelbergensis* in Africa, just mentioned above. The inference one can make is that this type of hominin was indeed evolving somewhere near the present east coast of Spain. This gives rise to the suspicion that the presence of *h. antecessor* there points to a significant entry point into Europe from Africa at the western end of the Mediterranean, a narrows now represented by the Straits of Gibraltar. Such a crossing would be facilitated by the area of water being less than it is now. It need not come as a surprise that there are signs of even earlier presence by men on the neighbouring coastlands of France to the east, namely by the artefacts found at Lézignan-la-Cèbe. *H. heidelbergensis* thus appears to be present in north-east Spain several hundred thousand years earlier than in East Africa. Apart from that, there are finds of *h. antecessor* and even earlier types of men in this general European location that are altogether absent from the mentioned African ones. The implication is that the early evolution of *h. heidelbergensis* occurred in the area around the eastern end of the Spanish-French border and probably on land to the south now covered by the waters of the Mediterranean Sea. From there, over a period of several hundred thousand years, examples of *h. heidelbergensis* carried this version of *homo* back into Africa and also penetrated deep into western Europe to the north when circumstances allowed.

From the above considerations it would appear that the *h. ergaster* form of *h. erectus* was evolving into *h. antecessor* at a time when migration was possible out of north-west Africa into Europe at the western end of the Mediterranean Sea from where he gained access to lands to the north, even including Britain. Later, from Spain and in the evolved form *h. heidelbergensis,* he again spread northwards through France and beyond. This form of *homo* was a hunter of the foothills and associated plains and adapted to more northern areas of Europe by learning how to overcome large animals of the temperate zone, and perhaps even sub-arctic ones. Absence of evidence of any earlier presence in north Africa could be due to loss of habitat as a result of increased aridity and being subsequently obscured by conversion of intervening areas to desert during glacial periods, these being historical changes to the Sahara. Yet despite lack of direct evidence, it would seem that over a long period folk closely related to the *h. heidelbergensis* found in western Europe had also managed to make their way - after some delay - from somewhere near Spain to Ethiopia. Any intervening evidence of this may subsequently have been lost beneath the waters of the Mediterranean or the ecological development of the Sahara, including its eventual covering of sand.

Since *h. heidelbergensis* did not abruptly become *h. sapiens*, it follows that one should expect to find intermediate forms between the two that one might dub *'homo pre-sapiens'* as a term of some convenience and the nature of whom will be addressed later.

■ Neanderthal Man

This phase of the story of humanity occurred in the later Pleistocene and with a main concentration in western Europe. Considerable remains have been found in Spain, France and Germany. Further east such presence is sparser, with anatomical finds occurring in places such as Italy, Greece, Israel, Iran, Uzbekistan and even as far east as the foothills of the Altai Mountains of Siberia.

While the popular conception of Neanderthal Man is usually of a powerful, brutish being who roamed frozen northern plains and hunted large animals such as mammoths and woolly rhinoceroses, evidence suggests rather that his more usual habitat was temperate woodland and that his favourite quarry there would be red deer, which were largely to be found in such an environment. However, the Siberian cave finds suggest a reduced population density at the eastern end of the range. These may have been adapted to exploit a more open habitat and used their increased mobility to explore such expanses that lay to the east of Europe. Here they could have survived by hunting steppe animals such as horses and bison. The signs are that Neanderthal Man became adaptable and that he could exploit other environments when there was a need to. This capability had been pioneered in western Europe by his predecessor there, Heidelberg Man, who had even reached the still attached Britain by c500 kya, e.g. the finds at Boxgrove, West Sussex. (Even earlier traces of humans have been made in East Anglia, i.e. dating from 700 kya and 950 kya. The earlier of these dates indicate that men even earlier than Heidelberg Man were involved - perhaps *h. antecessor* - and they were taking advantage of interglacial periods, while the ability to make fire can surely be assumed.) Later remains identified as Neanderthal have even been found in North Wales, namely a jawbone from Bontnewydd, near St. Asaph, and dating from 230 kya.

The term 'caveman' has been coined because finds from these early times have often been found in caves. This is because such provide the best conditions for organic remains to survive. Even so, although signs of activity on plains and in forests are less likely to be preserved, it seems reasonable to assume that open country at least did not provide an environment where such early men would like to dwell, at least those eking out a living towards the north. The existence of Neanderthal Man in the foothills of the Altai Mountains seems to be the end of a thin trail of sites on hilly terrain in northern Iran and Uzbekistan and thus indicating movement further east out of eastern Europe and mainly keeping to the north of the Carpathian Mountains.

The distribution of sites containing remains that have been identified as Neanderthal suggest that their preferred habitat was foothills and that they tended to avoid high mountain ranges and flat plains for their occupation sites. In Europe the massifs of the Alps and Carpathians are avoided, as are the Danube plains of Hungary and Romania. The type of zone favoured were the hills as found north of the Alps around the middle course of the Rhine, which is where the site-name Neanderthal actually originated (as did indeed Heidelberg). From such preferred habitats more prolonged hunting expeditions will sometimes have been found necessary in pursuit of prey such as mammoths, ibex, horses and bison.

The whole expansion of early populations from the Iberia/Catalonia zone may have been driven by the occasional need to find new areas and such pioneering would eventually be fueled by knowledge gained of suitable terrain elsewhere when searching for herding animals across bleak plains. In this way the plains of Hungary and Romania could eventually have been traversed and the Carpathians by-passed in the slow opportunistic drive towards the east. Having passed north of the Black Sea, where the Crimea seems to have offered conditions suitable for habitation, the mountainous areas of the Caucasus and northern Iran were reached where desirable associated hilly terrain was available. To the north of these were the vast plains known as the Steppes. The same trend eventually meant that even the distant mountains of the Altai were attained, which proved to contain some caves that were very suitable to provide for the needs of both the Neanderthalers and a distinct eastern type, the Denisovans.

During much of the Pleistocene, the earlier Neanderthalers inhabiting the variable northern fringes of the Mediterranean basin would as always be hindered from reaching further north by physical barriers even when the climate relented. Initially movement northwards into Europe from the shores of the Mediterranean was mainly blocked by mountains such as the Alps and some large seas at the eastern end, while that into Asia's interior to the north was mainly made difficult by even higher mountain ranges, e.g. the Himalayas.

Access to the interior was facilitated in places by river valleys and a particularly useful one was the Rhone in France, which flowed to the Mediterranean from the north on a course west of the Alps and whose upper basin eased access to the middle course of the Rhine and the headwaters of the Danube, in which area sites have been found, including the famous ones of Heidelberg and Neanderthal.

South of the main mountain spine, the northern shores of the Mediterranean were interrupted by the huge peninsulas of Italy and Greece, associated with which were gulfs such as the Adriatic and Aegean Seas, although during ice ages these would be smaller than now. Offshore were islands large and small, such as Sicily, Sardinia and Crete. Some of these were accessed in the course of the drive towards the east end of the Mediterranean, although when the ice advanced, this would be facilitated in some cases by them being attached to each other or to the mainland. Turkey is a mountainous country and, judging by anatomical finds, apparently avoided by the Neanderthalers, although an anomaly is provided by one site on the southern coast of Asia Minor.

At the east end of the Mediterranean are the Neanderthal sites in Israel. How did they get there? There is a choice between two possible routes. Were they the end result of expansion eastwards along the north shores of the Mediterranean, or had they made their way south from the Caucasus area? Of course either or both may have occurred, but the latter seems more likely. The scarcity of sites from the coast of Asia Minor and the elusiveness on Cyprus tell against approach from the west.

The presence in Israel suggests that these Neanderthalers there were the descendants of escapees from advancing ice when this made life difficult further to the north. Yet this is as far south as they got. There is no sign that they went on to cross the land bridge to enter Africa via Egypt.

Yet derived Neanderthal remains have also been found further west in north Africa, where the culture is known as Aterian, named after a site in Tunisia. The earliest examples come from Morocco, being dated 145 kya, while the type went out of existence c20 kya.

■ Later Neanderthalers: the Mousterians

The emergence of *h. sapiens* seems to be tied to the effect of the third glaciation (191-130 kya). The buildup of ice from the north caused a narrowing of the straits at Gibraltar, this allowing easier access between Iberia and north-west Africa (the Maghreb), while to the south of this zone aridity was increasing. The advanced culture that existed in Morocco (Jebel Irhoud, 350-280 kya) clearly preceded this stage by a long time, but any residual elements of it may have been threatened by a general deterioration of their environment due to climate changes caused by the advancing ice. One must also consider that their demise could also have been accelerated by the arrival of folk from beyond the straits to the north who were better suited to the changing climate of the Maghreb. These would be Neanderthalers from Iberia. While these may have become the dominant physical type in the area, the social contact with remnants of the Jebel Irhoud folk type may have led to improvements in their own culture and leading to groups who might be dubbed proto-Mousterians.

Eventual amelioration of the climate would then cause reflux of folk at this level and these carried the improvements back into Europe. Hence the presence of Mousterians there - with their Levallois features - was due to contact with residual elements related to the Jebel Irhoud culture in the Maghreb, which population had come some distance towards becoming *h. sapiens* as this progressive type of human was starting to pervade Africa. Artefacts of Mousterian type have been found along the north side of the Mediterranean. They appear in the

Balkans, apparently their eastern limit there, since they are absent from the bulk of Asia Minor.

The European Mousterian period covered the years160-40 kya. In north Africa it gave rise to the Aterian culture with its acquired Levallois technology as an iconic feature. The Mousterian culture in north Africa is considered to have begun c315 kya, with relevant artefacts having been found as far east as Cyrenaica (e.g. at Haua Fteah). However the period 145-20 kya is considered to be the Aterian period and it hence appears to have survived slightly longer than the Mousterians in Europe. (The type site of Le Moustier in the Dordogne region of south-west France is dated c45 kya, hence only a few thousand years before this breed of *homo* vanished as a distinct entity in Europe). The Mousterian period fell between the Acheulean period of the more primitive Neanderthalers with their hand axe culture and the Aurignacian period of populations of *homo sapiens*. These were destined to pervade and dominate Europe by intrusion from east to west, and by about 40 kya had apparently eliminated the Neanderthalers as a distinct component of the population, perhaps by a combination of competition, extermination and interbreeding.

In the meantime, further to the north, it was the primitive Neanderthal folk who persisted longest with their lifestyle. The evidence as reviewed above suggests that they were being gradually affected by folk from the south who were physically essentially the same as themselves, but whose lifestyle was Mousterian and using Levallois techniques ancestrally picked up in Africa. Yet they were eventually all doomed to elimination as discrete races due to the effect of those *homo sapiens* migrants who were drifting into western Europe from the east.

▉ The Origins of Homo Sapiens

Resulting from DNA analysis it is apparent that nearly all present human races retain vestiges of interbreeding with Neanderthalers, the significant exceptions being the various populations found in sub-Saharan Africa. This suggests that there was a block that caused a cessation to the possibility of return to the bulk of the great southern continent for northerners at this particular stage of human development.

In any case it would seem that the Neanderthalers were not good at crossing broad expanses of water, so the Mediterranean is an apparent barrier, while it was backed up by another obstruction and that was a great desert, namely the Sahara. Yet later on in the same period, some sub-Saharan folk were managing to get out of Africa. But this was not by the route from north-west Africa into Iberia via Gibraltar; they were either crossing the Red Sea (most probably at its southern end, the Bab el Mandeb Strait), or using the land bridge out of Egypt at the continent's north-eastern point, i.e. into the Levant. Eventually these north- or east-bound migrants came to consist of fully formed representatives of the species *homo sapiens*, but an important point of contact with the Neanderthalers, at an early stage of becoming this, would be in the Palestine/Israel zone, more specifically in the more northern parts of this tract. The suspicion is generated that this flow of migrants from the south in some way impeded entry into Africa by the Neanderthalers by way of Egypt. Indeed, the flow was to be reversed in that the home territories of the Neanderthalers in both Eurasia and north Africa were due to be impinged upon and eventually thoroughly penetrated and overwhelmed by *homo sapiens* perhaps from Asian stock already established further east, but ultimately from Africa by way of Egypt and the Levant.

The great difficulty with getting a grip on the status of early *h. sapiens* in the Near East, and subsequently further afield in zones such as north Africa, Asia and Europe, lies in their relationship with the Neanderthalers. There is not only evidence that they could be in social and sexual contact with each other, but they were also sharing each other's cultures. In such variable hybrid communities any hand axes of Acheulian origin were originally provided from the Neanderthal side, while flakes and blades of Levallois origin resulted from more gracile human constituents with roots further south, namely in Africa.

Hominins that were in the process of becoming modern humans were hence getting out of their core area at the Horn of Africa in at least three discernible directions: to the west to reach the Maghreb: to the north to reach Palestine: to the east to reach southern Asia. In the west they are evidenced by the finds at Jebel Irhoud, Morocco. These folk failed to cross over northwards into Iberia. Indeed, the signs are that they were eventually overwhelmed by Neanderthalers who at a later time were entering Africa by crossing southwards over narrows that have since changed to become the Straits of Gibraltar.

Early Homo in Africa Reviewed The emergence of modern humans in Africa needs to be revisited. It has already been mentioned that a version of *h. heidelbergensis* was found in Ethiopia (Bodo) from 600 kya. It has been claimed that this skull was already exhibiting some features that suggest early signs of *h. sapiens*. Down the east side of Africa, from Ethiopia to South Africa, sporadic finds have indeed been made that indicate stages in the evolution of *h. sapiens* from something more primitive. This all suggests an emergence of a type from a form of *h. heidelbergensis* that was destined to develop differently from the one that appeared in western Europe and which had led to *h. neanderthalensis,* as discussed above.

The image one has of *h. heidelbergensis* is of a burly man. As earlier versions of this hominin spread into western Europe from north-west Africa (as presented above), they were of a type that was not evolving towards *h. sapiens* and, once in Europe, they were moving into areas that lacked more primitive hominin presence, so that their successors were largely a population that on the whole comprised improved versions of themselves, and hence remained heavily built, namely Neanderthalers. But in Africa towards the south *h. heidelbergensis* would always be sharing the continent with remnants of earlier populations that were of more gracile build, these being residual elements that were derived discretely and in their own ways from *h. habilis/h. rudolfensis* through *h. erectus,* and hence were evolving separately and distinctly. They had retained a smaller stature and could be dominated by the humans returning from the north-west as 'Heidelbergers', but, by becoming specialists, still gained a living on less fertile land or in areas of extremely luxuriant growth.

However, despite differences in origins, appearance and culture, hybridization could occur with the results sometimes being more advanced physically - and eventually culturally - than the original types. Thus the evolving hybrids could derive greater height and bulk from the one, but become more gracile as an acquired trait from the other, although the proportions of such physical traits that were exhibited could vary. However, while the bigger hybrids would be dominant in the fringes of the more desirable savannah and grassland areas, the smaller components of the genus still hung on in the physically more difficult or remote places. Yet complete separation did not occur on a permanent basis. Continuity of hybridization, however infrequent, would ensure that overall fertility was retained and the integrity of the particular genus we have come to call *homo* was maintained no matter what differences in habitat, stature and lifestyle arose. Indeed, it can be claimed that it was only by retaining the ability to hybridize with other types of *homo* that the progressive form that survived into the distant future occurred.

Apart from the Ethiopian site at Bodo (600 kya), the earliest finds in Africa of this developing hybrid Heidelberger type are from Tanzania (Ndutu, 400 kya), which lies to the south of Ethiopia in the Horn of Africa, with Kenya in between. Folk in this line of evolution had reached Zambia (Kabwe/Broken Hill) c300 kya and South Africa by 260 kya (Florisbad) and 235 kya (Rising Star). There is another site in Tanzania (Ngaloba, 120 kya). The last can be regarded as representative of descendants of those left behind after a gradual migration southwards out of Tanzania later than 400 kya and –with some integration - reached South Africa before 260 kya. This distribution, thin though it is, suggests an early population of Heidelbergers in east Africa from before 400 kya that was starting to evolve towards *h. sapiens* by way of considerable hybridization with more elusive residual native populations of different and indeed more primitive stock.

So these progressive types eventually spread from Tanzania as far as South Africa, but earlier than this, movement from the Horn of Africa had been northwards, traces of which would be elusive if it were not for the finds from Ethiopia 195 kya (Omo Kibish) and 160-154 kya (Herto) and significantly later ones from Israel. From the anatomical details the remains from the first of these Ethiopian sites are regarded as the earliest representation of fully formed *h. sapiens* ever found anywhere.

There would seem to be no trace of the immediate source of the claimed movement out of north-west Africa across to Iberia which could give rise to *h. antecessor* and his successors in western Europe. This is all concomitant with the lack of evidence over the whole of western Africa of signs of early man such as this. For environmental reasons there may not indeed have been a presence over most of this area in the earliest times, but, even so, *h. antecessor* must have had origins in ancestral populations to be found in somewhere like Morocco or a lost adjacent littoral at the western end of the Mediterranean Sea. All available evidence suggests that these arrived there from a generally south-easterly direction, rather than from a more directly southerly one up the west coast of Africa.

Down to the Cameroons, the Guinea coast and the Niger basin little has turned up indicating early hominin activity. This is the classical homeland of the generally Bantu speaking negroid types. The signs are that these arrived in the area when already constituting a fully evolved example of *h. sapiens*. As such these folk were taking part in the eventual widespread dispersal of *h. sapiens* over Africa. These west Africans tended to be well built and dark skinned, but one particular feature they had marginally retained from Heidelberger ancestry was a greater prevalence of prognathism, i.e. a jutting forward of the jaw area of the face. This characteristic has generally earlier been reduced among the recent native inhabitants of Africa in other parts. The overall robustness of this feature suggests a greater dependence on vegetation as the main source of nutrition. This accords with this type being specifically associated with the more densely forested areas of Africa. Eventually their mainstay became mixed farming, in contrast to much of east Africa where pastoralism based on cattle became the norm, a lifestyle requiring plenty of grass. Prognathism is not a feature that has evolved; it has been retained from earlier forms, albeit to a limited extent. The lack of it in the bulk of the *h. sapiens* species is due to a more complete recession of the face as an evolutionary process, this resulting from nutrition being obtained from food that did not require the jaws to be so robust.

In modern times folk of west African descent are prominent in sports that require speed and explosive action, such as sprinting and boxing. The face is designed to exchange greater volumes of air and this facility is enabled by larger mouths and wider nostrils. Against this one can seek out the best endurance runners in the world; these are found in east Africa, near the Horn and the Rift Valley, where the retained features described above are less in evidence. Viewed historically, this suggests that the former primitively practised ambush hunting using hide and rush, whereas the latter are more derived from persistence hunters. The facial differences between the Bantu speaking Guinea Coast folk in the west and the populations of the Nile catchment and the Horn in the east are relatively slight: both types have a much lighter bone structure than their Heidelberger forebears. Indeed all the surviving types have generally lost the larger and jutting jaws found on the latter as well as on their *homo habilis, homo rudolfensis* and *homo erectus* ancesters. The evolutionary changes must have been largely brought about by a shared cultural acquirement, the ability to control fire. The heavier jaws of the past were no longer required once meat could be roasted and vegetable food cooked.

The one site in the west that delivered remains of suitable antiquity in this context was in Morocco (Jebel Irhoud, 300 kya). These finds showing strong *h. sapiens* traits seem more likely to represent a rare and isolated find site, which perhaps represented variant examples of descendants of *h. antecessor* (who had indeed earlier discretely found their way to Iberia and western Europe), rather than antecedents of the eventual negroid inhabitants of western equatorial Africa. While there is no evidence for *h. antecessor* in Morocco, the Jebel Irhoud finds suggest descent that had either been influenced by or replaced by developments that had earlier taken place further east in north Africa.

The above presentation based on anatomical remains suggests that the ancestors of *h. neanderthalensis* entered Europe out of north-west Africa (Morocco), whereas the route into Europe taken by *h. sapiens* was at a different time and from north-east Africa (Egypt) via Israel/Palestine and the Levant in general.

A problem arises in that the Jebel Irhoud finds showing early signs of *h. sapiens* before 280 kya are earlier than the presence of even the earliest Neanderthalers around the Mediterranean. These existed in Israel from 70-60 kya (and have been found in England and Siberia from a similar time). They spent two separate periods in the distantly eastern Denisova cave (100 and 40 kya). Sites in Spain and Italy have been dated to c120 kya. As suggested above, the Jebel Irhoud folk of the Maghreb were probably absorbed genetically by the incoming Neanderthalers, with their European origins northwards beyond Gibraltar: this would explain the improved culture that arose there among these and who have acquired the name Mousterians, this referring to their advanced cultural status.

The remains found at Bodo, Ethiopia are developed Heidelbergers showing the earliest signs of recognizable *h. sapiens* features and date from c600 kya. By 195 kya fully formed *h. sapiens* remains have been found at Omo Kibish in the same country. The inference one can make is to look either to the Horn of Africa for the origins of *h. sapiens* or to somewhere not so very far away from there. This would suggest that the upper catchment of the River Nile is involved. As already suggested above, the best explanation for the finds showing trends towards *h. sapiens* at Jebel Irhoud (Morocco, 300 kya) would be a drift westward across north Africa from their origins at the Horn and with any intervening evidence eventually being largely obscured by the Sahara or the inundation of adjacent littoral. There is no cause to believe that this human trend ever crossed water northwards into Europe over the Mediterranean Sea, although its effects were eventually felt through escape routes to both Iberia in the west and the Levant in the east.

The above scenarios are largely based on finds of anatomical remains. Sites without these, but consisting mainly of stone tools, are more common and might give a different overall picture. However, artefacts are only indirect evidence when it comes to the ethnicity of the makers and users and, being transferable both as objects and ideas, may be misleading, significantly so here with regards to the emergence of *h. sapiens.*

There would seem to be rough correlation between the change over from the fabrication and use of Acheulean hand axe culture (bifaces) to the core-and-flake Levallois culture that started to replace and eventually followed it in association with the successions *h. heidelbergensis > h. neanderthalis* and *h. heidelbergensis > h. pre-sapiens > h. sapiens.* In reality it is not quite as simple as this.

One finds that some early *h. sapiens* continued to use hand axes, while Levallois tools can be found on sites that can be regarded as of late or derivative *h. heidelbergensis* type. The Levallois culture often occurred in societies that were too early for *h. sapiens*. As a matter of fact the Levallois technique had already been developing over a long period before it was associated with modern humans. It is apparent that the development of flaking from a core was one of the changes of activity that were accompanying the evolutionary trends being followed among certain Heidelberger hybrids in Africa and were contributing to progress towards and eventual emergence of *h. sapiens*. So *h. sapiens* did not invent the Levallois technique; the evidence indicates that this was one of the cultural facets that was instrumental in the emergence of *h. sapiens* out of an earlier population developing in Africa and consisting of a range of hybridized *h. heidelbergensis* types that has been dubbed *h. pre-sapiens* above. The Levallois flaking technique from a prepared core was a refinement that had originated in the primitive knapping of stones, which modifications formed the basis of the Oldowan culture. The flakes themselves eventually became usable byproducts of this way of working the material. The biface culture of the Heidelbergers resulted from improvements on the core-based Oldowan culture, which tended to undervalue flakes, and as a result ended up as the beautiful hand axes produced by the culture known as Acheulian. In the fully blown Levallois culture the core was not developed as a tool, but was a skillfully prepared piece for the production of preconceived flakes.

The Critical Path Leading to Homo Sapiens

Homo erectus did not suddenly become *homo sapiens*. It follows that there have been intermediate and differing transient forms between these two identified types that could interbreed occasionally to provide divergent lines of descent, most of which were sooner or later doomed to failure. Under survival pressures such intercourse will have led to more advanced types at the cutting edge of such evolution. Such selectivity will first have led to developments such a, *h. ergaster, h. antecessor, h. heidelbergensis and h. neanderthalensis,* and eventually to the emergence of *homo sapiens.* However, the latter was not derived directly from the Neanderthalers, and was never a completely homogeneous group, but constituted a type exhibiting a variety of characters. These were selectively retained among sub-groups that exhibited the beneficial results of the specific genetic experience acquired from encounters that had led to intercourse that was still able to be sexual as well as cultural.

However, certain acquired characters did become specific to *homo sapiens* as known to us now and in the long run only those types of man that had them would survive. They were related to developments in the head to do with preferred food, ability to think abstractly, the development of speech and manipulation of objects. A case in point here is the arrival of *homo sapiens* in Europe. The area already had a long standing population of Europeans who had evolved on their own to become a hunting society exhibiting its own characteristics, both physical and cultural. Despite any presumed mutual hostility some interbreeding was possible. Even so, the arrival of the newcomers eventually resulted in the final disappearance of Neanderthal Man. (Yet it can be noted here that some bodies of opinion do indeed still consider the latter as being representative of a distinct race of *homo sapiens*, albeit at a more primitive stage.)

■ Distribution of Physical Types

The Spread and the Squeeze In the Pleistocene Epoch (from 2 mya+) migrants would be able to spread out into Eurasia during interglacials, but when the ice eventually advanced again they would be squeezed between these hostile cold conditions in the north and increased aridity that extended into the southern hemisphere. The latter state would apply in the west, but further east it was the Indian Ocean and the China Sea that prevented activity being extended into the far south, except where interrupted by the Indo-China peninsula and its extension formed by the then joined up islands of the East Indies (the so-called Sundaland). Beyond this eventual major archipelago beckoned myriads of lesser islands spreading out over the western Pacific Ocean, while further to the east and south still lay some detached and progressively increasing land masses, namely Sulawesi (aka Celebes) and a still unified New Guinea and Australia (the so-called Sahul), as well as a lot of smaller islands.

Homo was acquiring a capability to break out of any habitat of dwindling adequacy and penetrate distant harsh places because of his growing adaptability and mobility. Barren open spaces could be traversed because of his developed stamina, while rough or hilly terrain could be clambered over because of a lingering ability to climb that had its roots in his former arboreal existence. This latent adaptability, which imparted an ability to survive in a wide variety and often hostile range of surroundings, is the key to this one species - *h. sapiens* – now being found on every one of the World's continents, including now even Antarctica.

The Significance of Hair Types With regard to the situation with *h. sapiens* in the ancient Old World one can use a generalization from among present conditions to enable a simplified grasp to be made of such considerations. Taking head hair as a criterion, one can observe that there are three apparent zones that may be recognized now right across Eurasia. In the north this is straight, while in the south it is curly. In between there is a zone where both of these types may occur alongside a state that can be described as wavy. Of course these crude designations have their limitations; anomalies arise for reasons that lie in folk movement in the distant past as well as more recent displacements, which in some areas - together with recent hybridizations - have played havoc with the earlier distributions. Thankfully the existence of early photographs can sometimes ameliorate this problem.

In Africa one can consider that in the sub-Sahara zone tight curly hair is historically almost universal, although density of the locks varies considerably. Yet it is here that extremes of shortness of stature (Pygmies) and tallness (Nilotes) are found, all with curly hair. From this population there has been migration eastwards along the southern fringes of Asia, also incorporating ethnically distinct shorter and taller folk, but in present populations only with discrete elements consistently having curly hair. The question arises as to which type did the earlier emigrants belong, and were the curly people the first *h.* sapiens to leave Africa and gradually colonize the east? The problem then also arises as to when the straight haired people arrived in Asia: who were they and whence did they come?

It helps to postulate that the *homo erectus* types in Africa originally had dark, straight hair, and that the earlier ones who emigrated towards the Far East were all thus bedecked on their heads. This would then suggest that curliness originated well to the south in Africa and was one of the features that contributed to the emergence of *h. sapiens* there. So, how did it arise? It perhaps first appeared as a chance waviness among certain populations of *h. erectus* and was selected for since it was different from any other forms of primate encountered. Persistent differentiation led to curliness and it became widespread among later groups through the continuance of such preference and hybridization. The trend was so strong that straightness and even waviness were eliminated from Africa populations, except for the far north. Such selection led to curliness being characteristic of the type which constituted the various versions of *h. sapiens* that were emerging south of the Sahara and which would eventually become the only races of *homo* to survive there. The denizens of sub-Saharan Africa could hence have any kind hair they liked as long as it was black or dark brown: and curly.

The plausible solution to these difficult questions is that the straight haired people actually left first and continued as the basic constituents of emigrations well before curliness became so widespread and dominant in sub-Saharan Africa. It seems safe to say that the Heidelbergers who had earlier moved south-eastwards into the Horn of Africa from the Maghreb had straight hair that tended to be dark. The scenario that seems to work best is that their descendants in north-east Africa, the first fully evolved *h. sapiens*, still had straight hair until impinged upon by folk from further south who had evolved differently, including having acquired curly hair. The evidence is slight for the penetration of Asia by Heidelbergers from the African populations, but, even so, there are signs that their general type may have managed to reach as far as India (e.g. Narmada) and even China (e.g. Dali).

However, when *homo sapiens* emigrants ultimately from Africa started to encounter folk with some different characteristic they could find it sufficiently attractive to wish they had it. This could apply to hair colour. The northern distribution of the Neanderthalers had led to them having acquired pale colouration, including fair shades of hair. The dark haired *h. sapiens* who encountered them became dominant, but they were attracted to the lighter shades of hair as found among the Neanderthalers, especially the women and children, and it hence gradually became a widespread feature of their own populations as an aspect of sexual selection. In modern times the wish to have somebody else's hair type can be seen in the anomaly that - while European women until quite recently frequently used to curl their hair - it is common for women of African descent to have their hair straightened, while sometimes also being dyed blond. (A parallel anomaly is the wish of dark people for lighter skin, while light skinned folk, such as many Europeans - and despite the risk involved - like to lie in the Sun in order to obtain a darker skin hue.)

East Asia and the Denisovans As a distinct entity in Eurasia the Neanderthalers did not penetrate further east than the Altai Mountains. The equivalent type further east - or rather south-east - from there is represented by the Denisovans, whose presence was discovered much later by science and for whom direct evidence is slight. How did they reach the Far East, from out of groups that had originally left east Africa and presumably mainly by way of the land bridge at Suez?

For convenience of understanding, all the versions of *homo* from that earliest time - when exodus is recognizable from out of Africa - can be considered to be varieties of *homo erectus*. Thus *homo erectus* was the basic stock that entered Asia, while in the far west the one entering Europe from the Maghreb was the variant *homo ergaster*.

The exodus from Africa into Asia via the north-eastern 'gateway' at Suez had begun before the ancestors of the Neanderthalers first used the north-western 'gateway' from Africa into Europe. The finds at Dmanisi in Georgia date from 1.85 to 1.77 mya. They were from the *h. erectus* stock that has been noted in east Africa from 2.0 to 1.5 mya. Yet their stature was quite small, so they may have retained more *h. habilis* ancestry than did *h. ergasster*. The site is remarkably productive of both skeletal and cultural matter. Yet evidence that folk of this type and of this period ever penetrated further east into Asia is lacking, except perhaps for the dental finds at Yuanmou, Yunnan province in south-west China (1.7 mya), around which there is some dispute.

Less dubious are finds from later than 1.0 mya that have turned up near Beijing, China (Pekin man) and Indonesia (Java man), with date ranges of 750-300 kya for the former and 1.0-700 kya for the latter, these being evidence for *h. erectus* in these areas. Pekin man, as already mentioned, was vulnerable to advances of the ice and seems likely to have contributed to the hardy stock of the Denisovans, who seem to have been adapted to harsh conditions, including suitability for living at higher altitude.

Evidence for the stages that led from *h. erectus* to the Denisovans is much sparser than that leading to the Neanderthalers from *h. ergaster.* This may relate to the amount of ancient human presence in or entering into Asia being actually less, or being subject to less investigation: or perhaps a combination of the two. Was there indeed an equivalent 'Heidelberger' stage before the Denisovans? Why have more traces of these not been found?

A Denisovan jawbone found on the Tibetan plateau has been dated to 160 kya and is hence tens of thousands of years older than the bone and tooth at the Denisova Cave (48 kya at the earliest). From DNA it is suspected that the hair, eyes and skin of such folk were generally of darker colouration. The hair would be straight.

This contrasts with the Neanderthalers, who can generally be thought of as lighter coloured, as discussed above. Despite the Mediterranean type being considered to be darker on the whole than northern Europeans, there are surprising numbers of blond folk in Italy and Spain. One could attribute this to the eventual passing south through Iberia of Germanic folk such as the Vandals, while there was extensive Gothic settlement in both Italy and Spain. Yet this may mask the issue of earlier light coloured people having been there in the past. These were the Neanderthalers who eventually passed over into Africa (the Maghreb) and formed the major component of hybrids that led to the Mousterians, who eventually returned to western Europe, where they acquired this name, or else remained in Africa to become the Aterians. The modern folk native to the Maghreb are known to the outside world as the Berbers, and it is not beyond reason that they owe the lighter colouring found among them at least partly to Neanderthal/Mousterian/Aterian ancestry, rather than ascribing it to later incomers, such as the Vandals who dominated there in the 5th and 6th centuries AD.

The sequence of *h. sapiens* emigrants out of north-east Africa into Eurasia is difficult to determine. However, one can postulate that when conditions were easier during interglacials it was more likely that they would be derived from the Heidelbergers who had evolved since their arrival in the Horn of Africa zone, as indicated by the finds at Bodo, Ethiopia (600 kya). This evolution was being accelerated by hybridization with earlier residual populations in eastern and southern Africa, who would be *h. erectus* – and ultimately - *h. habilis/h. rudolfensis* derivatives.

CHAPTER 23

The Movements Leading to Modern Humans

▌ A Four-Armed "Hubbed Cross": the Initiating Migratory Pattern in the Evolution of Homo Sapiens

The Hub To define this concept and as a simplifying aid to understanding the evolutionary developments involved, the Horn of Africa and adjoining areas can be postulated as the 'hub' of a 'four-armed cross' and in the following each 'arm' is to be examined discretely.

The earliest presence of hominins at this 'hub' in east Africa occurs in deposits containing worked stones and marked bones. They have been found to date from 3.3 mya (Lomekwi, Kenya) and indicate Australopithecines. Until this discovery the earliest recognized limit of such had been 2.6 mya. This primitive type of stone tool is known as Oldowan.

As a matter of convenience the four sinuous arms leading from the 'hub' have been designated after the four cardinal points of the compass.

■ The Western Arm

Migration that is generally westwards out of east Africa is indicated by hominin finds in north-west Africa (the Maghreb), the earliest of these possibly being at Ain Boucherit in Algeria, with a given date range of 2.4-1.9 mya. A nearby site at Ain Hanech has produced both Oldowan and Acheulian artefacts and dated to c1.8 mya, while at another site not far away at Ternifine, discovered Acheulean tools accompanied by bones of *h. erectus* type have been dated to 700 kya. Finds of yet later dating made far to the east are Acheulean hand axes from 700-500 kya occurring at Abu Simbel and, also near the River Nile, at Thebes, Egypt, 255-90 kya.

The hand axe wielding *h. erectus* (here recognized as *h. ergaster*) crossed northwards into Iberia where he first evolved into *h. antecessor*, then *h. heidelbergensis,* and finally *h. neanderthalensis.* The latter pervaded Europe south of the Ice and both south and north of the Alps. It was the latter case that provided migration that reached as far east as the Altai Mountains of Siberia.

So, while it is proposed that *h. ergaster* evolved in east Africa as a descendent of more gracile types such as *h. habilis* and *h. rudolfensis,* it was the robuster form that was more able to roam and instigated emigration towards the north-west. This line of thought leads to it having been *h. ergaster* who provided the antecedents of those who crossed over into Europe at the western end of the Mediterranean. As referred to above, remains of their predecessors along this route have been found at Ain Boucherit and Ain Hanech in Algeria with the first being from as early as 2.4-2.2 mya. Found nearby at Ternifine has been found the aforementioned material that included bones identified as '*h. erectus*' and dating from 700 kya. This date is significant with regards to the circumstances - and especially the dating - of the return of *h. heidelbergensis* to east Africa. These reflux hominins had been preceded in western Europe by *h. antecessor,* which ancestral type was already present in Spain by at least 900 kya.

So this 'migration' route out of Africa into Europe, via the western end of the Mediterranean, can be seen in reverse with the eventual return of *h. heidelbergensis* from western Europe via the Maghreb to the eastern area of their origin, where they have emerged in the records in the form of the Heidelbergers that have been found in east Africa. There are no signs of their ancestral form *h. antecessor* in the Maghreb or eastwards from there in north Africa. This seems to indicate that it was environmental changes that eventually drove hominins at the *h. heidelbergensis* level southwards across the straits from Iberia and back into north Africa and the similar conditions kept them going until more suitable terrain was reached in east Africa between 700 and 600 kya. Such driven movements would leave little trace on the ground they traversed. Yet, as the finds at Ternifine, Abu Simbel and Thebes suggest, these Heidelbergers could possibly have encountered, at any position on the route back towards the 'hub', residual populations of *h. erectus*, in his *h. ergaster* form.

Over the millennia in Europe south of the ice, the descendants of *h. heidelbergensis* emerged as a population of hunters - i.e. the Neanderthalers - both north and south of the Alps. The northerners eventually reached at least as far eastwards as the Altai Mountains. It was probably out of this northerly spread that Palestine/Israel was reached to the south (probably by way of the Caucasus), and where encounter was made with *h. pre-sapiens* folk who were coming northwards out of Africa. These were constituents of the Northern Arm of the "hubbed cross", and they had originated discretely in the 'hub', which was located in east Africa (as considered below).

At the western end of the Mediterranean, European Neanderthalers also moved south into Africa, where contact was also made with the different and arguably more advanced culture that had originated more easterly on that continent. From the Maghreb these 'improved-by-contact' Neanderthalers possibly spread eastward along the north coast of Libya and Egypt to eventually contribute to the 'Neanderthal' populations of Palestine/Israel. (Yet none of these reached southwards towards the 'hub' at the Horn). At this point in their travels these north African Neanderthalers were forming hybrid cultures with folk of *h. pre-sapiens* type spreading northwards from the 'hub' and who were colonizing Africa's north coast from the east, while eventually reaching northwards from out of Egypt and into the east Mediterranean coastlands, the Levant. These circumstances created the proto-Mousterians

So these Neanderthalers leaving Spain for the Maghreb in Africa eventually initiated the Mousterian culture there due to contact with these populations of *h. pre-sapiens* who had already arrived in north-west Africa from the east. The earliest date for the improved newcomers (from Iberia to the north) is from Morocco (145 kya). The combined culture is now known as the Aterian, named after the type site Bir al-Atir, Tunisia, and has in the main been identified from discovered artefacts. This culture has left but few anatomical human remains. Some found in Morocco have given a radio carbon dating of 30 kya and it is claimed that the type had disappeared by 20 kya.

These Mousterians appear to have replaced or overwhelmed and absorbed the preceding *h. pre-sapiens* population, evidence for which is somewhat scarce, but the finds at Jebel Irhoud, near the Atlantic coast of Morocco (350-280 kya) can be indicated as early evidence for them. The Mousterians also appear to have replaced the *h. pre-sapiens* populations that had become established in the Levant. A site located between the Maghreb and the Levant is at Haua Fteah, a cave near the coast of eastern Libya (Cyrenaica). Finds from the earliest level and dating 80-65 kya comprised blade and flake artefacts which suggest the presence of *h. sapiens*, and these were overlain by deposits containing 'Levalloiso-Mousterian' flints with a date range of 73-40 kya. It is logical to conclude that the earlier 'modern humans' here certainly had come from the east, while the later Mousterians perhaps did the same (rather than from the Maghreb).

Finds at Sidi Abd el-Rahmane, Morocco have been identified as *h. erectus* and dated 200 kya. This surviving population - that was probably of *h. ergaster* type - appears to have held itself discrete from the *h. pre-sapiens* also present in the Maghreb (Jebel Irhoud, 350-280 kya). The latter apparently waited until the arrival of those Neanderthalers from Iberia before they mingled significantly.

■ The Southern Arm

From their origins in the western zone (that included Morocco), elements of the expanding Heidelberger strain eventually also made their way back to east Africa and it was the impact of this particular body of immigrants reaching the 'hub', and subsequently beyond to the south, that caused the crucial evolution to take place whereby human beings took the early steps towards becoming *h. sapiens*. While these included descendants of immigrant Heidelbergers from the lands surrounding the west end of the Mediterranean, they owed the particular form they eventually took to the degree of hybridization that took place and the variety of the *h. erectus* derivatives they encountered in eastern and southern Africa. These African Heidelbergers had reached Ethiopia before 600 kya (Bodo), where they found themselves in an environment that precluded them from becoming similar to the Neanderthalers. There was something about this area that led to a species that became smart due to a better brain, and with a lighter stature. One can well assume that playing a significant part in this process was the environment they had come to, in combination with - in this area - hybridization by these immigrants of north-westerly origin with residual populations that were already making evolutionary progress of their own, including having dark curly hair. Such hybridization could produce a range of combinations, but at this stage certain of them were getting ready to take over the world.

The Heidelbergers of east Africa provided elements that continued the southwards spread. From this start they sporadically made contact and hybridized with the developing *h. erectus* types surviving and evolving discretely down the eastern side all the way to the Cape of Good Hope. More gracile and culturally advanced hybrid types evolved from such unions, like the examples from Tanzania (Ndutu, 400 kya) and Zambia (Kabwe c300 kya). The process resulted in the predominance of hybrids of the *h. pre-sapiens* type in eastern and southern Africa, who were initiated before 600 kya, as indicated by the finds at Bodo, Ethiopia. It was in this area near the Horn that the most advanced types eventually emerged towards 200 kya, to culminate as *h. sapiens*, as found at Omo Kibish, Ethiopia (195 kya). These were due to migrate along the other three arms. However, before this happened, populations of *h. pre-sapiens* produced earlier in the east African 'hub' had already used these routes. One example has already been mentioned, namely involving movement along the Western Arm of the 'cross' to reach as far west as Morocco, i.e. the site at Jebel Irhoud (350-280 kya).

In the meantime all the hybridization improvements taking place along the southern arm (from the Horn to southern Africa) were eventually to result in the elimination of both *h. erectus* and *h. pre-sapiens* there, to be completely replaced by *h. sapiens*. However traces of the earlier populations were to remain as various races of *h. sapiens* due to variety in the constituents of the original hybrids and subsequent divergent evolution. Further subdivision also was an ongoing occurrence, with these smaller societies usually being distinguishable not necessarily by their physical nature, but rather by culture divisions of variable quality. They persist into the present and are known as tribes. However, whatever their present nature, they had all only emerged and survived by evolving collectively in the same direction, i.e. the one leading towards *h. sapiens*. This all occurred because of the fusing of two diverse derivatives of *h. erectus*, one of which had always remained in Africa, while the other had returned from distant areas in the north-west, where it had evolved over many years and in changing environments to become *h. heidelbergensis*.

■ The Northern Arm

The study of the 'Northern Arm' is made difficult by the amount of relevant ancient material that has been found in Israel and the fact that other areas in this direction are rather barren. It looks as though Israel has been a 'promised land' since long before the Biblical exodus of the Israelites from Egypt. One can address this problem by postulating that the apparently empty areas either represent a more fluid or nomadic phase of activity or that these areas lacked caves (or were generally unsuitable for human occupation). In any of such circumstances the chances of evidence of minimal cultural activity or habitation being preserved are reduced or have even been completely eliminated by later environmental conditions. This could account for the apparent paucity of Aterian material from Libya and Egypt, while evidence similar to this culture has turned up in Israel.

Yet it has been found: for example, a suitable cave at Haua Fteah in Cyrenaica produced relevant finds. Such Mousterian material can occur anywhere along the southern and eastern Mediterranean coasts where folk of relevant origin have found a place with a cave where they could dally indefinitely before continuing their wandering. However, by 80-70 kya, the early occupation dating for Haua Fteah, Lybia, may mean that access between Cyrenaica and the Maghreb became rather difficult, this barrier essentially involving the coastlands of the present Tripolitania. It would prevent contact with those earlier westernmost zones of activity of *h. pre-sapiens* and any 'Mousterians' there in the west would be separated and continue to evolve discretely as the Aterians of the Maghreb.

The Earlier Coming of Homo to Asia The earliest tool-users of the species, *h. habilis,* were not alone as early men in Africa. Some finds have been sufficiently different to be dubbed *h. rudolfensis*, being named after a lake in Kenya (now Lake Turkana). This may have been a subspecies of *h. habilis*, but one that was more advanced and of bigger build. So, at the very least, two versions of these tool using creatures appear to have existed side by side in Africa in the early Pleistocene; but which one provided the basis for later human development?

Finds of *homo* later than *h. habilis* all seem to have been more robust, with the genus continuing in Africa as *h. ergaster*, whose type physically emerged from an earlier presence that embraced *h. rudolfensis* and *h. habilis. H. ergaster* would seem to be a form taken in Africa by *h. erectus*. The latter was the form that migrated into Asia. Java man (in the East Indies) and Pekin man (in north-east China) resemble this designation, and existed from a period commencing c1.5 mya, with some surviving to 300,000 years ago (300 kya).

As with the rest of the world, the population of Asia now consists of one mutually fertile species, *h. sapiens*, although divided into discernable races, which state has normally come about due to earlier geographical separation interspersed with intermittent opportunities for hybridization. In view of the immense lengths of time involved, it is not surprising that there are no examples of earlier forms such as *h. erectus* surviving into recent times. So did all other species simply die out without a trace being left? Were they simply all overwhelmed and eliminated by new men coming out of Africa? Surely they were not. Genetic compatibility always remained a possibility within the genus *homo* along with probable intermittent yet persistent sexual encounters. Intercourse was a continuing, if sporadic, process. Selection of the best characters would determine the form such hybrids would take and which would survive. Such preferred characters would be contained in the genome of the most advanced type, and eventually this would lead to the ever improving forms of *h. pre-sapiens* that were being created in Africa and from there eventually spreading widely as ever improving versions and eventually - as *h. sapiens* proper – into the southern parts of the land mass to the north, i.e. Eurasia. One can understand that *h. sapiens* is just a label to suit any latest model, a flexible designation that would actually be fit to designate any stage of *h. pre-sapiens* when considering the form taken in its own particular 'present time'.

While considering the 'hubbed cross' concept with regard to the migrations of modern man out of Africa, it is evident that there were basically two distinct kinds of *h. pre-sapiens* being produced by the hub and moving out from there during discrete periods. They can provisionally be identified as being the 'mainly straight haired' who usually headed north (but less frequently east and west) and the 'mainly curly haired' who headed east, apparently exclusively, the latter being the major constituent of the 'eastern arm' (as treated below). The question arises as to which set of migrants left the 'hub' first? It is provisionally claimed that it was those who were moving north that left first, and this is the reason for considering the 'northern arm' before the 'eastern arm'. The earliest folk taking part in this movement towards the north were actually of a *h. pre-sapiens* type that embodied a greater degree of *h. heidelbergensis* constituency from the past and who were mainly making their presence felt in the Near East.

The Levant Perhaps the most productive site ever found with regards to the origins of *h. sapiens* in Eurasia is Tabun Cave, near the coast of Israel at Mount Carmel. This has produced evidence for occupation from 500-40 kya. In the lowest – and hence earliest – levels, the Acheulian culture was found in the form of hand axes. Later than this scrapers were added to the tool kit, after which hand axes were phased out. The scraper phase is known as the Yabrudian, which was eventually followed by the Amudian, a culture whose tools were based on long flakes known as blades and this culture elsewhere preceded the Aurignacian one that was firmly in the age of *h. sapiens*.

At Qesem Cave, also in the Mount Carmel area, the time of occupation was more restricted, all the finds being from the earlier period, known as the Acheulo-Yabrudian, followed by Amudian tools, and with a date range of 350-200 kya.

The number of Neanderthal, Mousterian and more modern human sites in Israel suggests that the Levant can be regarded as a sort of 'sub-hub' of human occupation, development and distribution during the Pleistocene. One relevant feature of this area at this time is the number of available caves. Cave dwelling is not a feature of the main hub at the Horn of Africa. This indicates that any *h. pre-sapiens* or *h. sapiens* migrants who moved north into the Levant were not drawn there by the presence of caves. However, it would seem that some of them were destined to become cave dwellers after arrival in these different circumstances. There has to be a reason for this. Well, they did come to share the zone with folk who were 'cavemen'.

Neanderthalers were cave dwellers and there is evidence for their presence in some of the Levant caves, such as those involved in the Acheulean hand axe culture as found at Tabun and Qesem. These had arrived from the north, probably by way of the Caucasus to the east of the Black Sea, as already mentioned. They were presumably driven south by the onset of a colder climate. They found what they wanted in the Levant, especially in the more northern parts of the present Israel, where there was hilly terrain with plenty of caves. Yet they failed to penetrate much further south than the latitude of Jerusalem. Apart from the terrain and the climate always being unsuitable for them – since they were adapted to survive on Eurasian flora and fauna – the fewer caves to the south were not within reasonable reach. A Neanderthal woman's skull from Tabun Cave has been dated to c120 kya.

The Neanderthalers would move south within the bulk of Eurasia when the ice tightened its grip. At the same time Africa would become more arid and this tended to cause *h. pre-sapiens* types there to drive northwards – with the River Nile being available as a corridor - and eventually find their way to the Levant (while presumably leaving the Horn area temporarily depopulated). The result would be that both populations would sometimes happen to be forced together on an east-west strip of territory at the latitude of the Levant. The enforced social conditions and associated intercourse would give rise to hybridization between *h. neanderthalensis* and *h. pre-sapiens.*

Away from Mount Carmel, to the east, a skull from 300-200 kya was found eastwards at Mugharet el-Zuttiyeh, Upper Galilee, Israel. Its heavy brow ridges suggest a Neanderthaler, but the lack of prognathism indicate some progress towards hybridization between *h. pre-sapiens* and *h. neanderthalensis* and this - together with the dating - membership of the Acheulo-Yabrudian culture.

Hominin remains of later date have been found in the Skhul Cave on Mount Carmel and the Qafzeh Cave inland to the east near Nazareth with a mixture of anatomically archaic and modern humans and with date range of c130-80 kya. Skulls with Neanderthal faces but modern braincases, - as present there - suggest hybrids of *h. pre-sapiens* type. The archaeologists found thousands of Mousterian stone tools in the sediments, and in the lowest layer they discovered ten burials of modern humans. An anomaly is evident in that the modern humans there (dated c130-80 kya) predate the Mousterians at Skhul, It would seem to be indicated here that earlier *h. sapiens* immigrants to the Levant moved on in some way to be replaced by beings who were of more primitive culture: such hybrids were Mousterians.

Mousterians such as these may have found their way northwards to the Kurdish area of north Iraq. Remains of 8 adults and 2 infants were found in the Shanidar Cave there, along with points, side scrapers and flakes and dating from 60-45 kya.

Returning to Israel and on Mount Carmel (i.e. not far from Skhul), Kebara Cave was occupied by Neanderthalers c60 kya and they left a skeleton from this time. These Mousterians used the cave until c47 kya and Levallois artefacts from the later part of this period established their Mousterian credentials. In the 2000 or so years after this the occupation then passed through Ahmarian and Aurignacian phases, the last indicating the presence of early *h. sapiens*.

The above discourse suggests that the Mousterian culture originated in two places, the Maghreb region of north-west Africa and the Levant region of the Near East. The common factor is provided by the *h. pre-sapiens* elements who moved northwards out of the Horn of Africa and occupied southern and eastern coastal areas of the Mediterranean Sea. In both zones they were eventually impinged upon by displaced Neanderthalers from the north, with whom they merged to form physical and cultural hybrids. A difference occurs because the migration from the south into the Levant was more persistent. Here the *h. pre-sapiens* elements became much more dominant and eventually provided the major flow of culturally advanced populations into Eurasia, which finally included examples of the latest model, this being *h. sapiens* proper.

Thus can one roughly sum up the sequences of immigration and occupation in the Levant as follows. During hard times, displaced *h. pre-sapiens* from the Hub at the Horn made their way northwards to reach the Eastern Mediterranean, but split near Suez. Some turned westwards towards the Maghreb where they were eventually overrun by Neanderthalers from Iberia to provide the Mousterian culture that eventually produced the Aterians. They were not the main progenitors of *h. sapiens*. The other line of *h. pre-sapiens* from the Hub continued northwards into the Levant. Here they were eventually joined by Neanderthalers from the north. This too provided a Mousterian culture that resulted from hybridization. In this case, however, it was the *h. pre-sapiens* immigrants who eventually became dominant and their strain overwhelmed the Neanderthal one in the course of time, so that the end result was a population of virtual *h. sapiens* with only vestiges of *h. neanderthalensis* in their genome. According to this reasoning the Levant was a major source of *h. sapiens* as he exists today and indicates one cause for the fact that, apart from sub-Saharan Africa, all other humans still have DNA with Neanderthaler traces.

The Neanderthal culture eventually became Mousterian: the hand axes were phased out and replaced by a Levallois style flake and blade culture. The 'sub-hub' in the Levant became a centre from which waves of improving examples of *h. pre-sapiens* emanated to penetrate areas of Eurasia to the north as and when they became accessible. This 'northern arm' activity resulted in a spread of populations across more northerly Eurasia whose hair was generally straight, but sometimes wavy and even occasionally curly, but rarely tightly curled as is historically the ubiquitous case in Africa south of the Sahara.

What did the folk of *h. pre-sapiens* type who moved north from the Horn look like before any hybridization took place with *h. neanderthalensis*? While the latter were lighter skinned with straight hair that tended to be fair or red, the former would be of more gracile build and darker in colour of both skin and hair, with the latter feature sometimes being wavy or curly. To various degrees it was these characteristics that featured more and more in the hybrids. It was the lighter hair colouring of the Neanderthalers that resulted in the fair and red-haired populations found later in northern and western Europe, while those of Western Asia and the Mediterranean were historically much darker.

Folk of *h. sapiens* type were eventually making their way along three of the arms of the hubbed cross, even if this occurred during discrete periods. Those of the Northern Arm are typified by Cro-Magnon Man, who we can understand as having *h. sapiens* credentials through him having direct roots in the Horn of Africa. These origins suggest that he would generally have sallow skin, dark hair and brown eyes. This was because he had retained some ancestral form that left the Horn due to limited contact both with *h. pre-sapiens* and Neanderthal populations strewn along the route that his ancestors had taken westwards into Europe. They formed one extreme of the Caucasian race of *h. sapiens*, while other populations also falling under this label had absorbed more primitive elements from previous inhabitants of Europe, with Neanderthal traits being the most extreme of all.

Others - who drifted eastwards across the wastes of central Asia - also at first had some input to their genomes from contact with Neanderthal derivatives, but eventually, at the far eastern extent of their range, their make-up was influenced by the descendants of that other ancient stock the Denisovans. These circumstances ended with populations who tended to have straight black hair and sallow skin, in other words the "Mongoloid" type. Yet like the "Caucasoids" of the west, they owed their *h. sapiens* credentials wholly to their distant ancestors who had moved north out of the Horn and passed through the sub-hub of human distribution in the Levant.

Europe and West Asia Early arrivals of *h. sapiens* in western Europe are usually identified as Cro-Magnon Man and the culture they brought with them was the Aurignacian. However, Aurignacian remains have been found near the Mediterranean coast of the Levant and they predate anything reaching this stage in Europe. It would seem that the widespread European Aurignacian culture has reached the far west subsequent to having originally been developed in the Levant. There is however a hiatus between the Levant and Europe in that the Aurignacian appears to be absent from Asia Minor and the Caucasus-Ukraine zone. It would appear that one or both of these routes could have been used to reach Europe, but, whatever the case, a restless period is suggested, with the migrants not settling down in any area until they reached territory north-west of the Black Sea.

Whatever else happened in these areas they were liable to encounter primitive Neanderthalers, while further penetration westward would involve contact with more advanced versions of these, namely Mousterians whose culture had its origins in the Maghreb of north Africa. The newcomers themselves probably had a little Levantine Neanderthal DNA in their genome and their Aurignacian culture had surely developed with some input from the Levantine Mousterian one. Culturally they were on the whole more advanced than any form of Neanderthaler they met as they spread westward across Europe and they surely considered themselves to be superior beings. Yet interbreeding leading to hybridization will have occurred when there were encounters, even if on a limited scale.

Cro-Magnon man appears to represent the eventual *h. sapiens* type from the Near East who penetrated Western Europe without signs of significant contact with Neanderthalers, or derivatives from them, at any stage since his kind left Africa. With others it was different: at stages during their progress westward they would have contact with the existing populations and contacts would be made that proved to be fruitful in that functioning hybrids could result.

An apparent anomaly between the historical inhabitants of northern Europe and northern Asia is that the former contains a considerable proportion of light haired people, whereas the latter consist almost entirely of dark haired folk, often absolutely black. One should surely expect that all populations in this similar latitude across Eurasia would have the same colouration. The reason has to be that the populations in the west are of different historical stock and were in place at an earlier date. This represents the conclusion that the Neanderthalers as such did not indeed physically penetrate further east than the Altai Mountains, even if their genetic presence became more widespread and is detectable in DNA samples from other areas lying further to the east and south.

Whenever the ice relented it provided opportunities for populations in the Levant to move northwards into latitudes beyond the Caucasus Mountains and the northern extremities of the Caspian and Black Seas. The land was already inhabited by a sparse population of Neanderthalers, as discussed earlier. These were to contribute to the genome of the incomers of *h. pre-sapiens* type, especially in the west, where the prospects for human evolution, both physical and cultural seem to have been greater at that time.

A stage in the drive westwards is apparent in the Bohunician culture that existed to the west and north-west of the Black Sea between Bulgaria and south Poland. This culture was at a stage intermediate between the Mousterian and Aurignacian. Relationship with the Levant is clear in that there are resemblances to the Ahmarian culture found there, this having evolved from the earlier one dubbed the Emiran. These were producing artefacts in the Levallois tradition, which later were being replaced by the Aurignacian. As already mentioned, this is well illustrated by the Kebara Cave on Mount Carmel, where Mousterian occupation (c60-48 kya) using Levallois technology eventually started producing artefacts resembling but predating the European Aurignacian. The Bohunician culture has been given a date range of c48-40 kya.

There were also those who turned east after crossing northwards through the Caucasus: they had dark hair that tended mainly to be straight. They are best considered alongside others, who comprised the Eastern Arm proper.

▓ The Eastern Arm

Mitochrondial DNA analysis has suggested that the earliest date for *h. sapiens* to leave Africa is 85 kya. While the western arm of the 'cross' was apparently not involved this early in such movements, the eastern arm certainly was. The departure could have been from the Horn to reach nearby Arabia by crossing the Bab el Mandeb Strait at the south end of the Red Sea. It is clear that these movements involved folk of darker hue, and one might perhaps initially surmise that they had curly hair, while also suspecting that they included the ancestors of the Negritos, those diminutive folk who still survive in some remote areas of south-east Asia. However, their moves along the Eastern Arm were surely later than those that were diversions from the Northern Arm and heading towards the orient along routes that lay further north, as implied at the end of the above Northern Arm presentation.

As already mentioned, *h. sapiens* had appeared in Ethiopia, Africa, by c200 kya. It was from about this time that the littoral started to be systematically exploited as a food source. Yet beachcombing would have been used as a way to find sustenance by hominins from a much earlier date. It is indeed currently observable as a way to find food by other mammals, such as some monkeys and bears; but such activity is on an "eat-as-you-find it" basis and leaves no lasting evidence that it has occurred. The change with humans was a cultural development whereby it came to form part of hunter-gatherer routines, with evidence being left that the consumption of such food could be delayed after being collected, transported, processed and finally eaten as a meal. The main evidence for this sort of activity eventually took the form of middens of discarded shells. This sort of organization originated with *h. sapiens*. An early manifestation of such (100 kya) has occurred in South Africa and apparently was originally an important component in the lifestyle of the African ancestors of the south-east Asian Negritos.

In the meantime hominins at the *h. heidelbergensis* stage using hand axes had been leaving Africa for Asia, as evidenced by finds from India, namely from Jwalapuram (300-200 kya), as well as a skull from Namada. But cultures using hand axes died out, having disappeared from Europe by 250 kya and never spreading as far as east Asia. This indicates that the *h. erectus* inhabitants of that region were never disturbed by hominins at the next stage of evolution and as represented elsewhere by the Heidelbergers. Thus in Indonesia did Java man survive between 1,000 and 700 kya and Pekin man in China between 750 and 300 kya. It would thus seem that folk of Heidelberger stock did not reach vast areas in east Asia at a time when they were spreading elsewhere across Eurasia, and this could explain the later dichotomy of the Neanderthalers being dominant in the west, while being substituted in the east by the Denisovans. The development of the former type has already been covered, but the latter remains to be considered.

The Denisovans These early inhabitants of east Asia, like all versions of *homo*, did not just magically appear there. Inevitably they had their roots somewhere in the west and their deepest ones in Africa. They are usually thought of as being at a similar stage of development as the Neanderthalers and it seems reasonable to assume that they were preceded by a stage similar to that of the Heidelbergers. Well perhaps they were, but the circumstances have somehow become obscure.

Evidence for the existence of the discrete Denisovan race first came to light with the scrap of bone and a tooth found in the Denisova Cave, in the Altai Mountains of Siberia. Initially this tends to generate a misleading impression that this northern area is where they were on home territory and living out their evolutionary story. But a Denisovan jawbone eventually found on the Tibetan plateau has been dated to 160 kya and is hence tens of thousands of years older than the bone and tooth at the Denisova Cave (considered 48 kya at the earliest). From DNA it is suspected that the hair, eyes and skin of such folk were generally of darker colouration. The hair would be straight.

Theoretically one would perhaps expect to find predecessors of the Denisovans in intermediate forms, as is the case with *h. accessor* and *h. heidelbergensis* preceding the Neanderthalers in the west. None has been found, or at least not recognized as such. Like the Neanderthalers, the Denisovans were designed to survive in a harsh climate, but perhaps even more so. Apart from being able to resist the cold, it is also thought that they could cope better with altitude. But in this context one needs always to bear in mind that actual evidence for the Denisovans, both anatomical and cultural, is minimal when compared to Neanderthal finds.

The Neanderthalers obtained light colouring as a result of spending much of their earlier formative time as dwellers in dense and shady temperate forests where they hunted red deer, while adapting to a damp climate with reduced sunlight. Later they spread into areas further north where they found themselves adaptive enough, both physically and culturally, for hunting the plains animals and eventual subarctic ones found nearer the ice. The necessity to wear heavy clothing against the cold when roaming about meant that the light coloured skin was covered up and incurred no serious disadvantage. The associated light hair colour was maintained as a preference. They liked having children with light coloured hair, and this would work as a form of sexual selection. The necessity to wear a complete covering against the cold while outside meant that northern Europeans preserved the earlier acquired light skin colour. Folk types originating east of the Caspian Sea and the Ural Mountains (basically Asia north of the main mountainous divide) also acquired lighter skins, but to a lesser degree. The ancient inhabitants of the Far East (Denisovans) had not had such a prolonged time inhabiting temperate forest so as to become as "white" as the Neanderthalers, while the latter were sparsely enough distributed so as to have reduced effect on the modern human types from the Levant and who were crossing Asia from the west. The resultant skin colour can be described as "sallow" and this is the one that was preserved as a result of being forced to cover up due to the cold.

During separate periods the Denisova Cave was occupied by both Neanderthalers and Denisovans. Early examples of the former were there 100kya, while the latter were present 48 kya. By 40 kya Neanderhalers were again in occupation. However the two types had earlier made contact in the area. Remains have been found of a hybrid whose mother was Neanderthal and father Denisovan (90 kya), but this is still not as early as that Denisovan jawbone discovered in Tibet (160 kya). This all suggests that the Denisovans originated somewhere south of the Denisova Cave. Over the millennia they drifted northwards to a zone where they might make contact with Neanderthalers whose origins lay westward across Eurasia.

While direct evidence is lacking, the inference one can make is that the Denisovans are the direct descendants of the *h. erectus* populations as found in east Asia (e.g. Pekin Man 750-300 kya) and south Asia (e.g. Java Man 1000-700 kya). Over the millennia they proved themselves capable of penetrating the mountainous interior of east Asia.

Homo Sapiens In south-east Asia the Denisovans have not been found as such, but like the Neanderthalers, they are present in the DNA of the folk of Australia and the many islands. The highest count of Denisovan DNA occurs among the Papuans, the native Melanesian inhabitants of New Guinea. But like all known modern humans all these folk are regarded as *h. sapiens*: DNA representing such a final form provides the major component in all these populations. Where have they come from? Well, since the possibility of *h. sapiens* having arisen independently in the east is virtually so unlikely as not to be worth considering, they must have originated in the west.

As already discussed, representatives of *h. sapiens* were leaving the Northern Arm sub-hub in the Levant and once north of the Caucasus some of them turned eastwards and traversed the territory already pioneered by Neanderthalers. Thus did early arrivals reach the Altai Mountains (c40 kya), i.e. near the Denisova Cave. Others penetrated even further to the east to leave traces near Lake Baikal. From such frontier posts interglacial conditions would subsequently allow them to investigate environments in the vast spaces to the north as far as the ice would allow them. But as the ice started to encroach southwards towards its maximum c19 kya, so would such populations be forced south in order to survive. In doing so they would tend to dislodge folk already there, who would then be likewise encouraged to seek refuge elsewhere, but generally in a southwards direction. A feature of the folk in the north of Asia is that they evolved so that they could better survive the cold conditions created by the presence of the ice and in doing so eventually developed the racial features that in the past have been referred to as Mongoloid.

The men of the far north tended to cope by developing paedomorphic features such as, less tendency to go bald, which could be accompanied by reduced facial and bodily hair and absence of overhanging brow ridges on the forehead. The cold also encouraged reduction in size of facial features such as nose and mouth. These became extreme features in the north, yet there were populations further south who never developed them fully, but only in more moderate versions. Advances in the ice would cause impingement on more southerly populations and explain how such extreme features could occur among them, but to a reduced extent. These forms of adaptation to survive the cold became the basis for the appearance of the so-called Mongoloid race. The more comfortably placed folk to the south were also liable to expand towards the north when interglacials set in and in doing so disrupt and displace the populations who were exhibiting the more extreme Mongoloid features.

Negritos and Other Eastern Curly Heads The Negritos are races of diminutive folk who are now found on Asiatic islands, namely the Andaman Islands in the Indian Ocean (off the coast of Burma) and the Philippines in the South China Sea. Such communities also survive in mainland locations in between, namely the border areas of Thailand and Malaya. As physical specimens they resemble closely the Pygmies of Africa. Like these, the question arises as to why they have survived in specific locations but not elsewhere. In the case of the Pygmies of the Congo forest, their absence from other apparently suitable forested areas in Africa is odd. They are clearly a relict population: they must once have been much more widespread and the Congo is their last stand, with survival being due to it being the most difficult of access for other kinds of folk.

The Negritos are not directly descended from Pygmies where found now, but one can postulate that they do represent former similar small folk from Africa, who originally depended on its east shores - rather than its forests - to sustain them. Like the Pygmies they now represent a relict population who only survive where they became isolated and thus protected from subsequent inhabitants who were different in appearance and stature; and with some confidence one might risk saying "culturally more advanced". One can indeed assert that the Negritos once also either inhabited - or at least traversed – all the Asiatic coastlands from the Horn of Africa to the Philippines. This would include intermediate coastal stretches of regions such as Arabia, Iran, India, Burma (Myanmar) and Borneo. They have only survived on islands that were remote (i.e. the Andamans), or in deep jungle (i.e. Malaya and Thailand), or both (i.e. the Philippines). Such terrain is now absent from most of the western end of their former range and they have been completely eradicated as separate social entities. Yet a partial survival of their type seems to have occurred on the island of Ceylon (Sri Lanka). A population of diminutive hunter-gatherers exists in the central forests and are known as the Vedda. Everything about them is redolent of the Negritos, except that their heads have lost their African traits. Their hair is usually straight and their features resemble those of other folk on Sri Lanka and in nearby India. It would seem that hybridization may have taken place here on a scale avoided by the Negritos.

The movements of Negritos eastwards out of east Africa would also have included others who were similar in appearance, but of greater stature and perhaps more advanced culture. But these too have all been eliminated as recognizable separate entities through extinction or being overwhelmed by amalgamation and hybridization.

It might be of some significance that both the Vedda and the Negritos all survive on a band of latitude shared with the Horn of Africa, which can be thought of as the "Negrito Band". Yet it seems odd that Negritos do not occur in the East Indies further to the south, on islands such as Sumatra, Java and Borneo where the habitat appears to be so suitable for them. The "Band" seems to have acted like a filter: those who passed through it changed. This effect has determined the populations of Australasia and Oceania, who have of necessity arrived from the land masses of Asia to the north. There has generally been a south-easterly movement out of Asia and originally consisting largely of Mongoloid derivation, but with some Caucasoid. These Asiatics have impinged upon and generally overwhelmed those populations of African derivation from the west who have come to occupy the Band. These Mongoloid traces of northern origin are still in evidence in most of the majority of inhabitants at the present time of the Negrito Band, as found in Indo-China and the Philippines, but are apparently completely absent from areas further to the south-east and now inhabited by such as the Melanesians and the Australians.

Southwards Movements in the Far East Far to the south, in Australia, one can note that curly hair is dominant, but not completely so. Among Australian Aboriginals waviness is also a common hair form, but with straight hair being rare. Far to the north-west, in Indo-China and including the Negrito Band, the curly heads did not remain equally dominant. With the lower sea level this region once extended further south to form a single entity - the so-called Sundaland - by incorporating some islands of the East Indies into its peninsular landmass towards the south such as the present Sumatra, Java and Borneo. To the east the clear water separating this from the Philippines was greatly reduced while New Guinea was joined to Australia. The island of Celebes (Sulawesi) lies between New Guinea and Borneo: it has been completely isolated from Borneo since long before *h. sapiens* activity reached the area.

Yet at the east end of the major land masses of the East Indies the presence of curly hair is so evident among the Melanesians. Indeed, curly immigrants must have been the prime source of the hair types found among the natives of Australia, who c45-40 kya made the crossing to this continent. There has been a straight haired addition to these, which was at the front edge of later arrivals from mainland Asia to the north-west (i.e. Sundaland) and which, because of mingling, modified the population, including the creation of a widespread wavy haired component.

To the north of Australia there now lies the large island of New Guinea (Papua), the inhabitants of which, together with those of the smaller archipelagos further eastwards as far as Fiji, are grouped together as the Melanesians. They resemble the Australians, but the incidence of curly hair is much greater, being virtually ubiquitous. An apparently anomalous feature is the presence of blond hair, which also occurs with less frequency among the Australians. The incidence of this appears to be greatest on the Solomon Islands, which lie immediately to the east of Papua.

The later straight haired folk apparently entered north-west Australia using the same route as the earlier curly haired ones, namely by way of the East Indian island-chain that now stretches from Sumatra through Java and Bali to Timor, but in the Pleistocene formed the southern edge of Sundaland. From there they pervaded the present extent of Australia to some degree, which, as already mentioned, led to the common wavy haired population elements, this being due to admixture with the curly haired folk already there. But these straight haired ones from the north-west did not reach the lands to the north east of Australia – at least in significant numbers - including the adjacent large island of Papua (aka New Guinea, which for some time in this period formed a single unit of land with Australia, an erstwhile entity that has been dubbed 'Sahul'). The curly haired populations there remained the main component, thus to become the Melanesians, together with the islands beyond to the east that stretched through the Solomons as far as Fiji.

Thus might it be asserted that curly-haired folk such as the first Australians and Melanesians were the earliest modern humans to reach the Far East, at least at its southern extremity, but that they have only survived there as distinct populations when not since overwhelmed by intruders from elsewhere. (Further north, in the Negrito Band, similar survival had also taken place of other curly heads.) This suggests that by the time that the straight-haired migrants arrived in north-west Australia from the north the straits between this continent and Papua had either grown too wide to cross, or they were disinclined to head northwards again, or both. Since the Torres islands still make convenient stepping stones from the northern tip of Australia (Cape York), it would seem that the second proposition is most likely.

So, in consideration of the ultimate origins of the curly heads, it might be observed that they are not much in evidence immediately eastwards from their ancestral African homelands, namely on the intervening coastlands of south-west Asia (e.g. Arabia), presumably because the evidence has been obscured; or it may be that they are difficult to recognize because of seldom occurrence now among other more numerous racial types and having thoroughly mingled with them as a minority presence. They have survived best as a type south of the Negrito Band. The Australians – when they have wavy hair, beetling brows and bulbous noses - can often be found to resemble some of the population of India: they are especially redolent of the image of a typical Hindu holy man, while there is also corresponding resemblance in physique.

It can be seen that the curly haired component found among the Melanesians has arrived from the west. These *homo* sapiens had come by way of India - where they do seem still to survive as a noticeable minority among some populations there – and Sundaland (now the western part of the East Indies archipelago). They originally formed a stage involving coast hugging migrants who were due to explore the very extremes of available land, as long as it did not involve crossing oceanic stretches of water. Hence New Zealand and the various archipelagos of small islands now known as Micronesia and Polynesia were beyond their reach. These particular *h. sapiens* folk had started to migrate out of east Africa and eastwards along the coastlands of Asia c70 kya. At first glance they appear to have been somewhat reluctant to penetrate the great hinterland to the north, as well as those coastlines of mainland China lying to the east and curving northwards.

Yet it will emerge that the eastern coastlands of Asia were reached by a distinct strain and remnants of this are recognizable in the Ainu at the north end of the Japanese archipelago and certain small areas adjacent to this. This is a problem to be addressed later.

It can be noticed that Mongoloid traits diminish in eastern Asia as one's attention is moved southwards. Yet not all features characteristic of Mongoloids should be associated with surviving in a very cold climate. The original inhabitants of China and neighbouring lands would not exhibit the more extreme ones, such as the protective skin formations around the eyes. The more generalized Mongoloid is descended from Denisovan ancestors who had had some contact with immigrants from the west, first Neanderthalers and eventually with *h. sapiens*, with the latter finally becoming the dominant strain.

The majority of the later established Mongoloid type has straight black hair and a pale sallow skin, while the characteristics induced by exposure to icy conditions increase when viewed as one travels north and away from the Pacific coastlands. The characters that do not necessarily exist because of this climatic cause are much more widespread and are generally aspects of paedomorphism in males, such as lack of brow ridges, retention of head hair and reduced facial and body hair. It can be noticed that the Chinese in general do not have beetling brows and that this trait continues down the eastern side of Indo–China, e.g. among the Vietnamese. On the west side the Burmese are however much more likely to have brows that overhang their eyes. This is something they share to some extent with inhabitants of the neighbouring Indian subcontinent to the west. When one considers the brows of the Australians and Melanesians it is apparent that the notion is correct that they are derived from earlier folk inhabiting regions that now include India and Burma. On the other hand it can be noticed that there is little sign of these Australasian populations ever being in contact with Mongoloids from the north. However there are others to be considered, those odd survivors, the Ainu.

The Ainu are a wavy haired folk who are mainly found on Hokkaido, the northernmost of the main islands of Japan, but with also a few on Russian territories to the north, i.e. the island of Sakhalin and the tip of the peninsula of Kamchatka. They can be seen as the remains of widespread populations whose type has been overwhelmed by the subsequent growth of the Mongoloid race and now only survive on extremities where the newcomers did not manage to reach or care to dwell. With their beetling brows the Ainu physically resemble the Australians and Melanesians, although something about their features betrays a modicum of interbreeding with Mongoloids, traits that are absent from those southern curly haired folk. Conversely, sexual contact with the ancestral Ainu – or related types - by the Japanese can be traced in that wavy hair is much more likely to be found among them than folk on the adjacent continent, such as the Chinese.

At one time the forebears of the Ainu and the Australoids must have formed a continuous population of curly haired folk. They have since been eliminated from intervening territory by southward migrating Mongoloids, a movement that ground to a halt at roughly the same latitude as the Negrito Band.

Oceania To the north and east of Australia are two extensive areas of the Pacific Ocean that are dotted with smaller island, some of them tiny. They are known as Micronesia and Polynesia respectively and are inhabited by folk who are generally of heavy build, have lighter skin than the Melanesians and their hair on the whole tends to be black and slightly wavy, rather than straight. The Polynesians have proved to be excellent seafarers and they eventually reached the waters far to the south-east of Australia, where they – now known as Maoris - colonized New Zealand in the 14th century AD. The question arises: where did they originally come from?

The facts of their appearance suggest an answer. Micronesia lies between Melanesia and the Philippines. The latter islands lie south of and not too distant from Taiwan, a fairly large island off the eastern coast of China. The sallow skinned, wavy haired Polynesians hence originated in China, but without showing much sign of Mongoloid features.

Any traces of the forebears of the Polynesians have been completely obliterated in China by emigration or absorption by overwhelming Mongoloid immigrants, who would themselves have been forced to move southwards because of climate change.

The former presence of the forebears of the Polynesians is still evident on Taiwan, where a distinct Aboriginal population survives. Once this notion of folk making this crossing from the Chinese coast to Taiwan is established, one can perceive the route taken that resulted in the far flung islands of Polynesia to be populated. From Taiwan they had moved south to the Philippines and continued from there to reach the north edge of Melanesia. They continued eastwards along this edge until Tonga was reached. This was just beyond Fiji, the easternmost extent of Melanesia. From this point, and with the seafaring skills they had developed on their travels, the island dotted expanses of the Pacific were open to exploration.

These movements explain the gap between analogous populations such as the Ainu in the north and the Melanesians in the south. The corresponding inhabitants of the Chinese and Vietnamese coastal areas have been eradicated by newcomers. Some of the earlier inhabitants escaped by crossing the strait to Taiwan. Those that remained on the mainland were eliminated as a recognizable entity, but through hybridization contributed to the genetic structure of folk in that area. The initial arrivals in Taiwan were not great seafarers. It was the more Mongoloid hybrids, who had displaced them on the mainland and who followed them to Taiwan and subsequently progressed to the Philippines, where they largely displaced a curly haired population who had arrived there by a more direct route from Africa. The Negritos are remnants of such. The dominant more recent arrivals can be thought of as proto-Polynesians, since it was they who continued their seafaring to conquer the Pacific Ocean.

Reflux Relatively recent migration of straight haired folk into India from the north can be seen to have occurred c4 kya after the Pleistocene in the form of the Indo-European speaking folk whose main line of entry was the valley of the River Indus at the west end of the Himalayas. Similar folk also pressed to the south-west to form the Persians and the Greeks, as well as westwards into Europe until they were halted at its extremities where it was fringed by the Atlantic Ocean. This spreading blanket only left patches representing earlier folk, one such being the Basques.

America This continent is properly thought of as being "in the west". However, in terms of human evolution it is beyond the Far East; it is indeed the Far Far East. It was not until the end of the 1ˢᵗ millennium AD that European men managed to cross the Atlantic. The Vikings did it. That it was an earlier accomplishment of the Irish St. Brendan is a rather dubious possibility. Another doubtful proposition is the crossing of the Atlantic from south-west Europe at the zenith of the Ice Age. This is based on the striking similarity between the flint points of the Solutrean culture with the Clovis points used by mammoth hunters in the USA. However, until more evidence turns up to link the two vastly distanced cultures and the time lag of some 3000 years between Solutrean and Clovis is explained, this has to be ascribed to coincidence.

Studies and comparisons of early populations have led to the certainty that the first men to reach America were *h. sapiens* and the probability that this twin continent anciently received migrants from north-eastern Asia in two waves. The second wave was certainly of Mongoloids and these formed the basis for this type as now found over nearly the whole continent. The first arrivals were c40 kya. More numerous immigrants used the land bridge c20 kya that existed in the Ice Age between Asia and Alaska. Eventual access to the distant southern tip of the landmass, known as Tierra del Fuego, was mainly made by progress along the west coast. Common characteristics of the native peoples are a sallow skin and straight black hair (with no trace of fair-heads). These are traits that their ancestors brought with them from Siberia.

CHAPTER 24

Review of the Origins and Development of Man

▌ Mankind: Evolutionary Experience and Geographic Expansion

Long before the earliest presence of creatures in Africa that can be designated as *homo*, certain early apes lived there in an arboreal habitat, as well as developing a predatory life-style and having diurnal habits. This led to binocular and colour vision being present among them. Under serious environmental pressure, these hominins descended from the trees, eventually to indulge in communal hunting of the game to be found in more open country at ground level. Here they evolved to become ever more successful at this by exploiting their emergent technology. Sticks and stones were prominent among the materials utilised as tools and weapons; but once tools started to be deliberately used to make other tools then these primates can be regarded as having become human beings in their earliest form.

The *homo* society became increasingly specialised on a sexual basis, with the males concentrating on being hunters, while the females became food gatherers. This was accompanied by a pronounced sexual dimorphism, which is also a noticeable trait among other higher primates and in which the male is characteristically bigger, stronger and provided with specific additional distinguishing features not directly connected with reproduction (secondary sexual characters). A specific human example is the hair, which developed as a mane and on an adult male came to grow on the face and jaw, where it is known as a beard.

The life away from the trees and the necessity to carry tools or weapons - or indeed other personal possessions - had encouraged ancestors of these hominins habitually to go bipedal. This led to the body being held in an erect posture, with the legs becoming proportionally longer and the feet losing their hand-like traits. The 'thumb' (i.e. big toe) of the foot in particular lost its 'opposability', and when standing the body weight was supported by what had originally been the whole mammalian lower leg, which among advanced primates had long been modified to provide the dedicated contact areas of the foot, involving heel, sole and toes.

The organization required to prepare for and successfully accomplish hunting expeditions - and the production of equipment associated with this - was at the forefront of man's transition from being just one more kind of animal to becoming a human being. The brain developed in accordance with the special requirements needed to overcome and slay animals that could be much larger or faster than himself, or both. There was also the need to cope with or avoid other predators that could be dangerously large and aggressive. Co-operation with one's fellows was a necessity to make these capabilities possible, and this was accompanied by improved communication within the group. A vital facet of this was the development of a structured sound system that surpassed every other means developed in the animal world to convey information. This 'language' induced improvements in the forebrain, which caused growth in the area of the forehead together with retraction of the hairline. The last change was relatively abrupt and resulted in the formation of the temple points which are still a feature of the hairline of each surviving member of the genus *homo*. The acquired ability to speak and then to think abstractly led to the situation where the young could be deliberately taught by their elders, rather than simply learning by watching their activities. The capacity of the brain to indulge in abstract thought resulted in the mind becoming exploratory and entering areas far beyond the requirements of everyday living. Upon reaching this stage man was well on the way to becoming *homo sapiens*.

The hunting abilities of man allowed him to go wherever there happened to be game and, in addition, to spreading eastwards out of Africa and throughout the southern regions of Asia. Mastery over fire having been gained, he was then able to wander further north and penetrate into colder zones that would have been impossible habitats for his forebears. A virtually global presence eventually ensued. In the course of such distribution of mankind, together with tendencies to perpetuate the culturally and geographically discrete societies that resulted, differences in both structure and appearance evolved further and provided the bases for the later races.

The area that seems to have provided the requirements to produce beings that were due to evolve from ape-men into *h. sapiens* included the Rift Valley and other parts near the Horn of Africa. The early type of hominin known as *h. erectus* originated thereabouts and in the course of time spread in various directions, including the Far East (e.g. as Pekin Man and Java Man), and, as *h. ergaster,* reached the Maghreb in North Africa at the west end of the Mediterranean Sea. From there they crossed over the straits into Iberia to the north. Eventually their successors pervaded much of western Europe as *h. antecessor,* who thereafter developed into the more widespread *h. heidelbergensis.*

In the course of time circumstances again changed, which then caused some *h. heidelbergensis* types to retrace the steps of their *h. ergaster* forebears eventually to arrive back in the Horn. The return to Africa enabled *h. heidelbergensis* to hybridize with residual hominins there, who were the evolved descendants of populations of the *h. erectus* type and the results tended to be of larger stature than these but more gracile than *h. heidelbergensis.* These folk were of darker colouration and normally had curly hair. Climate change sometimes led to migration northwards out of East Africa, some again heading westwards to the Maghreb, where they eventually became subsumed by Neanderthalers coming across the straits from Iberia to the north. These were the descendants of the Heidelbergers who had remained in Europe and they were also spreading widely over the eastward extent of Eurasia. Some of these Neanderthalers reached the Levant, where they occasionally made contact with darker migrants moving northwards out of Africa, using the available land bridge. Over time various hybridizations - both biological and cultural - took place that gave rise to versions that one might dub *homo pre-sapiens*, but in the end one type became absolutely dominant, i.e. *h. sapiens*, and all the rest were due to become extinct.

In the meantime African populations of *h. sapiens* - among whom the proportion of *h. heidelbergensis* in their constitution was minimal - were leaving that continent from situations near the Horn (as *h. erectus* earlier had done) and headed eastwards to colonize much of southern Asia that lay along this band of latitude. These dark skinned folk, small and curly haired, reached as far as the Philippine Islands, but eventually they were affected by climate change in that their territories were penetrated by persistent migrants from the north, who were of lighter colouring and straight haired. Various degrees of hybridization arose and in the course of time some of these modified northerners continued further to the south. Those earlier African types only survived in their original form in certain remote areas that the Asian types from the north avoided. These remnants still exist as the diminutive Negritos.

Those northerners who held more towards the west were *homo sapiens* with Neanderthal connections. They would mainly enter India by way of the River Indus and then reach the Bay of Bengal by following the River Ganges on its eastwards course. From there the west side of Indo-China could be explored (now the west coasts of Burma and Thailand) to reach Sundaland, this being the since inundated southernmost extension to the present landmass. This included at that time the peninsula of Malaya and the two main islands of the Indonesian island chain, Sumatra and Java. These earliest arrivals spent time in the Negrito Band before moving on. Due to selective hybridization they became curly haired and eventually found their way to north-west Australia, from where the continent was going to be pervaded. Behind this Neanderthal derived spearhead, the curly heads left behind were being overwhelmed by the straight haired ones from the north, and their type only survived as the Andaman Islanders and the Negritos near the Thailand-Malaya border. Otherwise their characters have been subsumed by those of the normally straight haired folk who now inhabit Indonesia and Malaysia.

A similar movement from the north also occurred on the east side of the extended Indo-China. They would also have some Neanderthal content in their genome, but they were basically straight haired Denisovan hybrids. They may have reached the Philippines by way of Borneo, but their main drift was south-eastwards towards New Guinea, where they made a considerable addition to the population of curly heads already there. This was the basis for the Melanesians, whose African origin is physically most in evidence now in the natives of Fiji, which island marks the easternmost extremity of this folk type. This Denisovan presence will be examined in greater detail later.

Paedomorphic Tendencies In the course of migrations, folk who were more child-like (paedomorphs) tended to be left aside to persist in harsh fringe areas of Africa and Asia. Two distinct paedomorphic developments can be noted. There is the African type that still exists on that continent in fairly distinct forms as the Pygmies of the Congo and the Bushmen of the Kalahari. This type had been of wider distribution and there has also once been a representation of similar small folk on the east coast, from where they became dislodged by circumstances and made their way - from shores near the Horn of Africa – eastwards along the coastlands of south Asia, to reach as far as the Philippine Islands. Fragments of such populations also remain in Indo-China near the Thailand-Malaya border and on the Andaman Islands off the west coast of this.

The other population of paedomorphs is represented by the Mongoloids of Asia, with extreme forms having been created to accommodate an icy climate, as found in the far north. But the bulk of the Mongoloids thrived further south in eastern Asia, and it was from among these that emigrants with less extreme characters made their way – when circumstances demanded – to areas to the south (i.e. Indonesia) by way of Indo-China, and eastwards to coastal areas and the Japanese archipelago beyond. More spectacularly this led to penetration of the northern parts of the Americas by way of a land bridge between Siberia and Alaska (since flooded by the sea to become the Bering Straits).

The Types of Homo A historical (but crude) distribution of earlier ascribed divisions of *homo sapiens* can be listed as:

Negroid in Africa south of the Sahara;

Caucasoid in Europe, the Mediterranean Basin and western Asia;

Australoid in Australia and Melanesia;

Mongoloid in eastern Asia and the Americas.

The detailed reasons for this situation have been explored on above pages, but the divisions can be used for ready understanding of how things have evolved and stand now. This is by considering the degree of historical paedomorphosis still discernible in the various races of mankind, and with particular reference to the effect of this on adult males. The distinctive features to be considered are the state of the hair – especially head hair – and the amount of overhang of the eyebrow area of the forehead.

In respect of this, the Mongoloids can be initially separated out from the other three categories. With them there is a reduced tendency to baldness, less facial hair and the brows overhang the eyes and the bridge of the nose to a lesser extent. They still inhabit the zone in which have been found anatomical traces of the Denisovans, namely in the Denisova Cave and on the Tibetan plateau. But, when it comes to the percentage of Denisovan DNA found in modern populations, it is not among Mongoloids, but among Australians and Melanesians that the greatest relevant content arises, these being folk types who are devoid of Mongoloid appearance. So it was not Mongoloids who carried Denisovan DNA into Australasia; they were held back at the Wallace Line, a difficult stretch of sea channel to the east of Borneo and Java that has existed from deep in primeval times. The question thus arises as to why Denisovan remains have been found in mainland Asia, while their DNA mainly persists in Australasia.

Study has revealed that the Denisovans split from the Neanderthalers at some point well after their common ancestor left the main evolutionary stem that was leading to *homo sapiens*. The implication that one can draw from the discourse in the above paragraph is that the divergence away from the mainline leading to *h. sapiens* took place while all proto-humans were still in Africa, and to postulate that the Neanderthal-Denisovan split is represented at its earliest stage by the difference between *h. ergaster* and *h. erectus*. The former took the north-western route out of Africa, via the Maghreb, and evolved to *h. neanderthalis* by way of *h. antecessor* and *h. heidelbergensis* as previously described. The latter took the north-eastern route, via the Levant, from where he spread eastwards over the southern coastlands of Asia to reach the Far East. Although evidence is lacking, it is necessary to assume a stage was reached during this expansion that was equivalent to the *h. heidelbergensis* one that preceded the Neanderthalers in Europe. The Denisovans had reached the form known to us by way of obscure stages equivalent to the Heidelbergers of the west. These had followed a route leading along the southern coastlands of Asia in the primitive form known as *h. erectus*. These reached the Far East, as evidenced by Pekin Man and Java Man, but it would be in the neighbourhood of Java that the final evolutionary step occurred to produce the eastern "Heidelbergers". Further evolution then led to the Denisovans and this probably mainly took place in that part of Indo-China that has been lost to the sea (Sundaland). This constituted the main body of the Denisovans who were to spread northwards from there into east Asia, where they made contact with folk at a similar stage of development, the Neanderthalers coming from the west along a route lying to the north of the Himalayas, as indicated by the finds in the Denisova Cave.

But it has been stated above that the forebears of both the Denisovans and the Neanderthalers left the main stem leading to *h. sapiens* before any of them left Africa. The implication of this is that this "main stem" were dark curly haired folk who were evolving in east Africa. However, these did not proceed to become *h. sapiens* until they were impinged upon and hybridized with Heidelbergers making their return from Europe. These hybrids, with their superior culture, again left east Africa near the Horn to once more make the transit along the southern coasts of Asia. Upon reaching the East Indies they had the opportunity to hybridize with the Denisovans already there. It was western folk such as these of modified *h. sapiens* type who were the first to have the ability to cross the Wallace Line and populate Australia and Melanesia, and thereby introduce Denisovan DNA into those regions. After this most traces of the Denisovans were removed from the populations found in Indo-China and Indonesia. This would be due to the inundation of Sundaland by the ocean and incursion of folk moving south from mainland Asia due to pressure generated by the activities of the Mongoloids to the north.

CHAPTER 25

Reflections on Some Aspects and Prospects of Human Life

■ The Bilateral Brain and the Mind

The central nervous system of man has the same historical and biological roots as those of other vertebrates; at one extremity of the body, the head (or rostral) end, such an advanced nervous system has developed to become a brain. When such had evolved to a point that appreciation of the passage of time was possible, then a memory had evolved and consciousness was present in the brain; at this point the hugely advanced mind had also been created and with it an enormous degree of intelligence.

The brain, like the rest of the nervous system, is a bilateral formation where information received from the various senses is co-ordinated, pending action. With more primitive mindless brains, where memory and intelligence are absent, both halves of the system are responsible for the same activities, although co-ordination is increased, not only by simple lateral links across the system, but also ultimately by the original two chords being welded into a single organ and signals being decussated, whereby they are then dealt with by the opposite side of the brain. In this way one side of an animal knows what the other side is doing.

Except for the higher primates, the mammalian brain is mainly concentrated in the cerebellum of the hindbrain. However, with apes and man the rearward parts of the olfactory lobes have been greatly developed to form the cerebral hemispheres of the forebrain. There was an evolutionary opportunity here that induced this development, but it was presumably made possible by a redundant capacity in the olfactory lobes due to the decline of the sense of smell in the higher primates. As is usually the case, obsolescent organs are either converted to other uses or shrink out of existence.

As with the cerebellum, the hemispheres of the cerebrum, originally separate, grew together in a way that was independent of the brain stem and were linked by the commissural tract known as the *corpus callosum*. Such linkage enabled the hemispheres to co-operate and thus allowed the equivalent sections in each brain half to diverge functionally and evolve independently, yet still keep in touch with overall activity. At this stage the mind was progressing beyond intelligence into a capacity for reason.

■ Activities and Achievements of the Mind

The brain builds up the capabilities of the mind by pioneering pathways through its own immensely complex cellular structure, which are recognised against precedence and perpetuated by the memory. When activities are linked together in space and time repetitively they tend to be consigned to the memory as a package, no matter how disparate in kind, through being naturally linked. This is how the experimentally defined "conditioned reflex" works, yet is the basis for all thought. On earlier pages the process has been dubbed for convenience the Conditioned Reflex Mechanism. Among primitive men this has been exploited to form the erroneous concept of "sympathetic magic". To put it more mundanely, it is a simple comparison of a current situation against precedent, coupled with an exploration of any previously recorded incidence of similar precedents.

Although sight is the most important sense centred on the head, it was a development of hearing that was to provide the vital breakaway of the human species from the rest of the animal kingdom. Sounds created by the larynx and modulated by the mouth were originally intended as indications of mood, but were developed so that they could be applied to specific items. As a result of memory, further development led to the appreciation of the passage of time, as well as other conceptual relationships. Then sounds were added that could co-ordinate all this and thereby lead to the creation of primitive diction. Thus was language born and men became conscious of themselves as individuals and of their own mortality as an undeniable and personally appreciated inevitability.

Time appeared short and drove man to a certain dissatisfaction with simply living off the provisions of Nature. He aspired to make the most of his transitory life by aiming to reach various goals and fulfil targeted achievements. The facility that lay behind such aspirations was the same one that made their realization possible and that was language. However, a relatively easy life-style that provided surpluses was necessary for rapid cultural development and groups that found themselves locked in a survival struggle against a hostile environment would inevitably reach a cultural plateau from which they were unlikely to escape.

The revolution that was to lead to more advanced cultures only really took off among those folk who succeeded in controlling and selectively developing certain plants and animals used mainly for food, but also as additional sources of useful materials. These achievements initially took place in south-west Asia, but human societies may or may not have had the latent potential for later independent parallel developments elsewhere and indeed even eventually in America. The cultural achievements of civilisation exploited capabilities of the human brain already present in more primitive man and early traits such as these can have had very little to do with a logical approach to the physical aspects of the World. In recent centuries, whenever civilised man has come into contact with 'primitive' man, the latter has always been found to have very complex beliefs relating to the metaphysical aspects of life and which can be referred to as magic or religion. The human brain has developed to cope with a whole host of language-born conceptions and misconceptions, but it is civilizing man who has been able to divert much of his growing capacity to think more rationally into mundane, practical and wealth-producing activities on a large scale, yet still leaving an underlying metaphysical level of the mind as a constant source of ideas.

Various mammals can be regarded as 'intelligent', but only man has 'reason' because this is a function of language. The basis for reason is as explored in Chapter 4 and owes its existence to the ability of man to share his knowledge and opinions with his fellows. Reason is hence a verbal non-instinctive and communal form of intelligence. It created a vast fund of knowledge, especially when writing was added as a visual means of disseminating information; but it was not infallible by any means, owing to the workings of the human mind still being puny when pitted against the problems to be solved.

The reasoning mind started by appreciating change, while thinking of it as 'time' and then came to handle other abstract concepts. This allowed problems to be tackled in advance by consciously attempting to project past and present change into the future, with any conclusions hopefully being justified by events to come, even if initially condemned by contemporary opinion. This capability of the mind is 'imagination' and enabled reason to be projected some way into the future. It is even more prone to diverge from any ultimate truth than is reason. However, its deliberate cultivation is the source of 'inspiration' and without it even the progress of reason would have been very laborious indeed.

The Hidden Mind

Everyone is aware of his mind because of self-consciousness, the reflexive result of language. The known mind depends for its existence on the continuous activity of the brain, backed up by that aspect of it called the memory and is particularly human in that it can be projected beyond mere intelligence into the realms of reason and imagination.

However, behind such a conscious mind there exists a capability of more primitive origin dealing with more elementary stages of consciousness, yet nevertheless now organised to a high degree of complexity to cope with the demands made on it by self-consciousness. This is the subconscious mind and is the realm of learned memories that are generally available to the conscious mind on a piecemeal basis.

More remote still from consciousness there lurks a capability of even more primitive origin, dealing with the most elementary stages of awareness, yet likewise highly evolved to cope with the demands made on it by consciousness. This is the unconscious mind and its workings are completely hidden from the conscious mind, except for when it chooses to reveal tiny glimpses of them under special circumstances. It is the bottom of the mind, the lower reflexive end, for impulses cannot pass it, but must either be returned aloft with solutions, or retained as mild irritations or, at worst, festering frustrations. The unconscious mind could well have a sign up – "The buck stops here".

At the bottom of the unconscious mind are the primary or inherited instincts, the prime movers of thought and activities. They constitute the launching pad from which are projected those impulses that unwittingly (and often unwillingly) are appreciated by the conscious mind in the form that can be labelled "primitive emotions". In psychoanalytical terms this is the 'id' at work.

There are also instincts that are secondary, being only acquired during the lifetime of the individual. They are mental reflexes that originate in the higher regions of the mind, but are consigned to the unconscious mind for future reference and are really the 'irritations' and 'frustrations' referred to above. These secondary instincts create their own secondary emotions by means of which they are appreciated in the conscious mind. As well as their emotional manifestations being just as incomprehensible as those alluded to just above, these instincts are unpredictable, for it is only the capability to acquire them that is heritable and they will vary greatly from person to person, thus making further contribution to differences in personality that otherwise have been inherited. In psychoanalytical terms this is the 'ego' at work.

Frustration based emotions can often be satisfied by daydreams, or even wish-fulfilment dreams during sleep, especially with children. However, when frustration is deep seated (i.e. repressed) such methods do no good. The mind can then self-consciously deal with the problem by using culturally acquired techniques such as meditation, psychology and hypnosis. But these are last resorts, after the mind's own self-healing method has failed; the brain can assuage a good deal of naturally and culturally induced 'stress' by means of dreams, whereby the frustration is redirected using the conditioned reflex mechanism, but there are limitations, particularly in the cultural field.

The Soul

Mammals in general may have souls, but cannot be consciously aware of them. Man has a soul because he repeatedly informs himself of the fact through the medium of language. But what is a soul?

The soul is in many ways like the mind, but differs essentially in that theoretically it can leave the body under special circumstances, and thereby enjoy a separate existence, while also being able to live on independently when the body dies. While folk will readily accept the existence of the mind, the reality of the soul always remains a matter of opinion. While many are convinced that everyone has one, to others its existence remains a dubious postulate, or even a complete fallacy. Belief in the soul as an entity rests on shared observations based on individual introspection of the workings of the mind. However, in the everyday conscious state the only way a man can be aware of his soul is by way of an inward ear that 'hears' words inside the head, although none have passed any lips. This aspect of the soul is the reflexive effect of language on the mind; the soul of a man can be understood by him, even if always keeping perfectly silent.

Among more primitive folk the belief has been found to exist that during dreams the soul leaves the body and roams about. A dream is chiefly a visual effect on the mind by the mind and is 'seen' by an inward eye. Although this can be taken as a manifestation of the soul, as proof of the soul's ability to leave the body it is unacceptable to the scientific or any other rational mind.

Dreams are not confined to human beings and one might suppose that any animal that can experience them has a soul. But this can hardly be the same as the immortal soul that is believed in by much of the human race, but rather the "animal soul" is a manifestation of the non-linguistic mind, a sensual rather than an intellectual creation.

The extra ingredient that creates the "human soul" is hence language and its presence is appreciated through the mind giving a silent verbal account of its own activities. The human soul, albeit dumb, can hence be 'understood', while the animal soul, albeit invisible, can be 'seen' and while being silent, odourless and without substance, can perhaps be 'heard', 'smelt' and 'felt'. The world of the soul is not the same one that we can appreciate through our senses and comprehend through our normal conscious transitory thought processes. Even as the unconscious mind is veiled from us, so is the soul.

Our comprehension of the nature of the soul seems to result from a coalescence of its two aspects as described above – the "animal soul" and the "human soul". While the former is just a way of referring to those aspects of the unconscious mind that deign to reveal themselves to the conscious mind, the latter is grafted onto this and results from the relevant revealed experience being compounded by acquiring a linguistic component, thus moving the concept of the human soul into the world of the intellect, rather than merely of the senses. However, the concept is thereby exposed to critical examination, both by imagination and reason. Imagination tends to accept the soul as an entity, but reason to reject it.

Many folk claim to have experienced 'ghosts' while being fully conscious. These spirits are manifestly of human form, although sometimes human pets may feature, such as dogs and cats. No one claims to have seen the ghost of a starfish. One would hardly expect such an experience since a starfish has neither brain nor mind, and hence no soul. Ghosts that are stumbled across and seen are never heard and, whatever their cause or true nature, must be regarded as manifestations of the animal soul alone. On the other hand poltergeists are heard but not seen, yet never speak, so once again can be regarded as formulations of the animal soul.

Those spirits that apparently refuse or are unable to communicate verbally are both incomprehensible and frightening and in order to overcome the problem of reaching human souls desirous of communication the concept called 'spiritualism' has been invented, whereby spirits may indeed 'speak' using a suitably gifted human medium, alive but in a trance. Results are apparently achieved, but whether they are genuine, or depend on deception or even deceit, inevitably boils down again to 'opinion'. This is the word we use to disguise the fact that ultimate truths are determined not to reveal themselves, neither to individual humans, nor to humanity as a whole.

■ Way Out There

This is not a work about ethics. Nor is it intended to be philosophical. One of its leading themes is to show how the abstractions that form human nature actually have physical origins.

The intellectual achievements of man, based on the reflexive and social powers of language, have allowed him to consider himself both as a physical and metaphysical internal being, but also to externalise the latter concept as a soul. He has also acquired the means to consider what exists outside of his physical limits, both here on Earth and out there in the Universe. The evidence that can now be collected by means of technology indicates that there is nothing beyond the Earth but the infinite uniform nothingness of space, the infinite variety of matter included in it and the interface activity occurring between the two. However, the human mind tends to insist that there is a metaphysical component that is as yet beyond comprehension. Such an entity may or may not exist, but if it does it "passes all understanding". Is it aware of mankind, or is this particular reservoir of life on Earth just an insignificant and unrecognised part of such an entity?

The other view is that there "must be" life (as we know it?) out there. If there is, and there has been no divine input, then it is quite impossible for it to resemble what is known as 'life' here on Earth. For life here to reach the present it has been subject to myriads of changes, each sequence of which has followed its own critical path through time. Once divergence has taken place there has been no chance for a return to any other path, although superficial resemblance to other life forms has sometimes been reached by coincidence. Hence, if life here did originate among lifeless inorganic matter, then the same concept must surely apply elsewhere. If it does then life out there in space must have evolved into totally different forms. But if it does not originate thus then it must be of a nature that will remain completely unintelligible to the human mind for some time to come.

In the meantime, life as found on Earth needs to keep going. Should it once stop completely there is no way that it can be recreated here as we know it and no chance that it will spontaneously ever start again; well at least not without supernatural help. And are we its supreme achievement? Well, there is much to support this view, at least in accordance with the evidence presented in all the above.

So can we create a heaven here on Earth? Our past makes this very unlikely, but at least we might move things more in that direction if only we could learn, both individually and collectively, to give life and the planet it inhabits the respect it increasingly needs.

GLOSSARY

Adenine -n - a compound which is one of the four constituent bases of nucleic acids.

Adeno- - relating to a gland or glands.

Adenosine -n - a nucleoside consisting of adenine combined with ribose.

Adenosine troposphere - n - a compound which by the breakdown in the body of adenosine triphosphate (ATP) to diphosphate, provides energy for physiological processes such as muscular contraction.

Alcohol - n - any organic compound containing a hydroxyl group -OH (e.g. C_2H_5OH, as found in liquors).

Allele – n – one of two or more alternative forms of a gene that arise by mutation and are found at the same place on a chromosome.

Allotrope - n - each of two or more physical forms in which an element can exist (e.g. graphite, charcoal & diamond as forms of carbon).

Amino acid - n - any of a class of about 20 simple organic compounds which form the basic constituents of protein & contain both a carboxyl (-COOH) & an amino ($-NH_2$) group.

Amoeba – n – a single-celled animal which catches food and moves about by extending finger-like projections of protoplasm.

Archaea - n (pl) - micro-organisms that are similar to bacteria in size and simplicity of structure but constitute an ancient group intermediate between the bacteria and eukaryotes.

Archenteron – n – the rudimentary alimentary cavity of an embryo at the gastrula stage.

Avicularium – n – (on some bryozoans) any of a number of modified zooids that take the form of a pair of snapping jaws resembling a bird's head and serving to prevent other organisms from settling on the colony.

Bacterium - n- a member of a large group of unicellular micro-organisms (prokaryotes) which have cell walls but lack an organized nucleus and other structures, and include numerous disease-causing forms

Base - n - a substance capable of reacting with an acid to form a salt and water, or of accepting or neutralizing hydrogen ions

Basic - a - containing or having the properties of a base, alkaline

Bicyclic – a - having two rings of atoms on its molecule

Blastopore – n – an opening into the archenteron during the embryonic stages of an organism.

Blastula – n – an animal embryo at the early stage of development when it is a hollow ball of cells.

Carbohydrate - n - any of a large group of compounds (including sugars, starch & cellulose) which contain carbon, hydrogen & oxygen, occur in food & living tissues, and can be broken down to release energy in the body.

Carbonic acid - n - a very weak acid formed when carbon dioxide dissolves in water (H_2CO_3)

Carboxyl - n - of or denoting the radical -COOH, present in organic acids

Carboxylase - n - an enzyme which promotes the addition of a carboxyl group to a compound

Chellate - n - a compound containing an organic ligand bonded to a central metal atom at two or more points.

Chlorophyll - n - a green pigment which is responsible for the absorption of light by plants to provide energy for photosynthesis.

Chloroplast - n - a structure in green plant cells which contains chlorophyll & in which photosynthesis takes place.

Chromoplast - n - a coloured plastid other than a chloroplast, typically containing a yellow or orange pigment.

Chromosome - n - a thread-like structure of nucleic acids & protein found in the nuclei of most living cells, carrying genetic info. in the form of genes.

Cilium – n – a short microscopic hair-like vibrating structure, occurring in large numbers on the surface of certain cells.

Cleavage – n – cell division, especially of a fertilized egg cell.

Coelom – n – the principle body cavity in most animals, located between the intestinal canal and the body wall.

Conjugation – n – the fusion of biological partners, such as gametes.

Coordinate - adj - denoting a covalent bond in which one atom provides both the shared electrons

Covalent - adj - relating to or denoting chemical bonds formed by the sharing of electrons between atoms. Often contrasted with ionic

Cyanobacteria - n - micro-organisms of a division comprising the blue-green algae, related to bacteria but capable of photosynthesis.

Cytoplasm – n – the material or protoplasm within a living cell, but excluding the nucleus.

Deoxyribose - n - a sugar derived from ribose by replacement of a hydroxyl group by hydrogen

Deoxyribonucleic acid - n - (DNA) a substance present in nearly all living organisms as the carrier of genetic info., and typically consisting of a very long double-stranded chain of sugar and phosphate groups cross-linked by pairs of organic bases

Deuterostome –n – a multicellular organism with a mouth that does not arise from the blastopore, such as a chordate, vertebrate or echinoderm

Diploid - adj - containing two sets of chromosomes, one from each parent.

Ectoparasite – n – a parasite that lives on the outside of its host.

Electron - n - a stable negatively charged subatomic particle with a mass 1,836 times less than that of a proton, found in all atoms and acting as the primary carrier of electricity in solids.

Endoparasite – n – a parasite that lives inside its host.

Enzyme - n - a substance produced by a living organism & acting as a catalyst to promote a specific biochemical reaction.

Eubacteria - n - a large group of bacteria with simple cells and rigid walls, comprising the 'true' bacteria and cyanobacteria as distinct from archaea

Eukaryote - n - an organism consisting of a cell or cells in which the genetic material is DNA in the form of chromosomes contained within a distinct nucleus (i.e. all living organisms other than the eubacteria & archaea).

Fission – n – reproduction by means of a cell or organism dividing into two or more new cells or organisms.

Flagellum – n – a slender thread-like structure, which enables many protozoans, bacteria, spermatozoa, etc. to swim

Gamete – n – a mature haploid male or female germ cell which is able to unite with another of the opposite sex in sexual reproduction to form a zygote

Gastrula – n – an embryo stage following the blastula, when it is a hollow cup-shaped structure having three layers of cells.

Gene - n - a unit of heredity which is transferred from a parent to offspring & is held to determine some characteristic of the offspring; in particular a distinct sequence of DNA forming part of a chromosome.

Genotype – n – the genetic constitution of an individual organism.

Glucose - n - a simple sugar which is an important energy source in living organisms and is a component of may carbohydrates.

Glucoside - n - a glycoside derived from glucose.

Glycoside - n - a compound formed from a simple sugar and another compound by the replacement of a hydroxyl group in the sugar molecule.

Group - n - a combination of atoms having a recognizable identity in a number of compounds.

Haploid - adj - having a single set of unpaired chromosomes.

Hermaphrodite – n – an animal having both male and female sex organs.

Heterocyclic - adj - denoting a compound whose molecule contains a ring of atoms of at least two elements (one of which is generally carbon)

Hominid - n - existing humans, their immediate fossil ancestry, and apes regarded as 'anthropoid', i.e. chimpanzees, bonobo, gorrillas and orang utans

Hominin – n – a primate of the tribe Hominini which comprises those species regarded as human, directly ancestral to humans, or very closely related to humans.

Hydroxide - n - a compound containing the hydroxide ion OH or the group -OH

Hydroxyl - n - of or denoting the radical -OH, present in alcohols and many other organic compounds

Infra-life – n – this specifically coined word describes the stage preceded proto-life, with life itself still being absent, but with the molecules and compounds being present that would enable life gradually to begin.

Ion - n - an atom or molecule with a net electric charge through loss or gain of electrons, either positive (a cation) or negative (an anion)

Ionic - adj - denoting chemical bonds formed by the electrostatic attraction of oppositely charged ions. Often contrasted with covalent.

Isotope - n - each of two or more forms of the same element that contain equal numbers of protons but different numbers of neutrons in their nuclei.

Karyo- - comb. form - denoting the nucleus of a cell

Ligand - n - an ion or molecule attached to a metal atom by coordinate bonding

Medusa – n – the free-swimming sexual form of a jellyfish or other coelenterate.

Meiosis -n - a type of cell division that results in daughter cells each with half the number of chromosomes of the parent cell.

Mitochondrion – n – a structure found in large numbers in most cells in which respiration and energy production occur.

Mitosis - n – cell division that differs from meiosis in that daughter cells have the same number and kind of chromosomes as the parent nucleus.

Monomer - n - a molecule that can be bonded to other identical molecules to form a polymer.

Neoteny n - the retention of juvenile features in the adult or the attainment of sexual maturity while still in the larval or other early state.

Nucleic acid - n - a complex organic substance, especially DNA or RNA, whose molecules consist of long chains of nucleotides.

Nucleoside - n - an organic compound consisting of a purine or pyrimidine base linked to a sugar, e.g. adenosine.

Nucleotide - n - a compound consisting of a nucleoside linked to a phosphate group, forming the basic structural unit of nuclei acids.

Organelle – n – any one of a number of organized or specialized structures within a living cell.

Oxidize - v - undergo or cause to undergo a reaction in which electrons are lost to another substance or molecule.

Paedomorphosis - n - Neoteny (q.v.)

Paramecium – n – a single celled fresh-water animal which has a characteristic slipper-like shape and is covered with cilia.

Parapod(ium) – n – one of the fleshy paired appendages of polychaete annelids that function in locomotion and breathing.

Peptide - n - a compound consisting of two or more amino acids linked in a chain.

Phenotype – n – the observable characteristics of an individual resulting from the interaction of its genotype with the environment.

Plasmid - n - a genetic structure in a cell that can replicate independently of the chromosomes, especially a small circular DNA strand in a bacterium or protozoan.

Plastid - n - any of a class of small organelles in the cytoplasm of plant cells, containing pigment or food.

Polymer - n - a substance with a molecular structure formed from many identical small molecules or other units bonded together.

Polyp – n – the sedentary form of a coelenterate, typically columnar with the mouth uppermost and surrounded by a ring of tentacles.

Polypeptide - n - a linear organic polymer consisting of a large number of amino acid residues, forming all or part of a protein molecule.

Porphyry - n - a pigment (e.g. haem or chlorophyll) whose molecule contains a flat ring of four linked heterocyclic groups

Prokaryote - n - a single celled organism with neither a distinct nucleus with a membrane nor other specialized structures (i.e. a bacterium of archaean)

Protein - n - any of a class of nitrogenous organic compounds forming structural components of body tissues and constituting an important part of the diet.

Protista - n. pl - a kingdom or large grouping of typically single-celled organisms including the protozoans, slime moulds and simple algae & fungi.

Proto-life – n – first life: life at its very beginning.

Proton - n- a stable subatomic particle occurring in all atomic nuclei, with a positive electric charge equal in magnitude to that of an electron.

Protostome – n – a multicellular organism whose mouth develops from a primary embryonic opening or blastopore, such as an annelid, mollusc or arthropod.

Protozoan – n – a single-celled microscopic animal such as an amoeba, a flagellate, ciliate or sporozoan.

Purine - n - a colourless crystalline bicyclic compound with basic properties.

Purine base – n - a substituted derivative of purine, especially adenine or guanine present in DNA.

Pyrimidine - n - a colourless crystalline heterocyclic compound with basic properties.

Pyrimidine base - n - a substituted derivative of pyramidine, especially thymine and cytosine present in DNA.

Reduce - v – to cause to combine chemically with hydrogen. …. To undergo or cause to undergo a reaction in which electrons are gained from another substance or molecule.

Retro-fusion – n – This specifically coined word describes a special situation that goes beyond the failure to separate and represents a degree of reintegration so that signs of former individuality are masked or eradicated. The term is more applicable to segmentation and combination of equals than to colony forming. It applies where segmentation was superseded by the advantages of cramming together the appendages of a number of segments into one mass, with such de-segmentation enabling them to specialize and co-operate more effectively.

RNase - n - an enzyme which breaks down RNA into smaller molecules.

Ribonucleic acid - n - (RNA) a substance in living cells which carries instructions from DNA for controlling the synthesis of proteins and in some viruses carries genetic information instead of DNA.

Ribose - n - a sugar which is a constituent of nucleosides and several vitamins and enzymes.

Ribosome - n - a minute particle of RNA and protein found in cells and involved in the synthesis of polypeptides and proteins.

Ribozyme - n - an RNA molecule capable of acting as an enzyme.

Spore - n - a minute typically single celled reproductive unit characteristic of the lower plants, fungi and protozoa and capable of giving rise to a new individual without sexual fusion

Sporangium - n - (in ferns and lower plants) a receptacle in which asexual spores are formed

Starch - n - an odourless, tasteless carbohydrate which is obtained chiefly from cereals and is an important constituent of the human diet.

Sucrose - n - a compound which is the chief component of cane or beet sugar.

Sugar - n - any of the class of soluble, crystalline, typically sweet-tasting carbohydrates found in living tissues and exemplified by glucose and sucrose.

Syncytium – n – a single cell or cytoplasmic mass containing several nuclei.

Transfer RNA - n- a folded form of RNA which transports amino acids from the cytoplasm of a cell to a ribosome.

Transcribe - v - synthesize (RNA) using a template of existing DNA (or vice versa) so that the genetic information is copied.

Transcript - n - a length of RNA or DNA that has been transcribed.

Transcriptase - n - an enzyme which catalyses the formation of RNA from a DNA template, or (reverse transcriptase) the formation of DNA from an RNA template.

Trochophore – n – a type of free-swimming planktonic marine larva with several bands of cilia.

Virus - n - a submicroscopic infective particle, typically consisting of nucleic acid (RNA or DNA) coated in protein, which is able to multiply within the cells of a host organism.

Zooid – n – an animal arising from another by budding or division, especially each of the individuals which make up a colonial organism.

Zygote – a cell resulting from the fusion of two gametes.